AF781911

INVENTOR'S HANDBOOK

PETER HORSLEY PUBLICATIONS

Revised February 1994.

COPYRIGHT © 1990; 1994;
PETER B. HORSLEY
115 - Riverbirch Crescent SE,
Calgary, Alberta, Canada
T2C 3M1

No part of this book may be reproduced, stored in a retrieval system, or transmitted, in any form by any means electronic, mechanical, photocopying, recording, or otherwise, without the prior written permission of the copyright owner.

ISBN 0--9698226--2--6

PRINTED IN CANADA

ACKNOWLEDGEMENT

The author and publisher gratefully acknowledge permission to use the antique drawings and cuts from the book

HANDBOOK OF EARLY ADVERTISING ART
by Clarence P. Hornung.

Published by DOVER PUBLICATIONS, INC.,
180 Varick Street,
New York, N.Y., 10014.

FOREWORD

If you happen to be sitting in the living room or family room of your home, or relaxing in the den at your apartment, raise your eyes and look around you. Every item upon which your gaze may fall was once a glint in the eye of an inventor. The stereo, the television set, the computer are all patented for the many inventions they contain. There are over a thousand new patents pending relating to improvements in computer technology alone.

Every device that we take for granted today was once the brainchild of an inventor. Every advance in technology and in civilization were brought to you from inventors throughout the history of mankind. He gave you comfortable homes, leisure time, entertainment and quick transportation. When you think about it.....we owe the inventor a lot.

There is another important aspect to inventions. Think about all of the jobs that the inventor created, all of the folks that had to be hired through the ages to make gadgets and run the machines of the world. The inventor is mankind's greatest benefactor. The contribution of the inventor to society is priceless and beyond measure.

Now, if you have an idea for an invention, you have the opportunity to join the ranks of the those who conceive and create. You can become an innovator and inventor yourself. The inventor's life is not an easy one. Very few inventors are successful and fewer yet strike it rich. But inventing things is an exhilarating hobby. You can have a lot of fun and greatly improve the quality of your life even if you never become rich and famous.

Every time some new fact or new principle is discovered about the physical world, it is not long before some inspired inventor applies the new knowledge for mankind's benefit. The key factor is the dispersal of the information. The inventor must have the facts. If the information becomes widely known, then the average inventor has the opportunity to locate it and assimilate it.

What happens if the new facts are imbedded within some obscure technical paper read by a few specialists? Then the inventor never gets a chance to see it and understand it. The inventions which might have sprung from that idea are lost.

Every scientific principle is the basis for hundreds of inventions. Consider the burgeoning technology that was spawned by electricity alone. Think about the hundreds of incredible inventions that derived from electronics and the magic of the silicon chip. Every principle of physics is the basic stuff that inventor's thrive on.

This is a book about inventions and the principles on which inventions are founded. It is a book that is full of the kind of information that inventors need. It begins with the first inkling of an idea for an invention and what to do about it, including how to train yourself to become a professional inventor.

This book also contains many innovative ideas for applying science and physics. It presents in illustrated detail the critical information that many inventors may have overlooked. New technology is explained and old technology is scrutinized for more novel applications.

In the early days of the 20th century there was a great revolution in scientific thought. New ideas about the nature of the atom had evolved. The foundations for relativity were being laid. The interferometer had placed the existence of the ether in serious doubt. Also, the apparent duality of a light ray made the scope of physics more confusing. Important landmark questions posed in those days have never been answered, and never have been reduced to basic principles. The details of what happened back then need to be illustrated in a way that inventors and innovators can grasp and assimilate.

Another facet of this book is the basic courses that it offers at the back along with a glossary of terms used in patent applications. Many inventors hesitate to proceed with an idea because they think that they lack a technical education. In this book the basic facts about electricity and electronics are shown in a simple illustrated form. To develop an invention that is really practical some knowledge of electricity or electronics is often necessary. For the average inventor, these courses are written in a style that makes comprehension easy and interesting. The objective is to provide a quick reference whenever the inventor is stuck for information or guidance.

The other objective of this book is to enlighten and to entertain. Every individual alive has at least one excellent invention in him (or her) waiting to be released into an idea hungry world. That's what makes invention and innovation such a fascinating subject. The author has endeavored to retain the flavor of excitement that surrounds the development of new inventions and ideas. The main objective of the book is to be useful to inventors.

Good inventions are the basis of many fortunes. Perhaps this book will start someone on the road to fame and fortune. Good luck and have fun.

Peter Horsley.

LIVING ON THE LEADING EDGE

Imagine that all knowledge and everything known to mankind could be represented by a disc-shaped graph, like a huge pancake. As new things are discovered, the pancake graph grows larger at the edges. Outside of the circle is the great unknown. On the very edge of the pancake will be found the very latest invention, the brand new concept and the most recent discovery.

At the very center of the pancake you will find the details of how man learned to use fire and shape stone tools. The very edge of the graph is strangely fuzzy as it wavers between the present and the future. Remember that when an inventor gets the very first inkling of a new idea, something that was never thought of before, he is actually gazing into the future. This is the definition of genuine creativity. The fuzzy edge of the pancake becomes solid as the new idea is formed in the mind of a human being.

So it is on the leading edge of the pancake where inventors do their work. On the leading edge, doctors define new ways to conquer diseases and scientists develop new theories. Chemists work on new compounds, like plastic as strong as steel but as light as wood. And there is new computer technology, new designs for silicon chips, new medicines to heal the feeble body of man, and everything else that a human being can originate.

This pancake representing all knowledge is immense. There is a place along the edge for anyone who is interested in working on something new. All of the inventors and discoverers and innovators in the technical world work there all along the edge. Some of these people are geniuses. Some are persistent plodders sustained by a belief in themselves. Some are there because they hope to strike it rich. Some are sustained by the hope of recognition. Others are there because a fantastic idea struck them like lightning. Anything can happen on the leading edge and it does.

All together, the leading edge is an exhiliarating and fun place to be. It is like being in orbit. It is an outlet for those whose minds need to soar away from the mundane tasks of daily living. For those who are tied to endless toil to sustain and nourish a growing family, a few hours on the leading edge provides a release from tedium. For those whose lives are full of boredom or stress, the leading edge provides a safe haven for finding calm and stability. And there is always the possibility of a great golden strike, but if it never happens, the leading edge is a great place to be just for itself. It is an ideal place to relax and look around at what's new.

CONTENTS

INVENTIONS FROM THE PAST

The month is October. The year is 1883. Two brothers named Tissandiers in France are on the test flight of the new electric balloon that they invented. The propeller was to be operated when needed. It is driven by an electric motor powered by batteries. The electric balloon rose a quarter mile above the surface of the earth and travelled 585 feet per hour on the initial run from Auteuil to Croissy-sur-Seine.

The balloon (not shown) was filled with hydrogen.

DEVELOPING AND PROTECTING YOUR INVENTION

WILD-EYED INVENTOR

The word inventor or the technical title "INVENTOR" will often convey a mental picture of a person whose ambitions and dreams are somewhat wierd. You may think an inventor is a funny little man wearing a smock and standing in the cluttered debris of an unkemp laboratory. He is holding up a beaker of acid to the light to see what color it has become and meanwhile is spilling a few acrid drops of smoking liquid on the leg of his trousers.

Personally, I love it when someone calls me an inventor. But the average university professor or doctor of any field will tend to shun the title. They usually prefer to be known by their proper name and letters of achievement. To be called an inventor may indicate that the person is not educated and does not hold a degree from an accredited university.

Who is the average inventor? Good ideas can come to any one. Housewives are just as apt to develop excellent ideas as well as engineers and mechanics. Many inventive fortunes have been created by persons with little in the way of formal education. The great Edison had almost no schooling but he had plenty of experience with science and technology. Much of the technology of Edison's day was self-generated. He was like a fountain of new ideas. He knew how to get innovations to perform and how to market them.

Regardless of what inventors call themselves, the truth is that inventors and innovators through the ages gave mankind the comfort and opulence of modern living. There was an inventor behind every new advancement in science, medicine, technology, and industry. So be proud to be known as an inventor. Have the courage to wear the title with flair and dignity. You are one of the special people.

WHERE DO IDEAS COME FROM

Ideas, innovations and inventions are a highly prized commodity in the modern technological world. There is no course of training that a person can take to become proficient as an inventor or as an idea man. It is apparent that some people more than others have an innate ability to produce worthwhile ideas. The training that such people have had is the kind of training that you get from experience.

At one time I thought it would be a good idea to become a cartoonist. It seemed that cartoonists had a lot of fun and made really big money at a funtime career. The first step was to develop some good ideas for cartoons. That proved to be the difficult part of cartooning. The ideas would not line themselves up as you might expect they would. Some ideas that I thought were good were promptly rejected. Thinking up ideas became the key effort. If there was no idea production, there were no cartoons. Meanwhile I had a wife and a growing family so I had to keep my nose to the grindstone to earn the money to sustain a household. The result was that the cartoonist career lapsed and gradually died.

The important point is that I started to get sudden flashes of inspiration for cartoon ideas. It seemed as if that during the time I was practicing thinking up ideas, something happened in my brain. My subconscious still believes I am a cartoonist. To this very day, many years later, I am still getting flashes of brilliance in the field of cartoon ideas.

IDEAS COME FROM PRACTICE

The real basic fact about ideas and inventions is that you have to exercise your brain. You have to concentrate on idea production for a period of time. During that time you have to convince your subconscious mind that your well-being, and yes, your very life, depends on your ability to think up good ideas. Your brain will respond. Results are guaranteed. You will start getting ideas almost immediately. Some of the ideas you get will be rather useless and some of them may have possibilities. In amongst the ideas you get there will be the odd gem.

Here is the important part. The longer you

work at thinking up ideas, the better you will get at it. That's right. The ideas will gradually come more often and they will show a gradual improvement in quality.

The other factor that is very important in idea production is that you must give yourself all the basic information to work with. You have to fill up your storehouse of knowledge with facts about physics, facts about science, facts about chemistry, facts about electronics, and facts about the technical world that is all around you. It's like asking a chef to prepare a special meal for guests when your cupboard is bare except for macaroni.

When an idea for an invention springs forth, it is often a clever combination of two or more previous ideas or known scientific principles. If your technical knowledge is full of facts, your brain will have plenty of combinations to sort through in the search for ideas.

WHAT INVENTIONS ARE NEEDED

The best inventions tend to make life more convenient and less of a hassle for the average person. You may have noticed that in the wintertime, some snow gets onto the car floor off your rubbers. It is warm in the car so the snow melts. Then you place your foot on the gas or brake and the cuff of your pants or slacks drags in the wet mud on the floor of the car. An inventor who noticed this nuisance developed a floor mat for cars which would permit water from rain or snow to lay in the bottom of a series of narrow grooves where it can evaporate without ever touching your clothes.

Watch for things that happen to people. Watch for things that are annoying and exasperating. If you can come up with a simple answer to such a problem, you have an invention of merit. Take the car door, for example. You buy a new car and it represents an investment of anywhere from ten to thirty thousand dollars. Then what happens. You park it in a parking lot for an hour and when you come back, somebody has thoughtlessly taken a gouge out of your beautiful paint by opening their car door against yours. For some reason that is hard to understand, many people seem to delight in marking new cars with their doors.

Okay, you inventors out there, get busy and design a car door that opens and moves sideways, staying close to the side of the car. You have to eliminate the kind of door that swings open wide, and replace it with doors that do not swing open at all. When the door is shut, it has to look just like the present doors----all tight and snug with no hardware showing on the outside. And on the inside of the car the mechanism must tuck away neatly without hitting anyones legs and not taking up space that is needed for other things like feet or luggage.

However, before you go to a lot of effort, arrange for a patent search of methods of opening car doors. The door we are looking for might already be in existence. If such a door does exist, one of the enterprising automobile manufacturers could make a killing by coming out with it first. I would be one of the first customers for a car with the kind of doors in question. Think of it----a car with doors that don't swing open. That is the kind of idea on which valuable inventions are founded.

THE INVENTION

Nobody knows when a good idea will strike. Often ideas arrive like a thunderbolt out of nowhere. When a really good idea strikes, the feeling you get is just the same as striking gold. It's a great feeling that only inventors and innovators share.

The wife of a friend was planting grass on a slope at the front of their yard. As is the custom, after the grass seed was planted, the slope was covered with wetted burlap sacks. This prevents the seed from being washed away by a sudden rain, and it protects the seed from being eaten by birds. The important thing is to remove the burlap before the grass begins to grow through it so as not to tear up the new grass shoots.

Thinking about this and wetting down the burlap with a hose, the lady suddenly got an idea. What if you made the burlap material (or a substitute) out of a fertilizer that would gradually dissolve away and fertilize the grass as it began to grow. That would save you all the trouble of taking up the dirty wet

Figure 1: Car doors designed to open and slide sideways close to the car, something like the side-doors on some vans, would be a great feature for a car. If you recently paid from twenty to thirty thousand dollars for a car, you would be dismayed to find that some careless or malicious person has damaged your door by swinging their car door into yours.

sacking material, and it would be beneficial to the grass.

The lady told her husband and they both became excited. They wanted me and my wife to witness the disclosure they had made. In the course of time they applied for a patent search and were greatly disillusioned. The patent search revealed that the idea had been patented before, not once, but several times with different novel ideas in each case.

GOOD IDEAS ARE SELDOM NEW

Our friends found out what many inventors discover to their dismay, that someone beat them to it. Sometimes what seems to be a fresh idea is as old as the hills. I once applied for a patent on a method for making 3-D jewellery and 3-D novelties. I was so sure it was a great idea that I didn't bother to get a search made of prior patents. I could hardly believe it when it turned out that the same idea and method exactly had been patented in 1882.

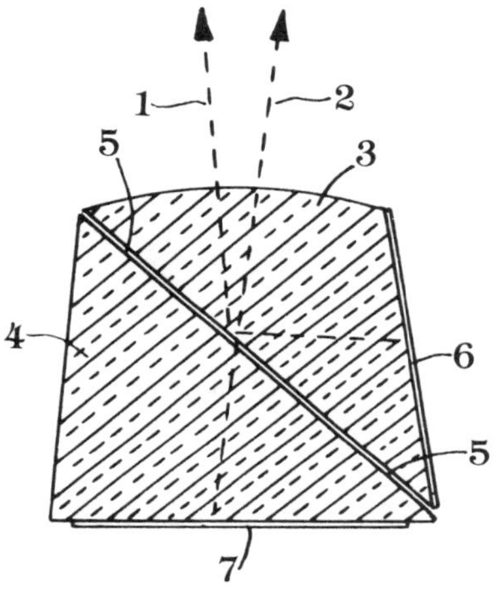

Figure 2: Cross section of a gem for 3-D jewellery. Light ray 1 is tracking to the right eye of the observer from one of the stereo pair of photos 6. Light ray 2 is tracking to the left eye of the observer from one of the stereo pair of photos 7. The air space 5 in the gem causes the light ray from 6 to be reflected due to the prism effect whereas the light ray 2 is not refelected due to the increased angle of the ray. Of the two pieces of the gem 3 and 4, 3 has a curved magnifying surface. This idea was patented in 1882.

JAMES WATT

When James Watt, inventor of the factory steam engine, applied for a patent on the engine, it was turned down. Watt was dismayed and chagrinned to discover that somebody else had patented the crank connecting rod. What the connecting rod does is to convert reciprocating motion (back and forth) into rotary motion. James Watt considered the problem and very soon came up with two ideas to circumvent the connecting rod. He invented the planetary gear. The two methods he devised are shown in the drawings. Of course the connecting rod is a much simpler and direct method of doing the job, so it was promptly adapted when it became public domain. The patent on it had run its course.

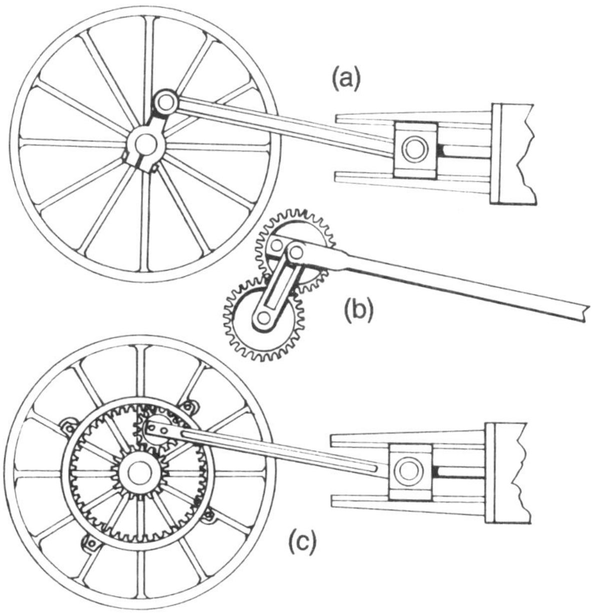

Figure 3: A connecting rod is illustrated in drawing (a). When James Watt discovered that the connecting rod had been patented, he quickly devised methods of circumventing the use of the connecting rod. Methods that he used for many years are shown at (b) and (c). Both methods use a planetary gear. In method (b) the "sun" gear is fixed to the main flywheel shaft. In (c) the "sun" gear is free to rotate on the shaft and the outer gear ring is bolted to the flywheel.

STEPS TO TAKE TO PROTECT YOUR INVENTION

The ultimate protection for an invention is to secure a strong patent. However, there are some steps you can take right away that will give you some protection in United States. When you get a sudden flash of inspiration and become excited about the commercial possibilities of your idea, perhaps the best thing to do is to relax and think it over carefully. Let it stew in your brain for a few hours, preferably overnight. This will give you a chance to see your idea in a conceptual way.

When you have examined your idea in detail, two things can happen. You may discover ways to improve on your idea or make it more practical. Or you may discover than the idea was not really important after all. It happens to everyone.

Sometimes after months of development work the inventor will discover that he is on a false trail. I once got the idea that I would make a personal gyroscopic compass. This would be a great advantage over magnetic compasses because a gyroscopic compass points to true north. I built several models, each one an improvement over the previous one. None of them worked the way they should. I talked it over with a physicist friend and showed him the models. I told him that the gyroscope did not seem to be effective. To make it too heavy would defeat the purpose of the invention. My friend did some research on the subject and later advised me that the gyroscope rotor would have to weigh at least 45 pounds and rotate at many thousands of RPM to be effective. That was the end of that invention.

After considering the implications of your invention for a few hours, if you are still convinced that it is an important idea, the next step is to make a complete disclosure. Make a drawing of it and a description, describing what the invention is supposed to do and how it works. Then seal the disclosure in an envelope and mail it to yourself by registered mail. The date stamp on the registration will establish a date of conception for your invention.

When you get the registered letter, don't open it. Write on the back of it what it contains and then file it away in a safe place. A safety deposit box is ideal, but it probably will be quite secure in your own filing cabinet or drawer.

The next step is to make another complete disclosure, including any improvements you might have thought of. Get a trusted friend to go over the disclosure with you. Explain how it works and make sure that he or she understands it. Now have your firiend sign an affiadavit that he or she was there and did see you make the disclosure on that particular date in a particular place (your postal address). When that process has been completed, again place the disclosure in an envelope and mail it to yourself by registered mail. This will establish a date of conception with a witness, a valuable document if your date of conception is ever contested in court.

GETTING READY TO PATENT YOUR INVENTION

The next step is to develop your idea into a patentable invention. You may want to make some detailed drawings of your invention and you might want to make a working model as well. It is not necessary to make a working model, although it often important to have a model so that you can refine and test your idea. At some point during the development of the invention you will decide that it is ready to patent. If you had not thought about it before, you now have to decide whether you will take out the patent yourself, or have a patent attorney do it for you.

THE PATENT APPLICATION

The preparation of the patent application is a very important process. The patent office advises that the preparation of a patent application should be done by a patent attorney. The value of the patent hinges on the ability of the person to word the claims in a manner that will give the patent the greatest immunity to a legal challenge. Unless you are skilled in the art of technical writing, drawing and legal matters, you would be wise to hire a professional.

There is another route that you can follow. That is to take a year off and spend the time studying patents and patent applications. Study the claims and you will begin to understand why they are worded so carefully. Become a patent expert and then, when you feel that you are ready, start the process of applying for a patent yourself.

THE PATENT SEARCH

There is a saying that there is nothing new under the sun. Of course that is a wild generalization, but the fact remains that to think of something new is not as easy as it appears. After you have sweated over your invention for many months and changed this or that aspect of it to make it better and better, it is hard to believe that somebody else might have already followed exactly the same route that you have taken and arrived at the same invention. The sheer novelty of your idea, without the inventive part, seems to preclude the possibility that somebody else might have the idea.

What you have to realize is that amongst the millions of people in the population of the technical world, hundreds of thousands of them are working on ideas and inventions. You must not be surprised if somebody else has anticipated you.

The time factor is also an adverse condition, because while you may have been working on your idea for a few months, other inventors have been working on ideas since the origin of the patent system over a hundred years ago. That means that you have several generations of idea men to contend with.

Regardless of the possibility of being anticipated by somebody else with a prior claim, there are still hundreds of new inventions being patented in a steady stream. Do not be discouraged by these considerations. Your idea could very well be the very first of its kind.

ANYONE CAN MAKE A PATENT SEARCH

A patent search is undertaken to find out if there are any prior patents that do anticipate your invention. The patent search takes place in the patent office. Anyone may search through all prior patents in the category of possible overlap. If you live within travelling distance of Washington or Ottawa, your best interests are served by making the patent search yourself. The patent files are arranged in a very orderly fashion so that it is not difficult to find your way to the category of invention that you are interested in.

If you cannot go to Ottawa or Washington to make the search, you can order the search by mail. Those people who make a living by knowing their way around the patent office advertise their search service in the technical magazines. The magazines with the word "Popular" in their titles usually have sections for advertising relating to inventions. Look in the classified section if you do not see it elsewhere in the magazines.

The problem with getting a search done by others is that you have no idea how reliable they are or how thorough they are. One other drawback is that nobody is permitted to search through pending patents. The important thing about pending patents is that all of the most recent patents are confined in that realm. Since it takes two or three years to get a patent, you can see that all of the most up-to-date patents on technical subjects are screened from public view.

Many professional inventors do not bother with a patent search unless they happen to live near the patent office. They prepare a patent application and proceed with the patent. Their rationale is that when the patent examiner does process the patent, the search for priority will be thorough and professional. An added bonus is that all pending patents will be searched as well.

There are two drawbacks to following this procedure. One is that it will cost you the total of the filing fee plus the examination fee, an amount that may exceed $350.00. Since the fee schedules are always under revision, you can never be sure of the exact amount you may have to pay. The other drawback is the time factor. If you order a patent search, it will be completed in a few weeks. But if you proceed with a patent application instead, it may be a year or more before your patent is examined.

EXCLUSIVE RIGHTS

A United States patent gives the inventor the exclusive right to make, use or sell his invention for a period of seventeen years. Originality is a prerequisite. During the processing of a patent application, the patent office makes a search through prior patents to determine the originality of your invention. In this process, there is no guarantee that your claims are clear and free of legal entanglement. One or more of your claims may infringe on some prior patent. If your claims do infringe on a prior patent, one of three things might happen:

(1) The fact of the infringement may never see the light of day and therefore will never be challenged.

(2) The other patentee does discover that your patent infringes on his but he doesn't care enough to do anything about it (his interests may have switched to something else meanwhile).

(3) The other patentee feels that he has been injured by your patent and decides to sue. In this case the decision will be made by the courts.

There is no law that forces the patent office to guarantee that your patent does not infringe on other patents. If there is a dispute, the courts are the only recourse. After a patent has been granted, the patent office no longer has any jurisdiction over it. The patent office is in no way interested in the problems that the inventor may have in regard to infringements or legal matters. Once the patent has been issued, the courts are the final arbitrator of any dispute.

THE PATENT INFRINGEMENT

The general feeling about patents is that once you have one stowed away in your safe or safety deposit box, your idea is protected for seventeen years. The facts are quite different. A patent is a legal document. All legal disputes are settled in a court of law. Therefore, what a patent does is give you the kind of evidence you need to legally face an adversary in court and win.

The average person would rather not go to court about anything. It is expensive and time consuming. The average person does not earn the kind of money that would be necessary to fight a legal battle successfully.

The usual practice in the industrial world is to test your position by close encounter. Suppose that you have a patent that is closely related to a patent held by a manufacturer. Perhaps your idea is somewhat better. The manufacturer orders a copy of your patent so that his high tech group can study it. They are trying to find out if they can incorporate your good idea into their patent without risking a legal challenge.

They decide that your patent is too strong for them to copy without infringement. The next step is a consultation with their attorneys and patent experts. They are asked to study your patent and state a position on the status of your patent. They want to know if they could successfully face a legal encounter in the courts.

At this point you might have some very unflattering names to call the manufacturer, but he is on safe legal ground. Don't expect any large company to be charitable. They answer only to their shareholders. What you have to try and do is put yourself in the position of the managers of the manufacturing company. They have a lot of prestige as well as their jobs on the line. They have to be prudent and take whatever steps are recommended by their attorneys.

This is the way they are thinking. If they infringe on your patent, nobody might ever notice. Inventors are notorious for abandoning their ideas and their patents after they have lost interest in them. Another aspect is that it might be two or three years before the inventor knows that his patent is being infringed on. So, their decision is to go ahead and copy your patent, incorporating your idea into their own design of the invention.

They commence a program to manufacture your invention and in the course of time you see one, perhaps months or years later. The first thing you ask yourself is how come they are manufacturing your invention when you have a perfectly good patent on it. You write them a letter asking them to explain their actions and they ignore you. You don't get an answer.

At this point you ask the advice of an attorney and he says by all means take it to court. This is what the manufacturer has been waiting for. The court challenge will permit his lawyers to test their position in a legal challenge. The evidence you have is that there is a clear infringement on your patent. The evidence they have is (1) a patent that is very similar to yours, and (2) two years of expertise, design expense, tooling up expense, production expense and distribution expense. All of this evidence is placed before the court and the overwhelming amount of evidence they have accumulated is a very real threat.

The court is permitted to weigh the value of the evidence in dollars and cents. At the same time, they cannot rule against an obvious infringement. You may win your case. What do you get? It is up to the judge and the legal committee he has chosen to decide what is in the best interests of the injured party. You might be paid a royalty for all units manufactured in the past and in the future, you will receive a royalty for all units produced.

Your attorney might decide that you should sue for a lump sum as well---a penalty against the manufacturer to deter others from infringing on valid patents. But if the attempt fails, you might be left with nothing except your patent. One way or another, the manufacturer has proved his point. It was a wise move to go ahead and infringe on your patent. It was fairly tested in court, and your patent was the winner. But in the game of manufacturing widgets, the manufacturer and the shareholders of the company were clearly the winners.

HYPOTHETICAL CASE

The foregoing example is just a hypothetical case. But it is not unlike many cases that are fought in the courts over patent rights. Sometimes the inventor loses his rights through no fault of his own. The important point to remember is that what a patent gives you is the right to challenge an adversary in court.

THE WINNERS

The wisest move that an inventor can make is to visit the public library and take out the books which pertain to inventors who made it big. What special knowledge or what special procedure did these successful inventors take? One very general rule is that many of the inventors who made a lot of money from their own invention, also did everything else to make the invention succeed. They organized financing and set up plants to make the products of their own invention, and then they promoted and marketed the products themselves.

INVENTOR'S FORTUNES

The inventors who became rich and famous from their inventions all followed the same, or similar, paths to riches. The common factors are most evident when the inventors like Henry Ford, Westinghouse, Edison, and in more recent times, Commodore Sperry and Edwin Land.

The steps they followed can be outlined as follows:

(1) Test your invention thoroughly to be sure that it does the job you expect of it.

(2) Study the marketing implications carefully. Make sure that it is a saleable product or device.

(3) Secure patents in the United States first.

(4) Plan to manufacture and distribute the product or device in the United States.

(5) Plan a method to secure financing that permits you to keep the major share in your invention.

(6) Acquire a factory or manufacturing facility in or near the heartland of industrial United States.

(7) Manufacture and distribute your product.

(8) Set up suitable sales agencies where sales conditions indicate their need.

(9) Keep improving your invention and secure patents on the improvements.

SOME OF THE INVENTORS WHO MADE FORTUNES

THOMAS A. EDISON: With as many as 1800 employees at one time working on his ideas, Edison was the most prolific inventor of all time. Everything that he worked on spawned new inventions. From the invention of the phonograph a dozen other inventions were developed. The dictating machine is an example.

GEORGES CLAUDE: Claude pioneered in liquid air and related inventions. When he invented the neon sign he recognized the commercial potential of illuminated letters. He eventually built up a chain of neon factories around the world before his patents became public domain in I928.

NICOLA TESLA: Tesla invented the a-c induction motor and alternating current electrical systems. He sold his many patents to George Westinghouse and then worked with Westinghouse to bring alternating current to the world. Together they built the first water-turbine electric generating station using alternating current.

GEORGE WESTINGHOUSE: Invented and manufactured the air brake and a pneumatic system for brakes on trains of cars. His air brake system was soon in use on every railroad in the world.

CYRUS McCORMICK: McCormick invented the reaper which harvested a field of grain faster than several men could do it by hand. He began to manufacture the reaper in his blacksmith shop but soon had to expand as the demand for the reaper grew. The blacksmith shop expanded and became an immense manufacturing corporation.

ALEXANDER GRAHAM BELL: Bell invented the telephone and saw the power of communication develop into a vital industry. The demand for telephone service grew faster than anyone could have anticipated. Bell held many other patents relating to sound transmission. He also invented novel types of aircraft and hydrofoil water craft.

ALFRED B NOBEL: Nobel invented dynamite which soon became the only practical method for mining and excavating rocky terrain. The world wide demand for dynamite quickly made Nobel a multi-millionaire. In his will he bequeathed millions of dollars to the world to be awarded to individuals for the greatest achievements in literature and the sciences.

EDWIN LAND: Edwin Land invented a polarizing film when he was still a young man. It was used in many novel ways including non-glare sunglasses and 3-D movies. Later he invented and manufactured the Polaroid camera and following that, the famous Land Camera. He also invented a new method of color photography based on light frequency ratios.

WILSON GREATBATCH: Inventor of the implantable pacemaker which he helped to develop and market. The need for pacemaker batteries with special characteristics gave him the incentive to invent a new type of lithium battery. He organized a manufacturing facility to manufacture batteries for pacemakers.

STEVE WOZNIAK: Wozniak invented the Apple computer and the Apple II computer. He and his friend, Jobbs, began manufacturing personal computers in a garage, and started the personal computer industry. The personal computer demand took off like a rocket and the PC industry became one of the fastest growing industries ever conceived.

MANY MILLIONAIRE INVENTORS

Only a very few of some of the best known inventors have been listed above. The novice inventor who is interested in reading about the great inventors and how they made fortunes with their ideas can find many reference books about inventors at your Public Library.

A good suggestion is to try and visit your Public Library on a regular basis. Whatever books you do take home, always take at least one book on applied science and look it over in your spare time. In this way you will gradually build up a knowledge of science and physics. When you are working on an invention, you will be surprised how your reading habits will affect the quality of your inventive output.

Books have a magical quality. It is not usually taught in schools, but it would be a good idea to make students aware of the magic that books can perform. The simple act of opening a book and reading a few lines daily will gradually transform the person who reads. Every book that a person reads works a subtle change in the nature of that person. Each book that is opened makes the reader have more of the qualities that are necessary for a full and happy life. Each book increases the basic knowledge and the intelligence quotient of the reader. Each book that is read makes the reader more understanding of human nature, more caring, more thoughtful and not least important, more able to earn a living.

WHO OWNS THE INVENTION

Engineers, physicists and chemists who work for manufacturing companies or corporations, never own their inventions. Although they are the legal inventors, the whole invention is owned by the corporation. In many cases a royalty is paid to the inventor as an incentive, but if he is being paid a salary or wage, whatever he invents is legally owned by the company.

If you have a regular job working for a company that is of a technical kind, you will be wise to refrain from working on your invention at any time during working hours. If you have your lunch at work, again do not work on any inventions that you want to keep for yourself. Your inventions must be developed in your own time at home, and should be, preferably, kept secret. It is a great temptation to tell everybody at work what a great idea you have for an invention. Control your enthusiasm.

The legal aspects of inventive ideas that are developed while you are working for somebody else is fraught with risk. If your invention happens to be related to your work, and if the idea is really great, then you are definitely at risk in the legal point of ownership. You would be wise to keep your ideas in strict secrecy and perhaps think about getting a job where the product is in no way related to your idea.

Sometimes you feel that your invention is so important that you just have to tell somebody about it. Don't blurt it out at work, but do go home and tell your wife about it. Discuss it with a trusted friend if you must. Best of all, try to keep it to yourself until you are well protected and out of reach of possible ownership claimants.

The situation is entirely different if you happen to work for somebody where you have a close friendly relationship. If you know that you will be treated in a fair , supportive way, by all means confide your idea to your boss. You should be able to expect a 50:50 arrangement after all expenses are paid out of the profits earned by your invention.

INVENTIONS FROM THE PAST

Transportation by water, road and rail, opened up the continent of North America to development and exploitation. The invention of the steam engine and the invention of the internal combustion engine were the driving forces behind this immense expansion of human endeavor.

TRAINING YOURSELF TO BE A PROFESSIONAL INVENTOR

CULTIVATE INVENTIVE ATTRIBUTES

If you read about the great inventors and innovators, the men who advanced science and industry by their ideas and persistence, you will find that there is a common thread of character shared by all. There are some basic abilities, interests and traits that are evident in the lives of inventors and innovators. The successful ones have some of these characteristics to an unusual degree. You can advance your own inventive carreer and increase your own ability to innovate by incorporating these attributes into your own sphere of existence. They are:

(1) The number one prerequisite is that you must have the strongest possible desire to succeed. You must be willing to work diligently at learning the things you have to know to be a successful inventor.

(2) You must have a natural love of physics and science, mechanics and chemistry, and a love of knowledge for itself. Most inventors have a keen interest in everything from an atom to a galaxy of stars. They have an overriding desire to know everything.

(3) Many successful inventors love tinkering and model making. They love repairing things that are broken and can often repair anything from a car to a TV set. They love the feel of metal and wood in the fingers. They love nuts and bolts and all of the hardware and tools that are part of the fix-it scene.

(4) Most successful inventors enjoy drawing the ideas that come to their minds. This often starts in childhood with the exclamation, "Look at the spaceship I drawed, Dad!" If you have not done so already, make or buy a drawing table. Get into the habit of drawing charts and diagrams about your ideas.

(5) Most inventors are keen observers. Inventors habitually look carefully at everything, especially everything that is manufactured. Observe everything little thing that is made, that grows, that flies, that walks. Somebody called it 3-D recording.

(6) Develop a keen memory for facts. Many inventors display an unusual ability to recall facts from earlier studies and investigations.

(7) An unusual persistence. Many inventors succeed on the basis of persistence alone. Edison said that invention is 10 percent inspiration and 90 percent perspiration. A continual probing at what seems to be a blind alley will eventually open secret doors. This is guaranteed from the experience of many innovators, including Newton and Einstein.

(8) A desire for excellence. The inventor wants everything to be the very best that it can be. Some important everyday items have been improved many times. Everything under the sun can be made to work better than it does now.

(9) A strong desire to be admired and to be rich and famous. Many inventors are driven by a desire to strike it rich in a unique way that will atract the attention of the world. Inventors not only want to strike it rich, but they also want others to know that they were successful.

CIRCUMSTANTIAL ADVANTAGES

If an inventor has some or all of the personal attributes enumerated in the above paragraphs, his chances of success will be greatly advanced. However, there are often circumstances in the life of an individual that helps that person to take certain steps which lead to success. An inventor's chances of success are greatly enhanced by any of the following circumstances:

(1) The person's father, uncle, or grandfather, was a successful inventor. If not entirely successful, one of the above mentioned relatives owns two or three patents on inventions and is actively engaged in working on another invention.

(2) The inventor is associated by family ties or by friends with a successful manufacturing facility. Or may be associated by career or job category with a manufacturing facility in which inventions are known to be a factor in the success of the plant.

(3) The inventor has family ties to a patent attorney or patent draftsman, or someone who works with

patents in industry or government.

(4) The inventor is a technical writer or biographer on subjects in which inventors and patents are of prime importance. Working with inventors and inventions often leads a person to become a successful innovator by emulation.

SYSTEM FOR CREATING IDEAS

There is a method that can be applied by anyone who wants to become a professional inventor. The method is simply to work hard at being an inventor. Practice it systematically and diligently. Things that you can do right now to get started are listed below:

(1) Buy a large size note-book and start a habit of keeping track of bits of information and ideas. Write down the thoughts that occur to you about innovative gadgets or notes about scientific principles that are new to you. Make observations about things that could be improved.

(2) Start on a project. Pick something that is not easy. One of the best ways to develop your inventive talent is to work on something that really looks impossible at the start. You will get lots of ideas about how it might be accomplished. Work on it.

(3) Keep abreast of the latest developments in science and mechanics. Read and study to keep your mind alert and interested. Later, when you need a concept to advance a project, your mind will respond with the information you have filed away. Your memory will bring the needed facts forward into your conscious mind.

(4) Practice effective mental control. To do this you must get in tune with the latent capabilities of your brain and in particular, your subconscious mind. As detailed in later paragraphs of this chapter, use your mind as a tool for solving problems.

DIFFERENT KINDS OF INVENTORS

Many inventors have very individual styles of working and different methods of innovating ideas and originating inventions. Some like to work on inventions in the middle of the night when everything is still and silent. The lonely effort is supposed to be conducive to clear thinking. In the middle of the night you are not likely to be interuped by telephones or by visitors. Many inventors work at regular jobs during the week and play around with inventions on the week end. A basement workshop or a small room with a drawing table in it are excellent places to work on ideas.

Some inventors get into the business by happenstance. A friend who is presently working on a patent application said that he never in his life had a single thought that would cause him to think of being an inventor. Then one day this fantastic idea just evolved in his mind within a few minutes. It came out of the blue and his life was changed forever by that one incredible idea. He said that as soon as he realized the potential of the idea, he knew right then that he was going to be a millionaire inventor.

One type of inventor likes to work in a laboratory environment, preferably at a college or university. This type of inventor likes to test ideas with actual experiments in a highly controlled setting. Sometimes they only have a vague idea as to what will be accomplished by an experiment, so they are not surprised by the result. One experiment is quickly followed by another, sometimes with only minute changes, and thorough notes are kept on each experiment. Inventors of this type sometimes create a useful invention by accident, or by planned accident. Biographers of the great Edison say that he often discovered an invention while working on something else.

There is another type of professional inventor who likes to develop ideas in the mind. These people are good at visualizing what they are thinking about and also good at making drawings. When a certain stage of refinement has been reached by mental processes, then a model or experiment may be necessary. A professional inventor who makes his living from his patents claims that he has never made models of his ideas. He works out the details in his mind and then makes careful drawings. When he is satisfied with the drawings, he then proceeds with the patent application. It should be mentioned that he is an expert and has an engineering degree.

Another class of inventors, notably engineers, like to plan out the invention in meticulous detail. When he or she is satisfied with the design and the drawings, then a complete working model is made to exact specifications. The model is then tested for its basic functioning, and if it works, then it goes into endurance trials. Meanwhile he is conducting a marketability strategy. When everthing checks out, the patent application is then contemplated. Inventors who work for large corporations often work in this mode.

This author was present during the development of one of the well known automatic bowling pin setters. The working model was carefully engineered and built to exact specifications. When they got the pinsetter working, it had to undergo an endurance test of one thousand repeat performances without a failure. That took several weeks. Finally when the machine did one thousand settings without a failure, then it was set to operate for a thousand hours. No machine is absolutely perfect (even cars break down) but the pinsetter went through these rigorous tests before it was ever offered to a potential buyer.

REDUCING TO PRACTICE

When you get a good idea for an invention, the process of refining and improvement begins immediately. It would be a miracle if you happened to think of the very best way to manufacture your invention at the same moment that you conceived the idea. Improvement on the original idea is an automatic process. The first improvements you make are the obvious ones. Then gradually your invention gets closer and closer to the way you want it to look and operate. This process of improvement is called "reducing to practice".

STATE OF THE ART

Another term that is imbedded in the inventor's language is "state of the art". This term means the most recent developments in a certain line of invention. If you want to know about the "state of the art" of personal computers, you are asking what is the very latest innovative technology in that field. The term can also be applied to mean the most recent events or demonstrations of working processes in any field. Lets suppose that you get a sudden inspiration about how to build a new kind of helicopter. What you want to know is; are the modern helicopters already doing the thing that you plan to achieve with your helicopter? What you are asking is what is the state of the art of the latest helicopters.

THE DRAWING TABLE

A standard drafting table is 5 feet by 3 feet. It usually has a means for sloping it to a comfortable angle, or it can be set level for regular use as a table. If you have a drawing table, it gives you a place to leave notes and jottings of things that you are working on. Then when you want to do some work you don't have to spend a lot of time getting ready. You can just sit down on a high swivel stool and get right to work. Have a large pad of drawing paper on the drawing board (or typewriter paper is satisfactory). Then you always have a place where you can quickly sketch ideas or write down items you don't want to forget.

TRAINING THE SUBCONSCIOUS MIND

Many technical people, especially inventors, report interesting examples of how their brains apparently solved a problem while they were sleeping. Such problem solving ability is usually attributed to the subconscious. The reason is that it is known that the subconscious mind is the independent part of the brain. Those who have studied the phenomenon of creativity theorize that the subconscious mind is influenced over a period of time by induction. The mental process termed induction is that process whereby the subconscious gradually builds up a sense of altertness of even urgency about a set of circumstances in your daily life. In the same way that many dreams appear to be designed to allay fears or worries that a dreamer may have when awake, the subconscious appears to work on a level where by solving a problem, the individual is thereby relieved of the stress or worry of working on that particular problem.

When the chemist Kekule reported the discovery of the benzene molecule (1865) he said it was in the shape of a hexagon with a double bond on alternate sides. Of course he had never seen a benzene molecule, but he saw the shape of the benzene molecule in a dream. When he saw the shape of it in his dream, he knew instantly what it was because the word benzene was in his mind when he made the discovery. The next day he began the experiments that confirmed it.

Many other scientists and inventors have reported that the solution to problems came to them in a dream, or on waking up. There is a correlation between dreaming a solution and waking up to a solution. Both methods appear to have the objective of relieving the person of the stress caused by the problem.

What is known about the problem solving capability of the subconscious is that it is greatly stimulated by stress. If a problem is such that it has a life-and-death importance to the person with the problem, the subconscious mind is then alerted to deal with it. The more stressful the person feels about the problem, the more rapidly the induction process works to "program" the subconscious what the problem is about.

Training the subconscious to work on your inventions is not difficult if the invention means a great deal to you. If it seems like a life-and-death matter, the subconscious will co-operate. Ordinarily the subconscious is not too interested in your daily activities. It takes the special "stress" condition to get the subconscious working quickly. On the other hand, if you work on the same problem for a long time, the induction into the subconscious gradually gets through. Without the stress factor, it may takes weeks or months before the subconsciuous will respond. But with the stress factor introduced, the response occurs within hours, and sometimes immediately.

Another factor of importance is that the subconscious often gives answers in symbols or images which are like the pieces of a puzzle. Sometimes it co-operates and labels the images in such a way that you can't miss the interpretation. On the other hand, sometimes there is no solution at all although you have a gut feeling that the answer is there if you could understand it. The only recourse is to try again.

Apparently your subconscious brain is keyed to work on the problems that you encounter during your daily struggle to survive. That is to say, the process is a survival mechanism, and almost everyone has that ability. You probably have had the experience yourself of waking up the morning and finding that a problem you had been worrying about was solved while you slept.

Of course, it doesn't always function perfectly.

I once awoke in the middle of the night and realized that I had the most incredible invention in my mind. So as not to lose such a fantastic idea, I got out of bed and found a notepad and pencil. I made a brief note and went back to bed. When I woke up in the morning and looked at the note pad the words that were written on it were, "molecules imitate atoms---". I never did find out what it meant.

RECEPTIVE MIND

The process of training the subconscious does not work for everyone. You may have to try a different technique to get results, or you may not ever get a response. However, don't think that it is just a hocus-pocus kind of thing. It is not like ESP or telepathy. This process is well documented. This is a natural brain strategem that is part of you, like seeing or hearing. Almost every well known inventor has reported this kind of mental experience at one time or another.

CREATING A STRESS FACTOR

The subconscious seems to respond to your great desire, concern, worry, and sense of urgency about a problem. The response occurs to alleviate a stress factor in your life. Therefore, you can learn to program your subconscious to respond. Your subconscious apparently does not know the difference between a real stress factor and an imagined one. This means that if you are a good actor or actress you can create the conditions for the stress factor in your life artificially. It is simply a matter of convincing your subconscious mind that you have a real problem that it is causing you much alarm and stress. Accordingly, your subconscious will respond to your requests.

PROGRAMMING THE SUBCONSCIOUS MIND

The experts who write books and articles on these subjects claim that the subconscious is most receptive when you are calm and relaxed. It is most receptive when you are asleep. The process that works is to lie down and relax. You should do this preferably at your normal time to retire. Concentrate on relaxing your muscles one by one. Start at the toes and gradually work up the body. You will feel your muscles relax as your field of concentration moves from muscle group to muscle group. Think of being warm and comfortable and think about going to sleep. When you feel in a really relaxed condition nearing drowsiness, start to think about the problem you want to solve.

COMMUNICATING WITH THE SUBCONSCIOUS

The only input that the subconscious gets is from the filtered information arriving from your sense organs, mainly the eyes, ears, nose and sense of touch. All input arriving at the subconscious uses the identical channels. For this reason, a story that is acted out or only imagined in the mind gets the same credence and equal status as does real information. Experts in mind control are said to exercise their hearts and lungs by imaginging that they are jogging. The lungs expand and take more air and the heart beats faster in response to an imaginary jogging session.

Taking these kind of factors into consideration, you can communicate effectively with the subconscious by imagining that you are talking into a microphone. The wires from the microphone are seen to be in contact with your subconscious brain. It is not necessary to talk out loud. If you imagine that you are talking into the imaginary mike, the input to the subconscious will be exactly the same as if the situation was the real thing.

Don't be embarrassed about talking to yourself. All of the great inventors talked to themselves. Just use ordinary words the same as if you are talking to a person. Don't forget to say these important lines, "I am talking to you, my subconscious mind. The reason I am talking to you directly is because we always work together to survive. When I win, you win. We are a survival team." Start off each communication session with a monologue of this type and then convey the reason why you are calling. Don't forget to communicate a sense of urgency about your problem. Include the words frequently, "This is very important to both of us and it is really important to our survival."

After the initial contact, thereafter at least once a day try to communicate a sense of urgency to your subconscious about the problem you want to solve. Perhaps your problem has a real sense of urgency about it, like figuring out ways to make the mortgage payment. If that is the case, then you know what a sense of urgency is supposed to convey.

Don't be disappointed if your problem is not solved immediately. If you do not regularly dream and have some memory about your dreams, you will be at a disadvantage. When you go to bed, and when you feel that you are about ready to drop off to sleep, tell yourself (and your subconscious mind) that you would like to have an interesting dream and that you would like to be able to remember it. You will soon get some response.

If you find it impossible to remember your dreams, then try for a wakening response. Tell yourself (and your subconscious) that any possible solutions to your problem will be on your mind when you awake at your normal time. Again, don't expect immediate results. The problem has to have a real solution before your subconscious can solve it. To summarize, get on a one-on-one relationship with your subconscious mind and within a few days or a few weeks you will notice some interesting results.

A WORD OF CAUTION

Some very special attributes have been claimed

for the subconscious mind by sensationalists. The facts are that the subconscious was once the conscious waking brain of the humanoid from which human beings evolved. Out of the evolutionary past we retain vestiges of dangerous drives and urges. Human beings retain instincts which direct them towards territorial claims, including the killing of rivals for individual supremacy and mating rights, and the instinct to persevere against all antagonists. There are vestiges of unreasonable urges, including jealousy and rage. If the subconscious is permitted to subvert and control the thinking reasoning brain, mental illnesses and personality problems of many kinds can result.

This paragraph is included as a precaution to the impressionable. The belief that the subconscious has some kind of magic power can lead only to delusion. It is a dead-end. The subconscious is only a very useful tool that can be used to solve problems.

STICK-TO-IT-IVENESS

One other characteristic that an inventor must have is a desire to accomplish what he sets out to do. The desire is called stick-to-it-iveness. All successful inventors have it, but in fact it is a characteristic of most successful people. When asked how he could create such useful inventions, Edison told his admirer that invention is 90 percent perspiration and only 10 percent inspiration. One idea may be the culmination of hundreds of hours of work.

PERSISTENCE IS THE KEY TO SUCCESS

The persistence and tenacity that is needed to carry an invention through every phase to final success is exemplified by Edwin Land. He was 27 at the time when he first got the idea that a polarizing sheet would be valuable.

People get headaches from bright lights and from driving into the setting sun. Driving into the glare of oncoming headlights at night is also a hazard. When Land noticed how much the bright lights bothered him, the idea began to form in his mind. He knew that a polarizing material had the effect of eliminating glare, but he didn't know how or why. As he thought over the possibilities, he realized that a polarizing sheet material would have many scientific applications as well. It appeared to be a marketable product. It soon became his ambition to produce a sheet material that would polarize light.

A normal ray of light has components vibrating randomly in every plane. A ray of light that has been polarized has been changed such that all of the wave components in the light ray are vibrating in the same plane. When a normal ray of light is reflected from a surface, the reflected ray is partly polarized. Most glare is from reflected light so that a polarizing film is effective for reducing glare. That is why polaroid sun-glasses are popular.

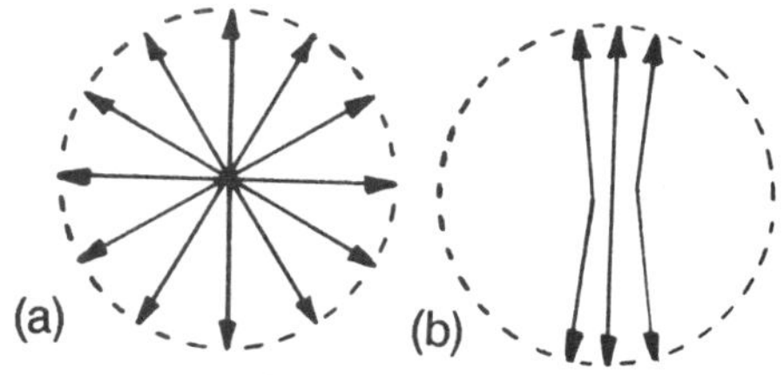

Figure 4: A normal light ray has wave components vibrating in every plane along the direction of motion. A cross section on the ray is shown at (a). When it is polarized, the wave components vibrate mainly in one plane, as shown at (b). A ray of light is partly polarized when it is reflected from any surface at a gentle angle.

Land began a study of polarized light and also the minerals that act as natural polarizers. Thin slices of tourmaline are natural light polarizers. A prism made out of calcite will select out polarized rays with a certain angular component.

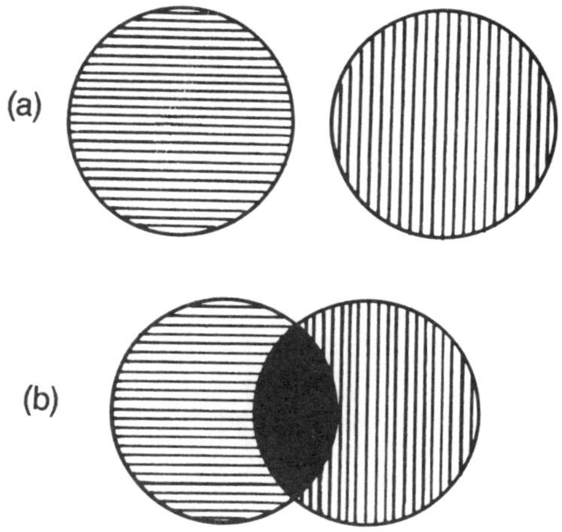

Figure 5: When a ray of light is passed through two polarizers which have been set at 90 degrees to each other, as at (a), the first lens polarizes the ray. When the ray passes into the second polarizing lens, it is completely extinguished, as shown at (b). When polarizing lenses are used as sun glasses, the polarizing attribute is set vertically. The glare from the sun bouncing off flat surfaces, such as a beach or table, is then reduced by the polarizers in the glasses.

Two years later, when he was thirty, Land had succeeded in making a sheet of transparent material that would polarize light rays. The one basic idea that was critical to the process of making a polarizer came to Land in a moment of intuition. He had been working with minerals that are natural polarizers. When the polarizing minerals are crushed up and finely ground, the individual crystals can be examined under a microscope. Land knew from his studies that the fine crystals of polarizing minerals were shaped like needles or toothpicks. When the tiny needle crystals were revealed under the microscope, their polarizing ability depended on their alignment. If they lay in a random way, there was no polarizing effect, but if they were lined up in the same direction, then light that

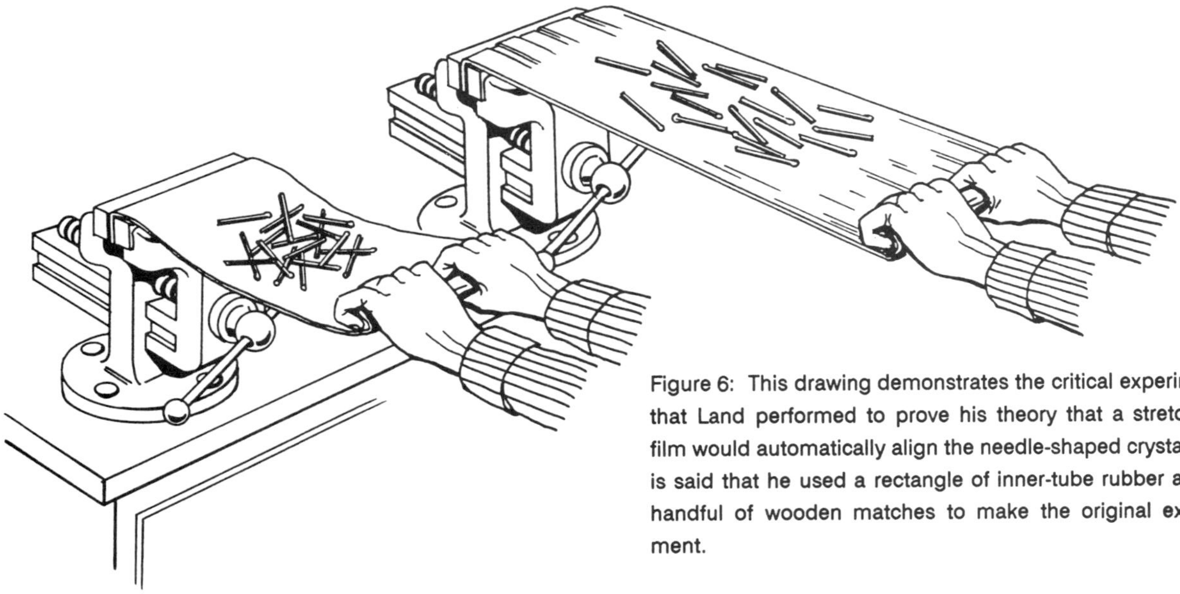

Figure 6: This drawing demonstrates the critical experiment that Land performed to prove his theory that a stretching film would automatically align the needle-shaped crystals. It is said that he used a rectangle of inner-tube rubber and a handful of wooden matches to make the original experiment.

passed through was polarized. The secret was in knowing how to make the microscopic particles line up. A magnetic field didn't do anything. An electric field didn't do anything. Centrifugal force wouldn't work either. But Land had figured out how to do it.

Land wanted experimental proof of his idea. It is reported that he went to the garbage bin of a tire shop and rescued an inner tube (the tubeless tire had not been invented). He took the inner tube to his work shop and cut out a large rectangular piece of rubber. He put one end of the rectangle in a vice. Then he emptied a box of wooden matches onto the piece of inner tube. Then he grabbed the rubber and stretched it out in one direction. As he expected, the wooden matches began to line themselves up in the direction of the stretch. The further he stretched it, the more matches lined up. That was the answer he had been looking for.

He went back to his lab and began to compound a clear plastic that could be stretched out into a sheet as it dried. When he had formulated the plastic, he floated the crushed mineral in it. As he had anticipated, the needle shaped crystals of the mineral behaved exactly like the wooden matches on the inner tube. On a microscopic level, the needles of the polarizing mineral lined up as the plastic was stretched. When the plastic was dry, polaroid sheet had been invented. It had been more than invented----the very first small sheet of it had been manufactured.

That is not the end of the story. Land had learned other attributes of polarizing materials during his years of effort. Now he wanted a material that would incorporate some of the important features of the ideal polarizing sheet. He began to formulate and test various compounds to see which ones exhibited the greatest degree of clarity and sharpness of detail desired by the scientific community. He found the answer to his quest in the crystals of herapathite, compounded from quinine and iodine.

When Land founded the "Polaroid" corporation in Cambridge, Mass., ten years had elapsed since the first idea about a polarizing sheet entered his mind. The "Polaroid" camera and the "Land Camera" were later inventions. It was not the camera that Land was noted for. Rather, it was polaroid lenses and filters, polaroid sunglasses, polaroid car windshields and a host of scientific inventions as well. All of them were based on the one idea of making a sheet material that would polarize a ray of light.

The lesson that Edwin Land teaches all of us is that persistance pays eventually. It often happens that the very first time you think of an invention the idea seems to be impractical. But after you have thought it over, you begin to see that it might not be impossible. The next time you think of the idea, you have made a bit of progress. Then you think it might be practical after all. That's when you start working on it with serious intent.

From the experience of Edwin Land, nothing is impossible until you have really studied it from every angle and thought about it for a long period of time. Can you imagine the response the very first time Land wondered aloud if it would be possible to make the film develop right inside the camera?

After many months of working on an invention and researching it and thinking about it, the knowledge that you gain and the character that you gain from the effort makes the invention a success even if you never make any money from it. The things that you learn from working on one invention usually lead indirectly to another invention, and then to a third. Once you get your mind on the inventive track, it becomes impossible not to invent things. Your brain gets into the habit of working on inventions and after that you just keep getting more ideas. At that stage you have become a professional.

RESEARCHING YOUR INVENTION

When you first conceive an idea, there are some moments of trepidation and doubt. You ask yourself, "What am I doing this for? I don't know anything about electricity or mechanics." At this moment the average person will tend to forget his good idea and let his thoughts return to the work-a-day world. He has enough problems just surviving and raising a family without having the added burden of trying to develop an invention in a field of which he (or she) knows little or nothing.

The inventor is not an average person. When he gets a good idea, he feels a sense of intense exhilaration. It is something akin to striking gold. When a really good idea strikes, there is an aura of excitement and the thrill of possible wealth. The inventor is gripped with the romance of discovery and a future that promises the kind of intense interest that makes life a song. Working on an invention changes the drabness of daily survival into a bright sequence of interesting events. Life becomes much more colorful and vital.

THE FIRST STEPS

When you decide that you are going to develop your good idea, the first step is to get a notebook and start a diary about your invention. If you went to the Library and took out a book on a subject you need to learn, make a note of the title and portion of the contents that you studied. If you search through your encyclopedia for details, make a note of the pages on which you found relevant material. What you now are doing is building up a field of knowledge relating to your invention. You are learning the "state of the art" into which your invention belongs. By making notes, you will have a reference handy when it is needed.

At first you might be lost for direction. How do you find out what you need to know? When Alexander Graham Bell was first thinking of a telephone, he had no technical experience. He wrote to his parents, "My experience in these matters are a great drawback." Bell was not thinking about the commercial possibilities of a telephone. His job was the teaching of deaf-mutes, and he wanted an instrument that would simplify his task. As he learned about using a magnet to make an iron string vibrate, Bell conceived a telephone in which every tone or frequency in the sound would have a separate sender and receiver. If that method had been used, the telephone would have been as big as a piano. It was a great step forward when Bell hit upon the idea of forced vibrations. Only one microphone was needed.

Bell consulted some of the great innovators of his time, notably Wheatstone and Helmholtz. He also got encouragement from Joseph Henry, famous for his electrical discoveries. He did not hesitate to discuss his idea with anyone who could guide him, because his aim was to help the deaf. The noble cause gave him the incentive to keep trying. Finally, he hired the assistance of Watson, a mechanic who could make the things that Bell visualized. Then in 1876, the words, "Mr. Watson, please come here," were transmitted from one room to another. The telephone worked.

KEEPING IT SECRET

Bell was fortunate in that his open quest for knowledge did not lead to disaster. Everyone knew that Bell was trying to make a telephone because Bell talked about it freely. Several other inventors were trying to patent the telephone at the same time that Bell applied for his patent. The result was litigation and the worry and concern over the outcome. Bell was lucky to win in the end.

Obviously, one good reason for keeping your invention secret is so that you will not be anticipated by somebody else. Good ideas are quickly picked up by those who recognize their worth. When xerox was invented, nobody wanted to buy the patent. But when xerox proved to be a good idea, everybody wanted to copy it. The inventor was forced to fight for his rights. Fortunately, he had strong patents.

The other reason for keeping your invention secret is to avoid the negative feedback from otherwise caring people. It is a fact that inventors as a class are poorly treated by society at large. This is no reason for not proceeding with your idea. The very first time I talked to a patent attorney, he told me politely to forget it. He told me, You need whatever resources you have to raise your family. You can't afford to get into this game where the odds for success are worse than a lottery. He felt sorry for me and adopted the role of father figure so that I might be saved the embarrassment of becoming an inventor.

There is another group of people who love to shoot down ideas, regardless of their value. They will give you lots of reasons to doubt that your idea will ever work. Or they tell you that they already saw it in the stores and you are too late. It is easy to get discouraged with friends like these helping you. The best thing to do is not discuss your idea with anyone. If it won't work, you will find out in the course of time. If it has already been invented, you will find out when you make your search.

EXISTING INVENTIONS

The patent office has records of thousands of inventions which have become public domain. These inventions are free for anyone to make and sell as they please. This great store house of ideas and invention represents mainly the dreams of inventors of bygone days. They had a great idea. They developed it and patented it, but nobody wanted it.

On the other hand, the records of all of the successful inventions are there as well. A patent has a lifetime of seventeen years, which is a rather short period considering that the invention has to be developed and marketed. The manufacturing plant has to be redesigned and retooled to produce the product. By the time the invention is actually earning money for the inventor and the manufacturer, several years might have elapsed. The remaining few years are hardly sufficient to permit the inventor to reap the rewards of his invention and to recoup the investment.

When the seventeen years has elapsed, the invention then belongs to everybody. Inventions that are producing good returns are then copied by every manufacturer in every country in the world. In 1928 when Claude's patents on the neon sign had run out, chains of neon plants sprung up all over United States and Canada. The neon sign enjoyed the greatest popularity in its history. It is still popular today, which shows that a good invention has staying power.

All of the records of the patent office have been arranged for the convenience of those who wish to use them. There are viewing rooms with comfortable equipment, similar to library reading rooms. Patent results are also published in the Canada Gazette and in the Official Gazette of the United States Patent Office. Patent information is available at most public libraries in Canada. In United States the Patent Office records can be viewed at the following Libraries:

Albany, N.Y.; University of the State of New York; Atlanta, Ga.; Georgia Tech Library; Boston, Ma.; Buffalo, N.Y.; Chicago, Il.; Cincinnati, Oh.; Cleveland, Oh.; Columbus, Oh.; Ohio State University; Detroit, Mi.; Kansas City, Mo.; Los Angeles, Ca.; Madison, Wi.; Milwaukee, Wi.; Minneapolis, Mn.; Newark, N.J.; New York City; Philadelphia, Pa.; Pittsburgh, Pa.; Providence, R.I.; St. Louis, Mo.; Toledo, Oh.

Publications pertaining to patents can be secured in Canada from the Department of Services and Supply, Ottawa, and in the United States there are several publications about patent matters supplied by the Superintendant of Documents, Washington, D.C.

CHANGING TECHNOLOGY

Technology is changing rapidly. Who would have believed a few years ago that the video cassette store would be in big thing in our times, or that there would be stores specializing only in home computers and accessories. So many things today were unheard of just a few years ago. This means that your idea has a good chance of being relevant if it is related to a new technology or a branch of the recording and entertainment field.

On the other hand, a good idea that is not developed and put out to market quickly may very well be too late for the changing technology of the day. For example, ordinary recordings are fast becoming obsolete as the new compact disc and specialized cassettes take over the industry. It takes so long to secure a patent that by the time the patent is granted, the world has already passed you by. You are ready to start thinking about marketing your product and the technical situation has made your product redundant or out-of-date.

To counteract these considerations, inventors must proceed with due diligence if their idea has a time-decay factor. The Canadian Patent Laws have been changed to take these factors into account. See the section on Patent Rules and Regulations in the following chapter.

HISTORIC PHONE CALL

Last week I received a phone call from an inventor. About 15 years ago (1975) he got the idea that it would be great to own a harmonica or mouth-organ that could be played in any key. Most mouth-organs play in one key only, and some can be played in two keys. But to play in 12 keys was asking too much. When he first got the idea, I thought to myself, "That's just too complicated!" It was one of those impossible dreams.

But he phoned me last week and told me to listen carefully. I could hear a harmonica playing in one key and then switching to another and another. Meanwhile the inventors voice was saying, "Key of B flat, now key of D----key of C-----do you want to hear some minor keys?"

After 15 years of working on it he had perfected the all-key harmonica. Now he had made a working model of it to prove to the patent attorney that it did work. Later he came and showed me the model. Except for the reeds, the whole unit was made out of clear plastic so that you could see the parts move when you changed key. It has 88 keys, like a piano. This inventor is my son, Douglas.

WAYS AND MEANS

If there is one near where you live, a Public Library is a good place to visit. In the Library will be found books about science and innovation. Books about machines and machinery are particularly interesting. Regular reading about the inventions and the inventors that you find interesting will build up your own mental store-house of ideas, facts and details about designing and manufacturing things. Ask the technician in the reference Library to help you find the books that you want to take home.

If you never plan to manufacture anything or even make anything during your lifetime, it is still necessary to know how things are made and to understand the problems of mass producing an invention. When you draw a specification for your idea, the manufacturing know-how shines through. It is the same with technical knowledge and engineering know-how. You do not have to be an engineer to be an inventor, but it is a great advantage if you have thoroughly read up on all of the engineering subjects that you find interesting. When that moment arrives when you must put pencil to paper and make a drawing of the invention that you are thinking about, all of the reading and know-how you have stored in your mind will act to guide your pencil along a profesional course.

Another important and interesting hobby is that of building up a library of books that an inventor needs. Start with the books that you find the most interesting and inspiring. A good place to look for books is at local garage sales. Never pass up a garage sale if they have books out at 25 cents or 50 cents each. Any book on science, chemistry, mathematics, physics, engineering, electricity, electronics, radio, television, mechanical drawing, innovation, how-to-do-it and many other topics are worth buying when the price is so reasonable. Magazines like Popular Science and Popular Mechanics are good sources of ideas. They have inspiring articles about innovations and inventions also.

One of the handiest reference tools available is an encyclopedia. Second hand sets of these books are often advertsied in the classified sections of your local newspapers. You can get like-new sets of books for one half or one third of the original price. An encyclopedia saves you a time-consuming trip to the library when you have to know some technical detail, or some pertinent thought that you cannot remember. An illustrated dictionary is also a useful reference tool.

THE MOST IMPORTANT ITEM

The most important thing that you can do with your spare time when regular chores are complete, is read. Read about inventors, read about science, read about anything technical that is interesting and that teaches you something new. Every time you (or anyone) read a book, you become a different person. The change may not be noticeable at once, but after you have read a few books, you will notice a difference. Your confidence increases. Your ability to converse increases. Your relationships with other people improve, and others will notice the difference, too.

EVENTS FROM THE PAST

The antique illustrations on this page and on several other pages in this book are reprinted from the HANDBOOK OF EARLY ADVERTISING ART by Clarence P. Hornung.

These illustrations were handmade by commercial artists. Some were prepared over a hundred years ago. The evolution of automobile design is evident in this group of vehicles.

PATENT OFFICE PRACTICES AND PROCEDURES

APPLYING FOR A PATENT

As requested by the United States Patent Office, all proceedings and communications with the patent office must be in a written form. In other words, you must sit down and write a letter to the patent office. All other types of communications will be ignored. And when you write to the patent office, all mail must be addressed to the Commissioner of Patents. This holds true even if you are writing to a specific person, such as a patent examiner.

When your patent application is received by the patent office, you will be sent an acknowledgement with a file number. The file number is important to you because it identifies you from many other inventors. After you have your file number, you can use it whenever you write to the patent office. When they receive your letter with a file number, your file can be opened immediately. Keep your own patent file up-to-date so that relevant information will be available if requested by the patent office.

The patent office also requests that whenever you write to the patent office, confine each letter to a single subject. If you are mailing in a fee, and you also want to order an abstract from a patent, separate individual letters should be sent.

COMMUNICATIONS BY OTHERS

If you have assigned your invention to a second party, then the assignee must conduct the correspondence with the Patent Office. Documents, such as amendments, requiring the signature of the inventor must also carry the signature of the assignee. The letter, however, will be mailed only to the inventor.

If you have appointed a Patent Attorney to conduct the affairs relating to your patent application, the correspondence must be confined to the attorney.

PATENT APPLICATIONS

In the United States Patent Rules, a patent is awarded to the first inventor of a particular invention. The problem with this system is that you can never be sure that at some later date an unknown inventor might step forward with proof that he was the original inventor, which is to say, he was the first inventor. Many of the Countries which belong to the International Patent Co-operative have the same system.

This system was also used in Canada until October of l989, at which time the method was changed so the patent would be granted to the first person who made the effort to patent his (or her) invention. At this moment, the process of establishing a date for your inventions is still valid. By making a full disclosure and sending it to yourself by registered mail, a date is thereby established that will give you a prior claim over any later invention of the same kind. On the other hand, such a disclosure would be of no particular advantage for a Canadian Patent. If you plan to take out a patent on an invention in Canada, the wisest policy is to move as quickly as is practical.

It would be foolhardy to rush your patent application before you have a chance to properly study the basic nature of your invention. It often happens that an inventor will mail his patent application to the Patent Office, and within a few days wish that he had it back again so that he could make some changes that seem to be important.

RESPONSE TO QUESTIONS

If you write to the patent office requesting certain specific information, by return mail you will receive only a copy of the Rules of Practice of the United States Patent Office. Every aspect of patent procedure is carefully detailed in this condensed handbook of the Rules of Practice. Then it is up to you to seek carefully through the handbook and find the answer to your specific question. If the answer to your question is not there, it means that you have asked a question to which there is no answer or which is contrary to the Rules of Practice.

PATENT ATTORNEYS AND AGENTS

The Patent Office makes an effort to advise inventors that the application for a patent must be prepared by a skilled person in order for the resulting patent to have any legal value. A patent that is not well executed will not stand up in court and therefore, may be worthless in cases of infringement. Infringement is when someone copies and manufactures your invention.

The powers of attorneys and the authorization of agents can be revoked at any time by the Commissioner of Patents if he believes that such person is not acting in the best interests of the inventor. In such rare cases, the Patent Office will then communicate directly with the applicant (the inventor or his assignee).

WHO CAN APPLY FOR A PATENT

Anyone who has invented something new and useful can apply for patent protection. The invention can be a machine or a chemical composition, or a new method of manufacturing a useful product. The invention can be a new material or it can be a new variety of plant other than tubers.

If the nature of the invention has been previously published or patented in a foreign country more than one year before the date of your application, the patent will not be allowed. This means that if you write about your invention in a publication, you must apply for a patent within a year. Otherwise the invention will be considered as public domain. The term public domain means that anyone can make, use or sell the invention without question.

If an inventor is killed accidently or dies from illness during the processing of his application, the executors of his estate may act to complete the patent proceedings. If an inventor becomes insane during the processing of his invention, the legal guardian or legal trustee of the inventor may proceed to complete the patent matters.

If an inventor really believes that he is the first one to make a certain invention, he will be granted a patent even if the invention was published or known about in a foreign country previously. However, if the invention had been patented previously in a foreign country, the new patent application would not be allowed.

APPLICATION FOR A PATENT

An application for a United States patent must be directed exclusively to the COMMISIONER OF PATENTS. Material arriving at the patent office which ignores this stipulation will be returned to the sender. The application must have the inventor's signature, or if the inventor cannot sign his or her name for any reason, the authorized representative may sign for the inventor.

An application comprises a Petition, a Deposition or Oath, a Specification including Claims, the Drawings, and the Fees. There is a fee for filing the application and a nominal fee for each claim. Each document or paper included in the application must be legibly printed or written in ink, and it must be in the English language. An application will not receive a filing number until all parts of the application have been received. The applicant for a patent will be advised when his application has been filed and assigned a number. The patent office will prefer if all parts of an application arrive in the same package, but if for any reason one or more parts of the application have to be revised or completed, there is a six month period during which all of the parts of the application must be completed and be ready for examination. If, after the six month period, all of the application is not received, the patent application will be considered to be abandoned.

THE PETITION

The petition consists of the full and correct names and postal address of the inventor. The title chosen for the invention is indicated and cannot be changed afterwards. The petition also includes the signature of the inventor. An example of the Petition is shown.

THE OATH

In the oath that the inventor must take, he must swear that he is the first inventor or discoverer of the matters in the application, to the full extent of his knowledge. He must state his citizenship and swear that he is the sole inventor to the best of his knowledge. The oath must be signed before a notary or any person authorized to administer depositions. An example Oath is illustrated.

THE SPECIFICATION

A complete description in detail of your invention is the main subject matter of the specification. The invention must be descibed in such form that any person skilled in the art can undertand it and put it into practice. Skilled in the art means a person who is qualified in the field to which your invention pertains, and who would be able to make or construct the invention from the information you have provided.

The specification must emphasize the exact principle on which the invention operates and describe how it is different or better than present day devices used for the same purpose. If the invention is merely an improvement on an existing invention, the exact structure of the improvement must be shown, as well as showing the old way that it was done. The inventor must be careful not to use language that might imply that parts of the invention are his own innovations when in fact they were previously existing as part of the old invention. A typical specification is shown.

THE CLAIMS

The claims are a series of numbered paragraphs describing exactly what your invention comprises. The claims emphasize structure and physical relationships. The first claim is as specific as is possi-

SPECIFICATION

My invention relates to improvements in meat-chopping machines in which vertically reciprocating knives operate in conjunction with a rotating chopping block; and the objects of my improvement are, first, to provide a continuously lubricated bearing for the block; second, to afford facilities for the proper adjustment of the knives independently of each other in respect to the face of the block; and third, to reduce the friction of the reciprocating rod which carries the knives.

I attain these objects by the mechanism illustrated in the accompanying drawing, in which-----

Figure 1 is a vertical section of the entire machine; Fig. 2, a plan view of the machine as it appears after the removal of the chopping block and the knives; Fig. 3, a vertical section of a part of the machine on the line 3 3, Fig. 2; and Fig. 4, a detailed view in perspective of the reciprocating crosshead and its knives.

Similar numerals refer to similar parts throughout the several views.

The table or plate 1, its legs or standards 2 2, and the hanger 3, secured to the underside of the table, constitute the framework of the machine. In the hanger 3 turns the shaft 4, carrying a flywheel 5, to the hub of which is attached a crank 6, and a crank-pin 7, connected by link 8, to a pin passing through the ctrosshead 9, and to the latter is secured a rod 10, having at its upper end a crosshead 11, carrying the adjustable chopping knives 12 12, referred to hereinafter.

The crosshead 9, reciprocated by the shaft 4, is provided with antifriction rollers 13 13, adapted to guides 14 14, secured to the underside of the table 1, so that the reciprocation of this crosshead may be accompanied with as little friction as possible.

To the underside of a wooden chopping block 15 is secured an annular rib 16, adapted to and bearing in an annular groove 17 in the table 1. (See Figs. 1 and 2). This annular groove or channel 17 is not of the same depth throughout, but communicates at one or more points (two in the present instance) with pockets or receptacles 18, 18, wider than the groove and containing supplies of oil, in contact with which the rib 16 rotates, so that the continuous lubrication of the groove and rib are assured. The rod 10 passses through and is guided by a central stand 19, secured to the table 1, and projecting through a central openingin the chopping block without being in contact therewith, the upper portion of the said stand being contained within a cover 20, which is secured to the block, and which prevents particles of meat from escaping through the central opening of the same.

The crosshead 11, previously referred to, and shown in perspective in Fig.4, is vertically adjustable on the rod 10, and can be retained by a set-screw 21, the upper end of the rod being threaded for the reception of nuts 22, which

Figure 7: The Specification, The Drawing, and The Claims, are illustrated on this and on the following pages. This particular example, the meat-chopping machine, is used by the United States Patent Office in their official patent guides. The patent drawings (next page) are excellent in the original size, with a border line 8 inches by 13 inches. The least amount of shading used gives a clearer view of the invention.

ble to make it. The following claims are progressively more general and the final claims may include variations as long as they remain within the scope of the actual invention. Some example claims are shown. The wording of the claims is very important. The ability of your patent to withstand a challenge in court is what makes it valuable. Therefore the wording of the claims is a critical factor. Note that the wording used has the objective of stripping away all detail except the actual structure of the invention and the relationship between the parts.

THE DRAWINGS

The objective of the patent drawings is to illustrate the invention clearly and fully. Anything that is mentioned in the claims must also appear in the drawings. If the invention is an improvement on an existing invention, the new part must be emphasized. A minimum amount of the old invention is shown. The objective is to show the old invention in relation to your new improvement.

The drawings must be black ink on pure white paper or bristol board. The black ink can be applied with drawing pens. Felt-tip pens can be used if more convenient. For the final copies a xerox machine or photolithographic printing can be used. The printing process often used is called "offset" printing for short.

The method that I employ is to use a drawing pencil and make the different views individually on any smooth white paper. I make the views one and a half size or even double size if the detail is fine. When I have done the pencil drawing for one view and I am satisfied with it, I lay a piece of tracing vellum over the drawing and tape the vellum in place. Then I use a new fine- point felt pen to go over the lines of the drawing. I ignore small mistakes as they can be corrected afterwards.

When one of the views is completely inked, I make corrections on it using a fine pointed camel hair brush. The correction fluid I use is graphic-arts white correction paint made especially for commercial artsists. The items you need for making the drawings are available at any art store.

The next step is to decide the final size that the view will be on the patent drawing. I then take the view to the xerox printer in my neighborhood and get it reduced to the final size. I follow the same procedure with all of the views needed. Then I take a piece

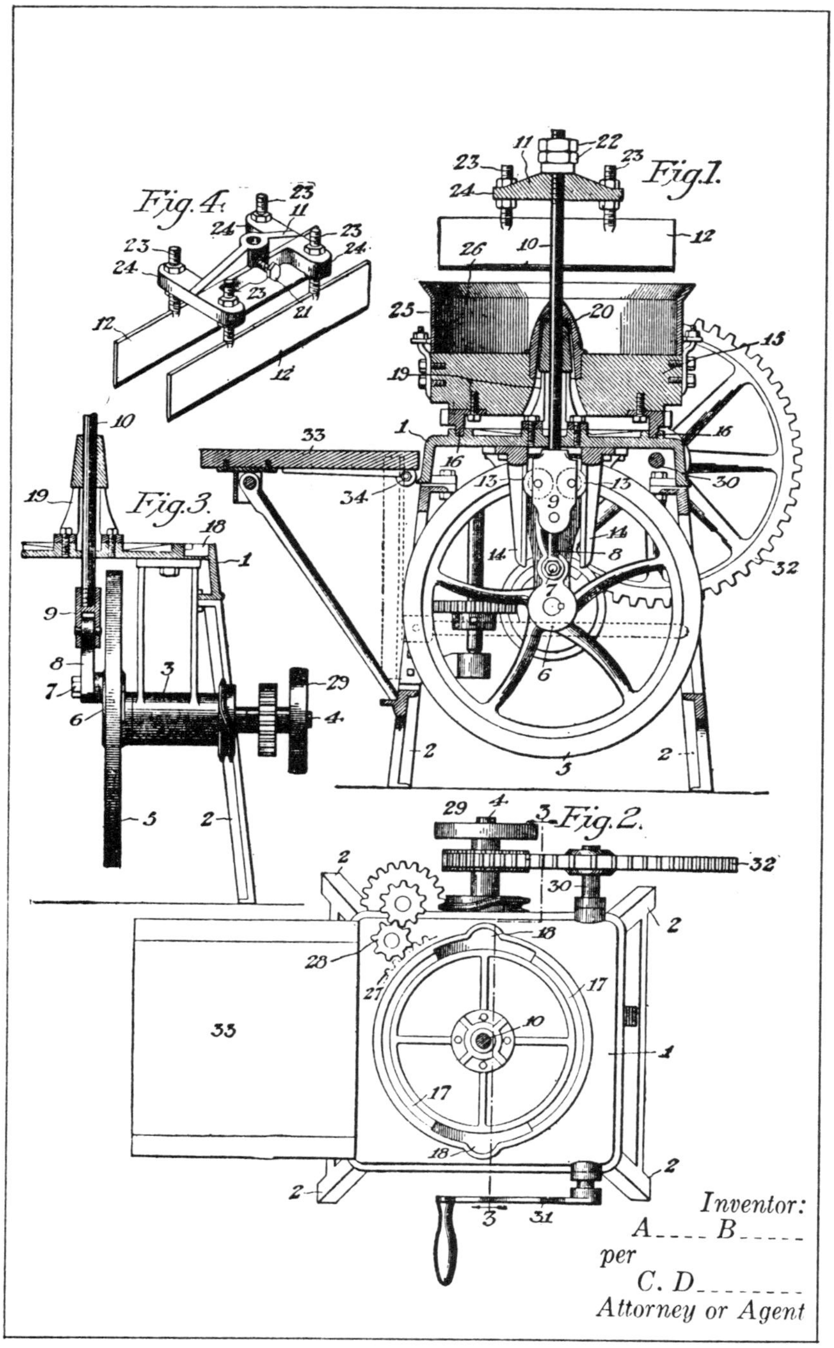

Figure 8: This example of patent drawings has been reduced in size for publication. Therefore, the lines used for shading and sections are too close together. These views are expertly done. Less shading would be an advantage in clarity.

of white drawing paper 15 inches by 10 inches and use the fine felt pen to carefully draw a border line making a rectangle exactly 8 inches by I3 inches. Within this space I arrange the views. If they won't all fit in nicely I make a second sheet for the extra views. The views are taped in position with clear tape, or fastened on the back using a glue-stick or rubber-cement for paper.

I use letraset (or equivalent stick-on letters and numerals) to stick on the letters "Figure 1", etc. You can also use stick-ons for the little numerals used to indicate the parts of the drawings. When the patent drawing is completed, with the different views fastened in place with clear tape or glue-stick, I then return to the Xerox machine and have a couple of good black copies made on strong white paper.

At this point there are two things that you can do. If you are in a hurry you can use these drawings in your patent application. The Patent Office will use them to make whatever drawings are necessary for the patent. The other route is to take your copy to an "offset" printer and have a few copies of each drawing made. I usually buy some bristol board and carefully cut it to the size for the patent drawings. I take the sized bristol board to the printer to use for the drawings. If you ask the printer to supply the board, he will usually come up with box board or something that is not white enough because of the poor quality.

The size of the bristol board or paper is 15 inches by 10 inches. I usually cut it half an inch to an inch larger and then trim it to the proper size after the printing is dry. In this way you can exactly center the drawing on your card board. The 10 inch width is the top, but a sheet can be laid sideways if necessary to properly illustrate the invention. A border line is drawn one inch from the edge, leaving a working space for the drawings of 13 inches by 8 inches. Different views in the drawings are consecutively marked; that is, Figure One, Figure Two, Etc. The parts of the drawing are numbered and a fine line is used to connect the number to the part so numbered. The same number must be used for the same part throughout.

Shading on cross sections of parts of your invention must conform to the code set forth in the patent rules. There are different types of shading to indicate metal, glass or porcelain, insulating material, fibrous material, etc.

The drawings must be clear rather than artistic or fancy. Professional drawings are an important part of a patent application and the inventor is advised to hire the services of a competent person. If the drawings supplied with the application are not considered suitable, the patent office will have the drawings made for you at cost.

resist the shocks imparted to the crosshead when the knives are brought into violent contact with the meat or the chopping-block.

The knives 12, 12 are adjustable independently of each other and of the said crosshead, so that the co-incidence of the cutting-edge of each knife with the face of the chopping-block may always be assured.

I prefer to carry out this feature of my invention in the manner shown in Fig. 4, where it will be seen that the two screw-rods 23, 23 rise vertically from the back of each knife and pass through lugs 24, 24 on the crosshead, each rod being furnished with two nuts, one above the other below the lug through which it passes. The most accurate adjustment of the knives can be effected by the manipulation of these nuts.

A circular casing 25 is secured to the chopping-block, so as to form on the same a trough 26 for keeping the meat within proper bounds; and on the edge of the annular rib 16, secured to the bottom of the block, are teeth 27, for receiving those of a pinion 28, which may be driven by the shaft 4 through the medium of any suitable system of gearing, that shown in the drawing forming no part of my present invention.

This shaft 4 may be driven by a belt passing round the pulleys 29, or it may be driven by hand from a shaft 30, furnished at one end with a handle 31, and at the other with a cog-wheel 32, gearing into a pinion on the said shaft 4.

A platform 33 may be hinged, as at 34, to one edge of the table 1, to suppoort a vessel in which the chopped meat can be deposited. The means by which it may be supported are shown in full lines, and the most convenient method of disposing of it when not in use is shown in dotted lines, in Fig.1.

I am aware that prior to my invention meat-chopping machines have been made with vertically reciprocating knives operating in conjunction with rotating chopping-blocks. Therefore I do not claim such a combination broadly; but

I claim:

1. The combination, in a meat-chopping machine, of a rotary chopping-block having an annular rib, with a table having an annular recess to receive said rib, and a pocket communicating with the said recess, all substantially as set forth.

2. In a meat-chopping machine, the combination of a rotary chopping-block with a reciprocating crosshead carrying knives, each of which is vertically adjustable on the said crosshead independently of the other, substantially as described.

3. A chopping knife having two screw rods projecting perpendicularly from its back and parallel with the sides of the knife.

4. A meat-chopping machine provided with a rod carrying chopping knives and adapted to be reciprocated, a crosshead secured to said rod, anti-friction rollers mounted on the crosshead, and guides with which the rollers co-operate, substantially as described.

Figure 9: This page shows the remainder of the specification for the example meat-chopping machine and the claims. The claims are the crux of a patent if a legal challenge occurs.

MODELS

There is no need for a model. In very special cases the Commissioner of Patents may request that a model should be supplied. If a model is made, it cannot be larger than 12 inches on any one side.

PATENT FORMAT

While the size of the drawings must be exact, the rules for paper size for the specification and other documents are more relaxed. Any legal size paper is accepted. The size can be from 14 inches by 8 and a half inches to 13 inches by 8 inches. A typewriter or other type of printer is acceptable if the type is black and clean. Use one side of the paper only. Leave margins of a suitable width, about 1 and a quarter inches on the left side of the page and about 1 inch on the right side. Top and bottom margins should be at least an inch. If the specification and claims occupy more than one page, number the pages at the bottom in the center of the page.

CANADIAN VARIATIONS

The foregoing applies to United States patent applications. There are some differences when applying for a Canadian patent. The new rules instituted in Canada in October 1989 are important in that the patent is now granted to the first person who applies for an invention. It used to be that the first inventor owned the patent if he stepped forward to claim it, and could prove that he was the first inventor. This system is still in force in most countries.

THE DRAWINGS

The size of the patent drawings for a Canadian patent application is different, in that the overall size of the bristol board is 13 inches by 8 inches. Note that this is the size of the border line in the United States method, which means that the same views with the same arrangement can be used. A space of 1 and a quarter inches must be left at the top of the drawings for a title to be added. This is the same as the drawings for a United States patent application.

THE SPECIFICATION

The other papers included with a Canadian application must be on legal size paper 13 inches by 8 inches. The typewritten sheets must be double spaced, and the claims can be triple spaced. A wider margin is left at the top of each page for a title to be added.

The Canadian specification comprises 4 sections, dealing with the following matters:

(1) The first paragraph starts with the words, "This invention relates to........" and in the paragraph the type of invention is described. See the example.

SPECIFICATION

This invention relates to a manually operable tool for driving posts into the ground.

It is common in devices for driving piles and posts to pull up a weight or hammer, e.g. by a cable and overhead pulley arrangement, and drop it onto the end of the pile or post. It is, of course, necessary that the hammer strike the pile or post squarely, and it has been proposed to provide the hammer with a depending guide which freely embraces and may slide up and down on the post to be driven. Tools of this type are, however, inefficient because the rebound of the hammer results in a loss of energy and a tendency to split the end of the post. They are, moreover, unsatisfactory for manual operation, because the hammer must be heavy to be effective, and the power of the operator is used only in raising the heavy hammer.

I have found that these disadvantages may be overcome by providing a number of handles secured to the guide and extending lengthwise along the outside of it. Such handles permit the use of a lighter hammer and the elimination of the overhead arrangement, secure a greater effect for a given amount of energy, and reduce the splitting of the post, since the power of the operator of the device is used not only to raise the hammer but also to bring it down on the post and hold it against rebound.

In drawings which illustrate embodiments of the invention,
Figure 1 is an elevation partly in section of one embodiment,
Figure 2 is a top plan view of this embodiment,
Figure 3 is a section on the line III----III of Figure 1, and
Figure 4 is a plan view of another embodiment having only two handles.

The tool illustrated comprises a guide 1 which is adapted freely to embrace and slide up and down on a post A which is to be driven. It may be of any suitable cross section, but in the form shown, is a cylinder open at the bottom and closed by a plug 2 at the top which may be the top of the device. The plug 2, which acts as a hammer, fits within the cylinder 1 and flanged at its edge so as to lie flush with the outer wall of the cylinder. Extending lengthwise of the guide 1 are handles 3 which may be formed from metal tubes, as shown, or may, if desired, be made from rods or bars covered with wood facings.

Figure 10: A typical specification for a Canadian patent application is this example that is used in the Patent Office literature as a guide for inventors. The drawings are shown on the next page.

(2) The second paragraph deals with the nature of the existing inventions of that class and explains why they do not function as well as should be expected of a useful machine or process.

(3) The third paragraph has for its subject matter a description of the way in which your invention overcomes the drawbacks of the class of invention, and in the way that it is a more useful machine or process.

(4) The fourth paragraph is a full description of the best way that you know of making your invention work and putting it into operation. This paragraph starts with the drawings and continues on to a full description of the invention, as shown in the example.

(5) The fifth paragraph is for the claims. The claims are separately numbered and may be referred to each other by number.

THE ABSTRACT

At the top of a separate sheet the Canadian application must include an abstract from the patent. The abstract is an exact description of the invention in the fewest possible words. In the abstract you are allowed up to 250 words or less. An example of an abstract is shown.

TIME LIMIT

The length of time it takes to get a patent is detrimental to the purpose of the Patent System. The technological changes occur so rapidly these days that an invention might be out-of-date by the time the patent is granted. The main objective of the patent system is to get good ideas out into the open and into the commercial main stream. Although the patent holder has the right to exclude others from making or using his invention, he cannot stop others from studying his invention and improving on it. This is one of the objectives of the patent system.

The new rules in Canada are set to release the details of the invention to the public within a shorter period of time. Once a patent is pending, and it is known that the inventor will get his patent in due time, the patent application will be made available for public inspection within 18 months of its Canadian filing date or within 18 months of its convention priority date. This gives the applicant up to a year to apply for a patent in a foreign country and obtain the same filing date as his original date in his own country.

MAINTENANCE FEES

In the new Canadian rules an inventor with a patent must pay a yearly fee to keep his patent in good standing. This means that if the maintenance

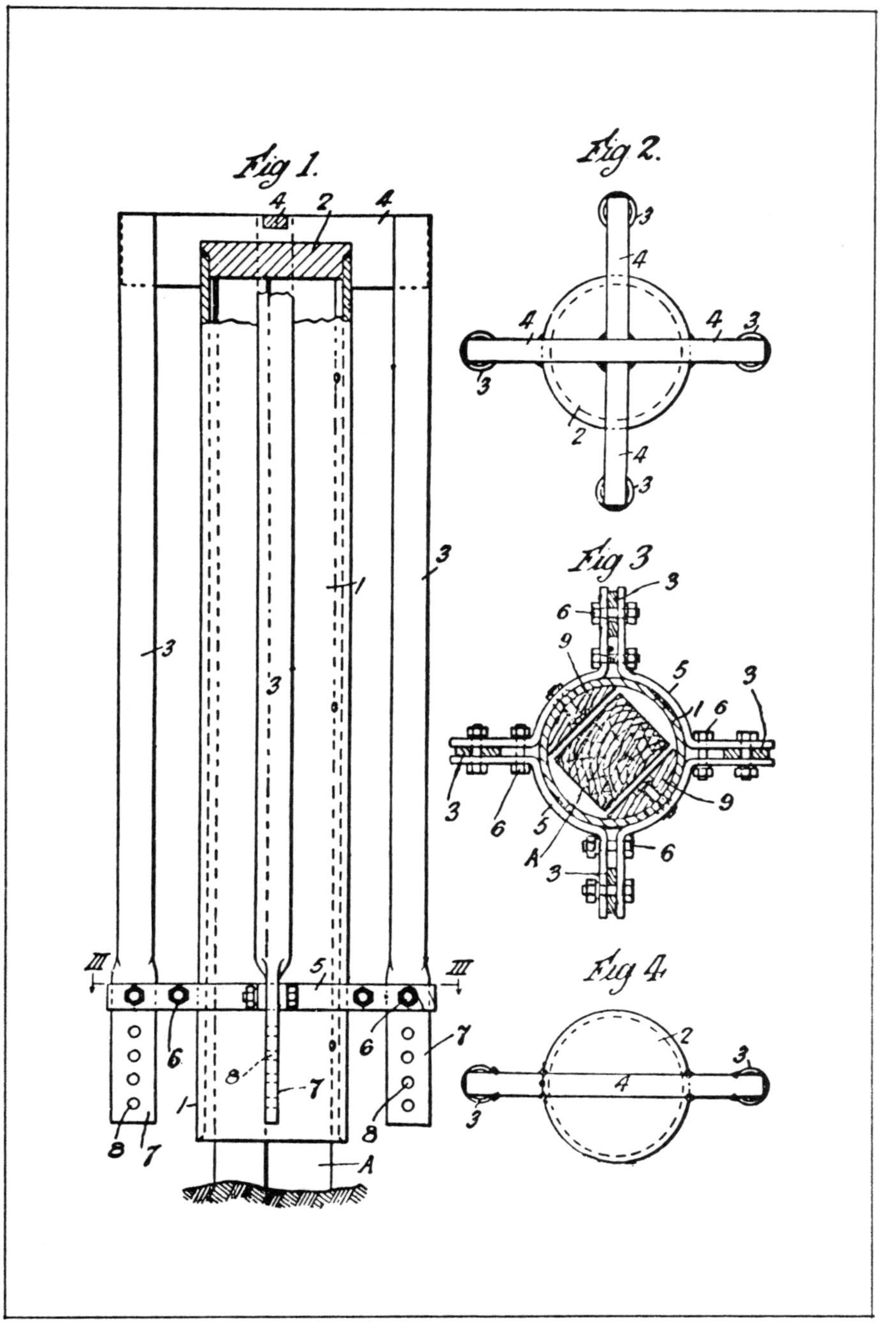

Figure 11: The patent drawings used by the patent office to illustrate the type of drawings preferred. Very little in the way of shading is used.

DOING SOMETHING ABOUT YOUR PATENT

After you have obtained your patent, you have to decide the best direction to take. Your objective is to convert a legal document, your patent, which has only a potential value, into hard cash. There are five ways that an inventor can go into the commercial world.

(1) You can advertise and sell your invention outright for cash. The main disadvantage of doing this is that you have not had the opportunity to prove the real worth of your invention. If someone wants to pay you cash for it, then it obviously must have some visible worth and saleable attribute. Assuming that you have found an interested buyer, the probability is that you will receive only a small fraction of what your invention is really worth.

(2) You can sell a part of your invention and retain a share interest. The advantage of this strategem is that the sale of a portion of the invention will give you some cash in pocket. However, if the invention becomes a big item in the markets of the nation, you will also share in its success. If you have retained a ten percent share of the invention, then you will receive a ten percent share of the actual profits from the invention.

fees are not paid, the patent will become public property. Again, the purpose of this system is to prevent good ideas from being wasted by abandonment or lack of interest.

NO MORE CAVEATS

For a small sum of money (it used to be 10 dollars) a Canadian inventor used to be able to send a disclosure to the Patent Office and secure a date of disclosure. Since the date of disclosure no longer has a legal standing in the Canadian Patent Office, the system of Caveats has been discontinued.

PATENT MARKINGS

Since November 1987, it is not necessary to stamp products with a Canadian patent number or to mark them Patented.

(3) You can license your invention to a manufacturer for a royalty. In this case you have no control over what is done with the invention after you have signed it away. Unless you have stipulated a time frame for the marketing of the invention, the licensee can take whatever steps that appear to be in his interests. If the invention is marketable, the licensee will most likely get it to market with all due diligence.

(4) Manufacture your invention yourself. If you have the courage and the energy, you can go it alone. You can start small in a garage or basement room. If you want to go on a larger scale, you can incorporate and sell shares. You can raise capital by selling outright small parcels of shares to friends or businessmen who have money to invest. I once advertised an invention share for sale and was soon ap-

The handles 3 are secured at their upper ends to bridge pieces 4, e.g. by welding, and the bridge pieces 4 are secured as by welding to the plug 2. At their lower ends the handles 3 are flattened for engagement between two arms of a sectional clamping ring 5 fitting around the guide 1 and clamped to it by bolts 6. The lower ends of the handles are extended below the clamping ring, as indicated at 7, for the attachment of extension members (not shown) and, for this purpose, bolt holes 8 are provided in the extensions 7.

In order to adapt a guide of circular internal cross section to a square post, segmental filling pieces 9 having their flat faces facing inwards may be secured inside it, the distance between opposed flat faces being slightly greater than the thickness of the post. Two filling pieces may be used as shown in Figure 3, but four may be used if desired.

In the embodiment shown in Figure 4 there are only two lateral extending handles instead of four as in Figure 1--3, but otherwise the construction may be the same as that described above.

CLAIMS

The embodiments of the invention in which an exclusive property or privelege is claimed are defined as follows:

1. A manually operable tool for driving posts into the ground, comprising a hammer, a depending guide adapted freely to embrace and slide up and down on the post to be driven, and handles extending lengthwise outside of the guide and rigidly secured thereto.

2. A tool as defined in claim 1, in which the guide is a cylinder closed at the top by the hammer.

3. A tool as defined in claim 1 or claim 2 , in which the guide has filling pieces secured to it in order to adapt its internal cross section to the cross section of the post to be driven into the ground.

ABSTRACT

In a tool for driving posts, it is known to have a guide depending from the hammer to freely embrace the post and slide longitudinally on it. In this invention, handles are secured to the guide such that they extend lengthwise along the outside of it. The tool with the handles may have a lighter hammer and thus be manually operated since the handles enable the operator to use his own strength to bring the hammer down on the post and hold it against rebound. The guide may have filling pieces secured to the inside to adapt its cross section to the cross section of the post being driven.

Figure 12: The Claims and the Abstract should each be on a fresh page of the patent application. They are shown here overlapped in order to save space. The top part is the remainder of the specification for the post driving tool.

proached by a businessman. He had a check for $5000 in his hand and when we first met, he said casually, this check is for you. But if you accept it I want a 50 percent share in your invention. I had to turn it down because I was certain that my invention was worth a million (I would have accepted $50,000 for a half share).

(5) The fifth thing that you can do about your invention is abandon it. This is what happens to most inventions. At the beginning of the abandonment, the invention gets placed on a shelf temporarily. The inventor usually intends to do something with it when he gets some money or time. Time goes by. What money does come through the inventor's hands has to go for other things, like shoes for the baby. Fifty or sixty inventions that I worked on over the years, only a very few ever became patents, and fewer yet were developed into a marketable product.

The other cause of abandoning an invention is something that happens to everyone. Other interests, especially those that are actually earning you money, take over your time and energy. You don't intend to abandon the invention, but neglect over a long period of time amounts to the same thing. When I was a young man, quite a few years ago, I was working on 3-D illustrations for books and magazines. The hologram had not yet been invented, but the idea that I had would have produced a similar result. I worked on that invention for several years and made some progress, but the effect was always marginal. If an invention doesn't work in the way that you want it, you have to keep trying. Meanwhile I started a sign shop and that took up my time. As the sign shop became a money maker, the inventions I had been working on became abandoned.

Now, a historic moment occurred this week. In the mail I received a postage stamp in 3-D. It is beautiful----a scene of space with the space-lab floating on by and the earth far, far below. The moon is in the background sky and right in the foreground there is a spaceman with a jet-pack. The fact is that the hologram and the laser made my idea a practical reality. Now it is a good feeling to know that what you were working on in the past was not impossible after all.

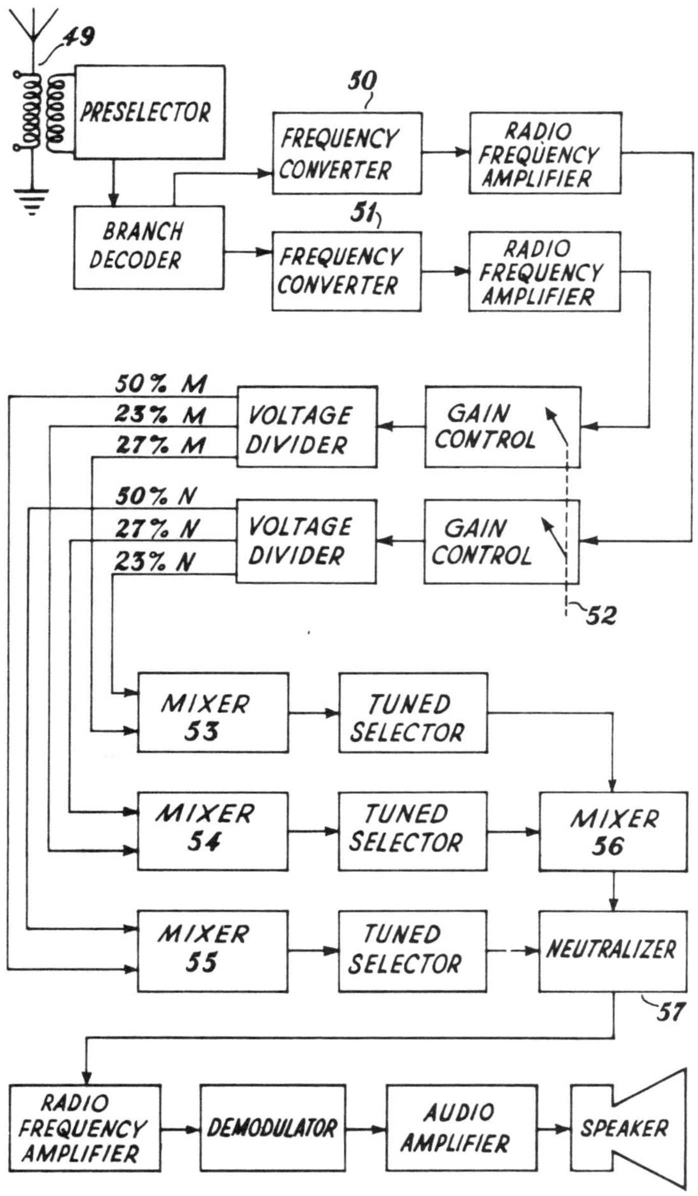

Fig. 15.

FIGURE 13: This example of a patent drawing shows the use of boxes with the words imprinted on the box to indicate what the box is. This is an acceptable strategem in patent drawings as long as the item is commonly known. The box marked neutralizer would be explained in the specification, and its exact function detailed, because a neutralizer is not a commonly used device in radio circuits. This drawing is from a method of selecting signals out of a common carrier frequency when the signals have become intermixed.

THE EXAMINATION

When your patent application has been completed and has been accepted by the Patent Office, it is given a priority number and sent to the examiners division for the examination. It will not be examined immediately, but will be taken in order when its date of examination has arrived.

The examiner will study the patent application to determine (a) if it meets the legal requirements of the Patent Office, and (b) to determine if it is an invention when compared to all other patents within the "state of the art" or in the class of the invention.

NOTIFICATION

The applicant for the patent is then notified in writing. This is called an "action". This action by the examiner will advise the inventor in detail if there is a rejection of any kind. The exact reasons for the rejection are given. Some claims may be allowed, but not all. Again, the reasons for rejection are stated.

The inventor must respond and pay the required fees within six months. During that interval the patent application can be amended to conform to the examiners opinion. On the other hand, if the inventor thinks that the examiner was wrong, an appeal can be made for reconsideration. In either case, the onus is on the inventor to point out the exact nature of the changes (if an amendment) or the exact nature of the claim to priority (if an appeal). References must be given, such as the titles and dates of publications cited, the patent numbers and claim numbers of claims cited, the names of pertinent inventions and the inventors. If the class of invention cited is different that the class of the present invention, the class must also be stated.

When the applicant's response is received by the Patent Office, the application will be re-examined. If the applicant has not responded in full, the application will be rejected again. To respond fully, the inventor must point out in detail the examiner's supposed error, and must respond to every ground of objection or rejection. To simply state that the examiner has made a mistake will not be accepted as reasonable grounds for a re-examination. During this process, the inventor is afforded every consideration as these matters are conducted on a level of polite decorum.

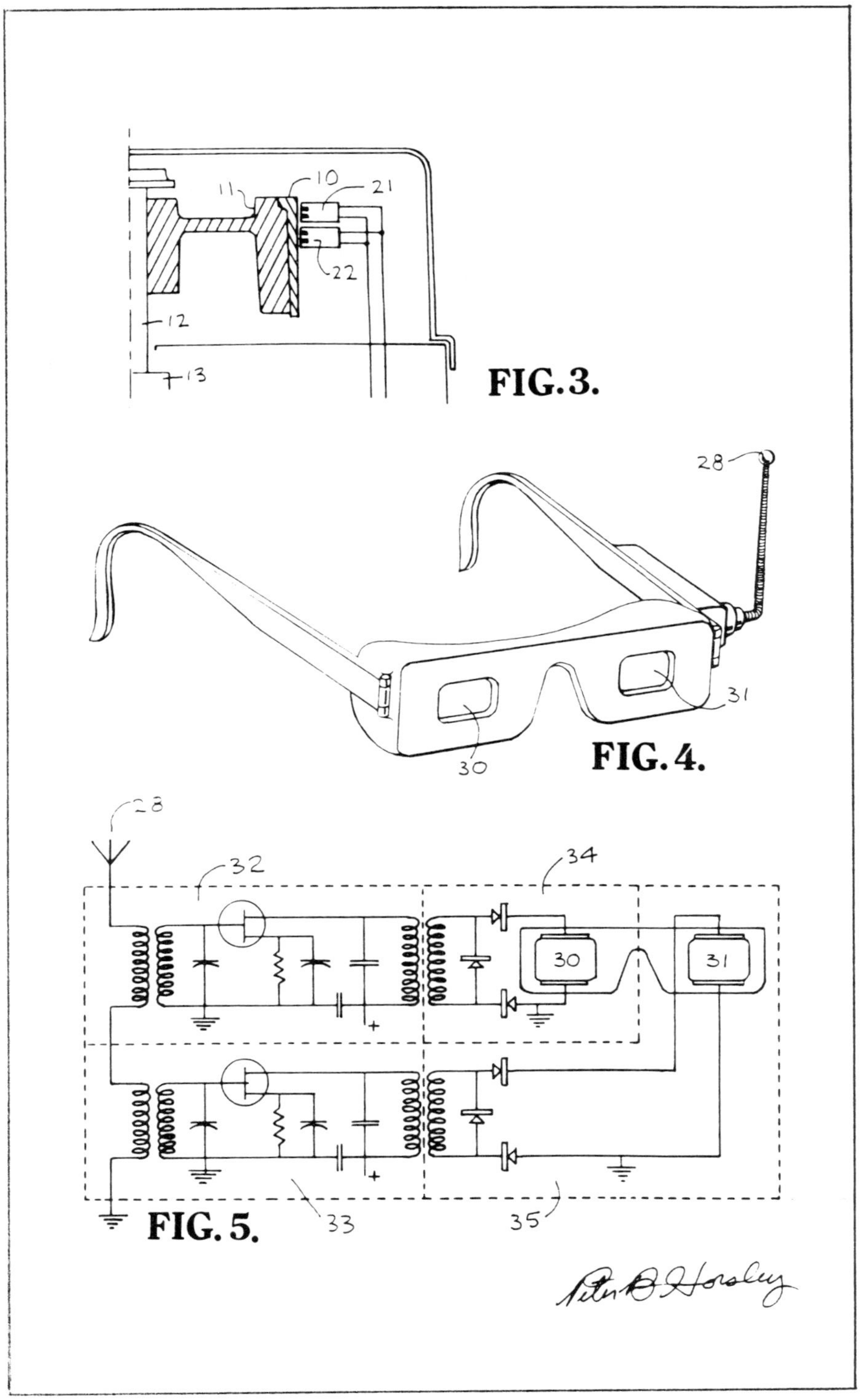

Figure 14: Example of a patent drawing. Note the lack of shading lines in the perspective view of the spectacles.

APPEALS TO THE BOARD OF APPEALS

If the application is again rejected by the examiner, and if the inventor is secure in his knowledge and is fully convinced that the examiner is in error, the inventor can make an appeal to the Board of Appeals. The appeals must follow a set format in order to comply with the legal requirements of the Patent Office, and they must be signed by the inventor. The set fee as required for such an action must accompany the documents.

ACCEPTABLE AMENDMENT

During the process of amending a patent application, the inventor has the task of circumventing the examiner's objections. It is important to study the objections carefully, because if the amendment changes the nature of the invention, then it is considered to be a new invention.

When amending the patent application, the drawings, the specification and the claims all must be changed to comply with the amended portion. The changes must reasonably circumvent the examiner's objections. However, if the actual invention becomes changed in the process of amendment, then the inventor must start from scratch and prepare a new patent application. This means that your application will go back to the bottom of the list where priority is a factor.

TIME LIMIT EXTENDED

In the event that the inventor is unable to get his amended application back to the Patent Office within the six month period stipulated, the application will be assumed to be abandoned and no longer pending. However, if the inventor can show that the delay was unavoidable because of reasonable circumstances, the application can be reinstated by the Commissioner of Patents.

ISSUE OF ALLOWED PATENT

After examination, if the patent application is found to be allowable, the inventor is sent a notice in writing that his patent application has been approved and that his patent is now pending. The inventor now has six months during which time he must respond and pay the stipulated fee. When the payment has been received by the Patent Office, a copy of the patent is then delivered by mail to the inventor (or his agent).

On the same day that the patent is delivered, copies of the patent can be purchased by anyone and the patent is placed in the records which are open to public inspection.

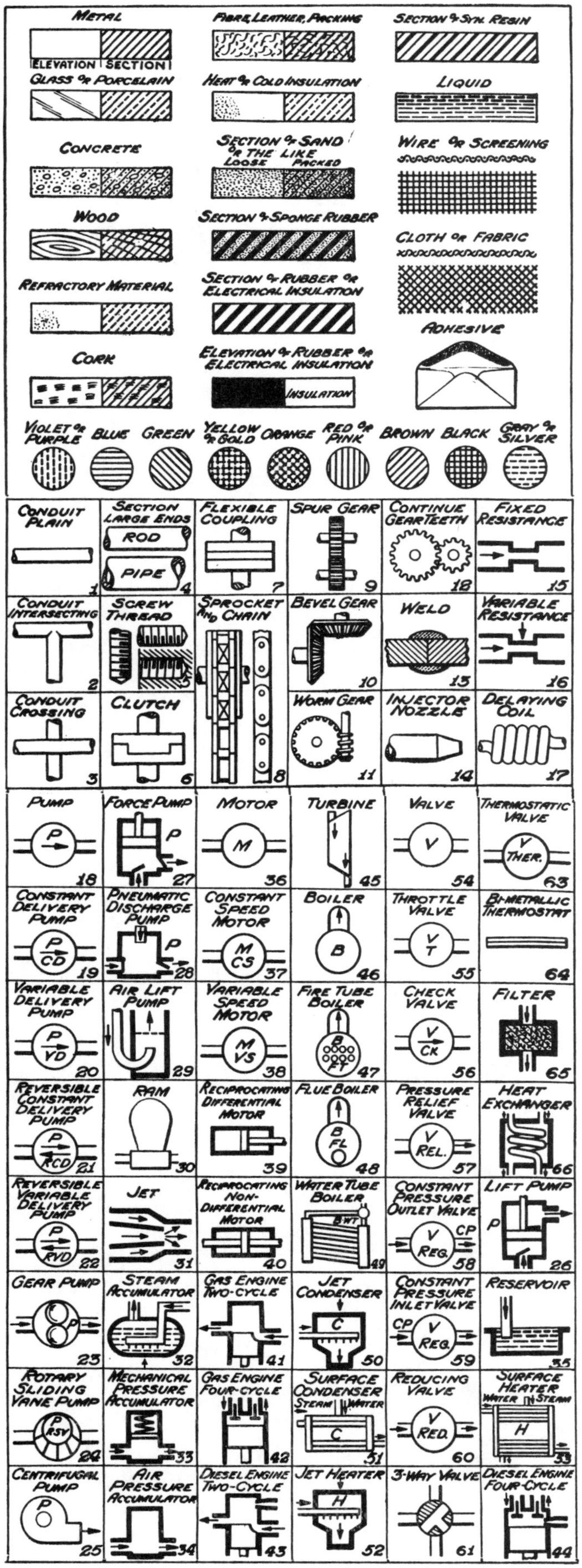

Figure 15: Symbols and substitute diagrams that are acceptable to the Patent Office on patent drawings. These conventional symbols are often much easier to draw and insert into patent drawings that the real thing. There are more conventional symbols continued onto the next page.

AMENDMENTS TO RULES

As time passes the rules and regulations governing patents and patent office procedure are changed to suit the changing technical world. This is true in United States and Canada, as well as other countries of the Patent Convention. It is therefore important for the inventor to be aware of the latest publications dealing with patents and patent procedure.

The Patent Record, an official publication of the Canadian Patent Office, contains the most recent information about patents that have been granted. Other publications are available that detail and explain all of the rules and regulations. This is important if you plan to assign a portion of your invention to a financial supporter, because the procedure set down is specific and should be referred to.

A good general knowledge of the operation of the Patent Office, as well as the publications issued, gives any inventor an advantage. If you have not already done so, send off some letters to the appropriate departments and start to build up your inventor's file.

THE PATENT AS A PROPERTY

A patent can be sold, leased, divided, assigned, partly assigned, etc., the same as a piece of property. It can belong to the sole inventor or to his heirs, or to anyone he may designate in his will. Whatever instrument (document) is used to change the legal status of the patent, the same must be recorded by the Patent Office within 3 months from the date. Any document relating to a patent must give the serial number, the title, the date of filing, and the name of the inventor. Of course, the particulars of the status of the patent must also be set forth, such as an assignment or an assignment of part interest. There is a fee that must be paid when the document is received by the Patent Office for the records. If an assignment is recorded prior to the granting of a patent, the assignee will be granted the patent. If an assignment of part interest is recorded prior to the granting of a patent, both the inventor and the assignee will be named as co-owners of the patent.

COPYRIGHT

Many inventors have a mistaken belief that if they copyright the information about their invention, that the invention will somehow then be protected. A copyright is intended specifically for written material. So the effect of a copyright would be to prevent anyone from publishing information about your invention. This would be no hindrance to anyone who had the intention of using or making your invention for profit.

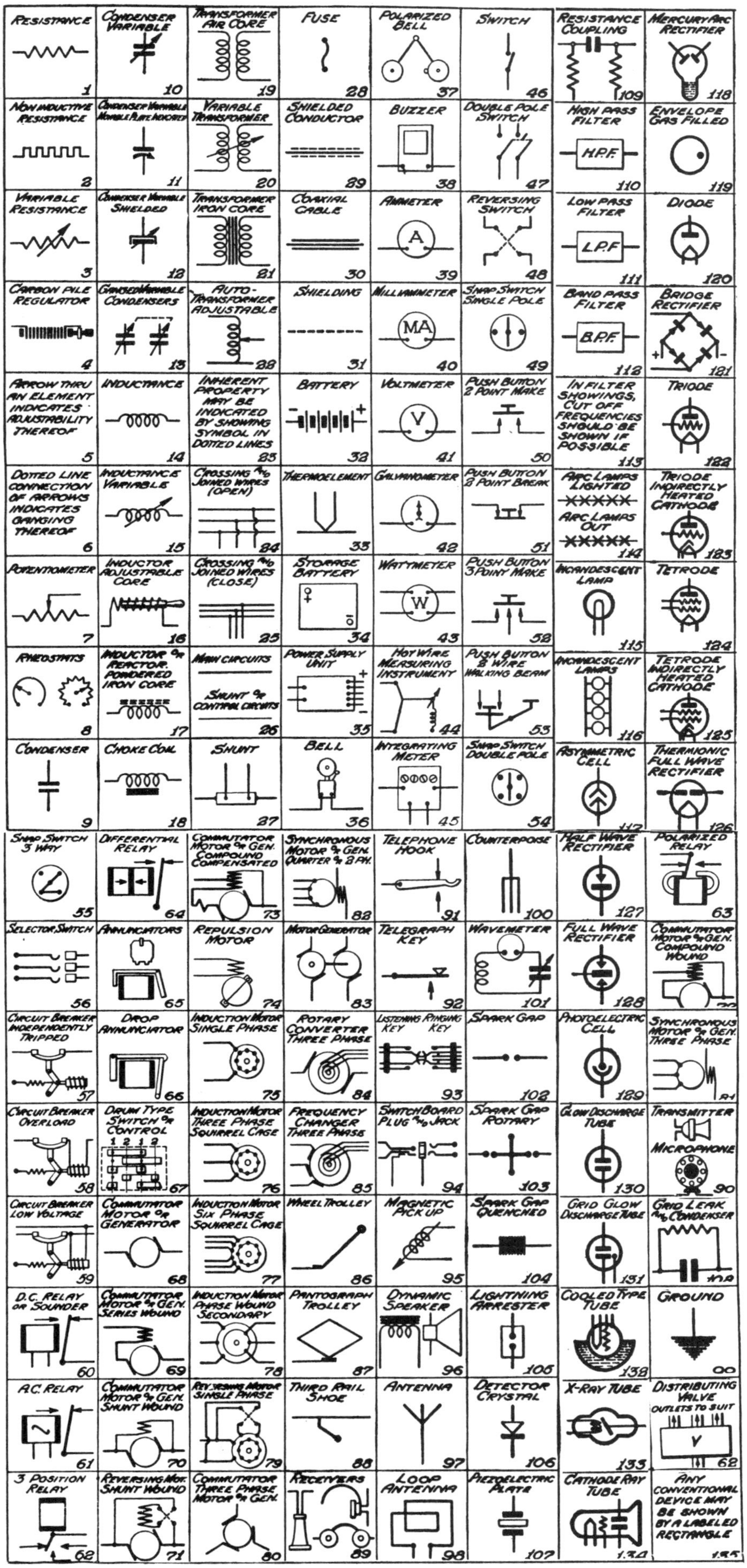

Figure 16: Conventional symbols that are acceptable on patent drawings to represent common devices and electrical components.

EXAMPLES OF CLAIMS

The examples of claims for a device are shown on page 24. The following examples show the form that the claims should take for a process or for a static (non-mechanical) invention. These examples are the ones used by the Canadian Patent Office in their advisory literature as a guide to inventors.

In the case of a process:

1. A process for cleaning the surface of a metal, which comprises converting contaminating matter by chemical attack to a residual film which is readily removable by anodic treatment, and removing the formed film by connecting the metal as an anode in an electrolytic system.

2. A process as defined in claim 1, in which the metal to be cleaned is iron or steel and the chemical attack consists of treatment of the metal surface with a strongly oxidizing acid.

3. A process as claimed in claim 2, in which the residual film is removed in an electrolyte comprising one or more acids or salts thereof.

Example of the claims for an article:

1. An insulated electric conductor comprising a metal sheath, at least one conducting core and, between the core and the sheath, highly compacted mineral insulation constituted by a mixture of two or more pulverulent mineral insulating materials at least one of which will, on exposure to the atmosphere, cause the formation over the exposed area of a skin or layer which is substantially impervious to moisture.

2. An insulated electric conductor as defined in claim 1, in which the insulating materials are calcium oxide and magnesia.

3. An insulated electric conductor as claimed in claim 2, in which the proportion of calcium oxide in the mixture is between 25 per cent and 40 per cent.

PETITION (Canadian wording)

The Petition of (full name of inventor) whose full post office address is (address of inventor),

SHEWETH:

That your Petitioner made the invention entitled (title of invention) which is described and claimed in the specification submitted herewith.

That your Petitioner verily believes that (he or she) is entitled to a patent for the said invention having regard to the provisions of the Patent Act.

That your Petitioner verily believes that (he or she) is entitled to pay a filing fee as a small entity as defined in section 2 of the Patent Rules.

Your Petitioner therefore prays that a patent may be granted to (himor her) for the said invention.

Signed at (City or Town), Canada, on this (number) day of (month), (year).

________Signature of Inventor

________Witness if desired.

PETITION (US wording)

To the Commissioner of Patents:

Your Petitioner, a citizen of (Country) and a resident of (City of Town), in the county of (County), in the State of (State), whose post office address is (full address of inventor), prays that letters patent may be granted to (him or her) for the improvement in (title of invention), set forth in the annexed specification.

Signed at (City or Town), in the county of (County), in the State of (State), this (number) day of (month), (year).

________Signature of Inventor

________Witness if desired.

OATH (US patent)

State of (State),

County of (County),

(Full name of inventor), being sworn, deposes and says that (he or she) is a citizen of United States of America and a resident of (full postal address of inventor), that (he or she) verily believes (himself or herself) to be the original, first and sole inventor of the improvement in (title of invention) described and claimed in the annexed specification; that (he or she) does not know and does not believe that the same was ever known or used before (his or her) invention and discovery thereof, or patented or described in any printed publication in any country before (his or her) invention or discovery thereof, or more than one year prior to this application, or in public use or on sale in United States for more than one year prior to this application, that the said invention has not been patented in any country foreign to the United States on an application filed by (her or him) or (his or her) legal representatives or assigns more than twelve months prior to this application; and that no application for patent on said improvement has been filed by (her or him) or by (his or her) legal representatives or assigns in any country foreign to the United States, except as follows (if any).

Inventors full name and Signature.

Sworn and subscribed before me this (number) day of (month), (year).

Seal and Signature of person administering the oath.

Figure 17: At the left examples of the Petition are shown. The petition is slightly different in Canada. The oath to accompany a United States patent application is also shown. In each case, the Petition or Oath should be on a new sheet of legal size paper, printed or typewritten on one side only.

4. An insulated electric conductor as defined in claim 1, 2, or 3, in which the limit insulation resistance, as herein defined, per free end of the insulation on exposure to the atmosphere, is not less than 250,000 ohms for an insulation thickness of 1.5 millimeters.

The following examples are claims for a composition. This example is from the Rules of Practice in the United States Patent Office, as a guide to inventors:

Claims for a composition:

1. A plastic composition adapted to form a light weight, fireproof, and waterproof lagging for steam heated surfaces, comprising a vegetable fibrous material, a mineral filler in powdered form and a mineral cementing substance.

2. A plastic composition adapted to form a lagging for steam pipes and the like comprising forty to sixty pounds of paper-pulp, forty-five to fifty-five pounds of powdered soapstone, and two to four quarts of a 30 degree Baumé to 35 degree Baumé solution of soluble glass.

3. A plastic composition consisting of a vegetable fibrous material, a powdered mineral filler, a mineral fibrous material and a mineral cementing substance substantially as described.

4. A plastic composition consisting of fifty pounds of paper-pulp, fifty pounds of powdered soapstone, twenty-five pounds of asbestos fiber and three quarts of a 33 degree Baumé solution of soluble glass.

TERMINOLOGY OF THE CLAIMS

(SEE PAGE 178)

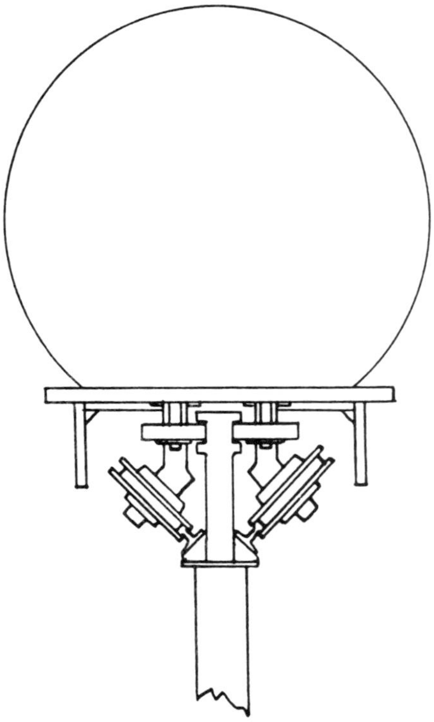

INVENTIONS FROM THE PAST

This artist's realistic rendition of a commuter train appeared in the news in 1886. The train is driven by an enclosed steam engine. A feature of the invention was that it took up very little space and could be mounted on individual pylons, which was an innovation in itself. It was designed by Joseph Meigs of Lowell, Mass. It was tested on a one mile track, but it was decided that electricity was more efficient and cleaner.

FINANCING AND MANUFACTURING YOUR PRODUCT

FINANCING YOUR INVENTION

Listed on page 23 are the five things that an inventor can do to proceed with his invention. These are:

(1) Sell the invention outright for a lump sum payment.

(2) Sell a share of your invention and retain a share for yourself.

(3) License your invention to a manufacturer and get paid by royalty.

(4) Manufacture and market your invention yourself, retaining control.

(5) Abandon your invention due to lack of funds or lack of interest.

VISIBLE VALUE AND INTRINSIC WORTH

If your invention is important, it has the potential to earn a lot of money for you or for any interested party who may want to invest in it. If your invention is clearly valuable, it is not difficult to find someone who will back it financially. It is only necessary to show drawings or a model to potential investors to judge the visible value of your idea. If it gets an enthusiastic response, you know it will not be difficult to finance.

SELECTING A PARTNER

On one occasion I was offered financial help by a business man who was a complete stranger to me. He was very enthusiastic and his wife was pleasant and enthusiastic also. It seemed like a good match. He was willing to advance money as I needed it for models, drawings, materials, or whatever expense that came up with regard to the invention. However, I would be expected to attend to my own living expenses, which meant that I would have to keep working at a regular job and work on the invention in my spare time. With this arrangement he expected a 50 percent share, or half the invention.

I was anxious to get going on a model, so I was very interested in the proposal. I asked the gentleman for references and he reacted cooly. So I decided to investigate him as thoroughly as possible before I committed myself to a deal. By doing some detective work I located his business and I introduced myself to some of his empoyees. I asked his foreman if he would recommend the deal that had been proposed. The man refused to say anything. However, later two of the employees phoned me at home and remained anonymous, but told me for godsake stay away from that man, that he was a real crook.

Perhaps the arrangement would have worked out fine. I would have insisted that everything should be in writing. However, my potential partner must have found out about my investigation because he lost interest and backed off.

Later I accepted a similar deal with another gentleman except that for 50 percent of the invention he was willing to pay my living expenses as well as pay for parts and materials. That partnership was very successful and we still retain a working friendship that has lasted for many years.

RETAINING CONTROL

Learning from experience, it is important for the inventor's interests and mental health to retain control over your invention if possible. There is no reason why you must release 50 per cent to someone else if it is obvious that the invention is going to be a big money maker. There are classes of people with money to invest who will be willing to buy into your invention for a small percentage.

Doctors and dentists are very good prospects. Sometimes a group of doctors who are known to each other will each buy a small share. They enjoy doing things together and it also gives them the feeling of confidence in numbers. If one of them makes it big, all of them will share in the success. If one loses, all of them will lose together.

You should try to retain 55 percent or preferably 60 percent of your invention. This will give you an edge and a strong position to guide the invention along the channels that will be to your advantage.

The big winners in the game of inventions are those who kept control over their patent rights. They directed the developement and manufacture of their inventions, built plants and factories, set-up distribution and marketing systems, and stayed in the driver's seat. Of course it takes a special kind of person to take control and make things hum. But often it is as simple as hiring the right people and choosing the right partners. If you are in charge of a good group of people and the aim of the whole group is along the same path, the chances of success are enormously improved. Persistence and energy of a group are self-sustaining whereas an individual might flag with fatigue.

VENTURE CAPITAL

Banks and regular channels of financing are always tightly closed to any kind of a deal where there is an element of risk. Most lending institutions must, by law, have secure equity to cover loans. As the saying goes, you can't borrow money unless you can show that you don't need it.

There are companies whose sole purpose of existence is to find worthwhile "deals" and finance them. They are willing to take a calculated risk based on the information that you can provide about your idea and your plans.

In order to catch their interest, you must have a worthwhile idea and a sound business plan. You must be able to show a realistic expectation of sales and profits based on a sampling of the market place. And you must have good people around you....the kind of people that will instill confidence in an investor.

They will want to see a complete plan for your product right down to the last detail. Your plan must be based on factual information and not just what you hope will work out. If they are satisfied, they will be willing to advance sufficient money to follow your plan through to success. The money may not be offered as a lump sum, but in installments as needed for the success of the company.

It is in the interests of the venture capitalist to make sure that you succeed. For this reason they may have one or more of their own people working right along with you to make sure that you have enough funding and to make sure that you are following the plan as agreed to.

SELLING SHARES

When it becomes apparent that your company is going to be a big success, but that a lot of funding is needed, a public stock offering may be suggested. Your venture capitalist is thoroughly experienced in this process and will guide you into it at the critical and proper time.

Selling shares to the public is a closely regulated business in which government agencies and government regulations are stringently applied. However, at this stage your personal involvement will be mainly as overseer. If a corporation has not already been formed, now is the time that it will be accomplished. The actual process will be directed by people who know exactly what to do to get the needed cash rolling in.

DEVELOPMENT COSTS

The development of an invention pertains to everything that needs to be done to get the invention ready for market. In the initial stage, making and testing models is the main activity. The costs of this process can run from a couple of thousand dollars up to a hundred thousand, depending on the complexity of the invention and the degree of sophistication of the tests.

If the invention is an operating machine performing a mechanical function in the processing of material, it must be tested for a long enough period of time to reveal weaknesses or periodic failures. When the invention fails to function, then it must be improved and new models or parts made. This step by step trial-and-error method can be very costly. It should be possible to have financing for development by releasing from 20 to 45 percent of the invention to a friend with money to invest.

A young inventor who asked for my advice told me that a man who was known to be a wealthy investor wanted to put up some money to help him develop his invention. He said the man was known to his family and had a good reputation. I told him to go ahead and accept the financing that it was a good idea. But the young fellow was still worried about the situation. He asked me, "How will I ever pay him back if the invention fails?" Then I realized he was afraid that he would never be able to repay his generous friend.

He was surprised to learn that when someone puts up money for a share in an invention, then his risk is of the same quality as that of the inventor. The inventor risks his time and energy to make the invention a success. It takes time and energy to earn money. So they both have equal value.

TESTING SALES POTENTIAL

When you have reached the stage where you know that your invention works, you now want to find out if anyone will buy it. Before you spend thousands of dollars on factory space and machinery to produce your invention in quantity, it is a great advantage if you have (a) an order book already filled with orders, or (b) a brief-case full of letters from people who want to know how much your invention costs and where they can buy it.

Mail order response is an important indicator of the sales potential of an idea because your ad will reach a complete cross section of the population. You can find out what the response will be to your product by placing one or two ads in selected advertising media. The first thing you will discover is that advertising is extremely expensive. In your mind you might be thinking that a two-inch ad in Popular Mechanics magazine may cost a hundred dollars. The actual costs are from 1500 to 2000 dollars depending on its position in the magazine.

The important factor is that it is a small investment to find out the things that you must know. Will people respond? Will your idea sell? You need these answers and it costs money to get answers.

Some magazines, Popular Mechanics among them, will not place an ad unless you send them a sample of the product for examination. For this purpose you can use one of your best models along with some good drawings of the device in use, or a written explanation how it works. However, you must be ready to accept orders that you may get, even if they are placed on a waitlng list. People who order goods must be advised how long it will take before the product is available. You must refund money to people who send money and then place them on your waiting list.

From the number of responses that you get over a period of two months, you can tell if it will be a viable exercise to proceed with the manufacture of the invention. Keep in mind that a mail order response of the type we have been talking about is only one facet of a wide range of ways to promote and sell your product.

You may have decided that your product will sell best in department stores and chain stores. In order to test this market it is necessary to produce a batch of your product, say ten thousand units. To get the kind of money that it takes for this step, you will need the backing of people with money. You may have to relinquish some partnerships in your company to raise capital.

You may also want to condsider selling shares to the public at this stage of your operation. At this time you might decide to seek the advice and help of a venture capitalist. It is a good idea to talk to a stock broker and see what kind of advice is offered. The money is always available for a worthwhile venture. It is only necessary to find the right channels and the right people to make your invention a success.

PERSISTENCE PAYS OFF

During the early days of a promotion, there is a period of time when the world of sales seems to be in a stagnant condition. Absolutely nobody seems to be interestd in what you have to say. They are too busy or just not sufficiently interested to watch a demonstration. Key people that you were depending on to show interest are away on holiday or they are in the hospital having an operation. The high expectations that you had for your promotion have gradually melted down and now the world is a bleak place indeed. At this stage the amateurs quit, the weak hearted go home and the tired and frail go to bed.

If you have ever tried to be a salesman for any kind of a product you will know the feeling. But it is not the end of the world of course. People who have worked in sales and promotions for many years know that the response is there. It is simply a matter of persistence. Don't give up. Experienced salesmen work on the principal of "percentage-of-calls", which is based on the idea that if you make one hundred calls, seven of them (or five or whatever) will place an order. In other words, orders will not fall into your lap without effort (although sometimes they do), but you can be guaranteed that if you make a persistent effort, you will eventually get the kind of response that you want.

MANUFACTURING YOUR INVENTION

With the ambitions you have for your invention, the ideal situation is to choose a good location for factory space, install the machinery needed, hire some good people and start the wheels rolling. Unfortunately, the kind of money needed for this step may be out of reach. From all of the sources that are available for financing, there may be no doors open to you. Perhaps the particular product you want to manufacture is just not interesting enough to attract the kind of money needed.

There are alternatives that you can consider. Very often a start can be made in a small way by having a sample quantity of parts made. If your product is made with one or two plastic parts and three or four metal parts, you can get one hundred of each made by different companies. Then assemble the parts yourself or hire a couple of housewives to do the assembly for you. In this way, by investing a few thousand dollars, you can actually have a hundred units on the shelf to sell and promote. It it very unlikely that you will show a profit at this stage, but you will learn the critical factors about your product that you need to know. Also, because the dies and jigs that had to be made for a small quantity are now available, a second batch can be produced at a reduced cost.

By applying some careful cost analysis and projecting your sales into the future, you will now know for certain whether you will be able to make money with your product. This kind of information is extremely valuable. You will be able judge that if you made a thousand units (for example) you would start to show a profit by the time you have sold 876 of them. With this knowledge you can generate the confidence you need to proceed on your own or to look for greater financing from others.

SETTING UP YOUR COMPANY

The owner of a company that manufactures a special kind of smoke and gas detector told me that he started his company on the kitchen table at home. His wife complained about the mess he made in the kitchen so he bought a space heater to heat the basement and moved his operations there. After a few months and an important success, he decided to rent a small office and work room. Within a year that had become a small factory and that is where he is working today.

Another successful entrepreneur who manufactures a fireproof electric motor got his start by becoming a sales agent for electric motor manufacturers. In this way he made his living doing something that he loved, and he was also able to detect the weaknesses and strengths of the electric motors that he sold and installed for the other makers.

When the time came to manufacture his own invention, he knew exactly what was required to make a successful electric motor. Furthermore, without investing a lot of money for equipment, he was able to order the exact parts that he needed from the manufacturers that he represented. Since many of the parts were standard, his costs for parts was minimal. With a small investment he set up an assembly line to put together his own electric motor invention. Meanwhile he continued to represent other manufacturers and his sales of their products paid his way.

These examples illustrate methods that can be used to become successful step by step. It is a great advantage if you are already familiar with a market and a product. It is a greater advantage yet if you are actively engaged in promoting and selling the same kind of product of the category into which your invention falls.

QUALITY CONTROL

There was a long period of difficulty for Japanese manufacturers when they at first tried to enter world markets. The Japanese products were avoided because of poor quality and workmanship. The change came gradually as they realized that the secret of getting the respect of world markets was quality control. Now Japanese products are rated among the best in the world.

Quality control can be done visually in a small shop. It is usually done automatically by machines and devices that check dimensions, check finish, check color, etc., in large factory operations. Quality control is important if you are having small lots of parts made by different manufacturers for later assembly. One employee who is new or careless can make a batch of parts useless.

MANUFACTURING PROCESSES

Some of the common production methods include:

Foundry for casting rough parts in metal.
Injection molding machinery for casting fine parts.
Metal forming and shaping machines.
Punch presses for cutting out shapes of parts.
Punch presses for forming and shaping parts.
Machine shop with production lathes.
Grinding and finishing machines.
Spot-welding and tack-welding machines.

Before you decide exactly how you are going to manufacture your product, it is a good idea to visit some factories where similar products are turned out in quantity. Such visits can usually be arranged by phoning and asking for the public relations manager. Visitors are usually treated with enthusiasm and respect.

The public relations manager and the owner are often the same person so that a tour through the production floor of the plant is a top echelon experience. If you get a guided tour by management, it can be a useful exercise. Observe, make mental notes and learn all you can.

You will see different kinds of production machines in actual operation. You will see the manufacturing steps that each machine is set to accomplish. You will see assembly line techniques and the ways in which space is used for assembly and packaging. You will see the number and type of employees used on the assembly line. These will be important factors when you start manufacturing your own product.

WHAT IS THE SELLING PRICE

Deciding what to charge for a product is something like peering into a crystal ball and trying to determine what the future holds. There are many elements and factors involved in the selling price of a product. You have to amortize all of the development, promotion and start up costs. This can be done in a series of payments over a period of time. Also, the backers are waiting for a share. If their expectations are too high, it would be easy to price the product right out of the marketplace.

There are two simple rules that will guarantee success. They are: (1) Sell a sufficient volume of your product per month, and (2) charge enough for your product to show a profit. Both of these rules are a challenge to put into practice. Rule one is saying that your price must be low enough to sell a large volume of product. But rule two is saying that the price must be reasonably high in order to guarantee a profit. A reasonable and workable balance must be established between the two opposing demands.

This is a very critical stage in the life of your invention and your company. All of the effort that you invested over the months and perhaps years in your invention, and all of the development costs, now must begin to pay off. All of the effort put into promotion and marketing strategy now must show a favorable result. Everything hinges on setting a price for your product that will be competitive and yet still show a profit.

Fortunately, the selling price of your product can be correctly set by adhering to a formula. This is a formula that has been successfully used by manufacturers for many years. But in order to use an effective pricing system you must have all of your invoices and costs of all kinds available. All of the money that you spent in various ways to make your invention a possibility now must be accounted for.

COST BREAKDOWN

The first step in setting the price for your product is to make a list of all of your costs. The costs can then be broken down into three categories of expense. These are (1) Labor, (2) Material, and (3) Overhead. All expense items are separated into these three groups:

(a) Labor includes all money paid out to employees in the form of wages, bonuses and vacation pay.

(b) Material includes all items that you have to buy to manufacture your product. Only materials that end up as a part of the product are included in this category.

(c) Overhead includes every other expense that you have to pay to make the business go. Into overhead goes expenses like rent and phone, insurance, freight and storage, etc.

Overhead is often separated into two kinds of expenses with all normal expense in one group and all government expense in another group.

Typical expenses in the overhead group will be:

(1) Rent paid for factory space.
(2) Telephone, computer interfacing, fax and telex.
(3) Advertising and directory expenses.
(4) Electric light and power.
(5) Gas or oil heating.
(6) Maintaining tools and machinery.
(7) Freight and shipping.
(8) Office supplies.
(9) Accounting services, bookkeepping.
(10) Insurance on equipment.
(11) Shop cleaning and maintenance.
(12) Bank charges; interest on loans.

To this list the government expenses must be added. These are licences, taxes, permits, cash paid out for compulsory auditing, and some others.

FACTORY COST

When LABOR, MATERIAL AND OVERHEAD are added together, the total is called the "factory cost". In this total all of the costs normally associated with manufacturing a product are included.

MARK UP

After you have carefully noted all of the items that must be included in factory cost, there are some items that have to be added on. These come under the general heading of mark-up. The mark-up numbers are rated as a percentage of factory cost. The factory cost is therefore set at 100 percent. The mark-ups include:

(1) A mark-up for a profit to the company. In large well-financed corporations that have been in business for many years and are accustomed to competing in the national market place, a mark-up of 5 percent to 10 percent is usual. In smaller, less stable or newer concerns, the mark-up for profit is about 15 percent. In non-competitive situations the mark-up can run from 20 to 50 percent.

(2) A mark-up for realizing the cash for repaying the financial backers. For this purpose a mark-up of from 10 percent to 20 percent is added.

(3) An amount for a pension fund for employees. This mark-up is usually from 3 to 5 percent.

(4) A small mark-up for wasted materials and estimating failures. The mark-up for materials that actually go into the waste barrels instead of into a product, the mark-up is from 1 to 1.5 percent. A similar mark-up will cover estimating failures. An estimating failure happens when you estimate that a thousand parts will cost you $800 and the actual cost charged to you is $1200. This happens when a supplier puts his prices up without warning, or when his costs go up unexpectedly. It can also happen when a manufacturer warns you in advance but there was a snarl up in your own communications, and you never got the message.

(5) A mark-up to cover non-collectable accounts. At the end of one year of operation, total up all of the checks that were no good and the accounts that never got paid due to bankruptcy and business failure. A mark-up of about 2 percent is needed to cover these losses.

(6) Depreciation of equipment and machinery is an important factor in a profitable operation. The amount of the depreciation is determined for each item by dividing the total cost of that item by the number of years that it will produce a product. That gives you the yearly amount of the depreciation. This amount is converted to a percentage mark-up on the selling price. The money realized for depreciation is put into a special account and allowed to earn interest. When a machine is worn out and will no longer produce, it is sold for scrap and the money added to the depreciation fund. New machinery is purchased out of the fund. If the depreciation fund is properly managed, you will always have enough money in the fund to buy whatever new machinery and equipment is needed as it wears out or has to be replaced due to obsolescence. The costs of repairing production machinery can also be charged to the depreciation fund if a careful record is kept of what was done.

UNIT PRICE FACTORY COST CONTROL

PRODUCT NUMBER A7-46 ESTIMATOR AL JOHNSON DATE MAY 17/90

OPERATOR	"Q"	PRICE	RATE	HRS	MATERIAL	LABOR
STATION A			8.80	7		61.60
Set-up main frame and install parts I6A-3	1000	.40			400.00	
and 22B-7 1000 units	1000	.32			320.00	
STATION B			8.80	4½		39.60
Install parts 18A-4	1000	.16			160.00	
and 18B-3 Check alignment.	1000	.31			310.00	
1000 units.			8.80	2		17.60
STATION C						
Install parts 17C-42	1000	.14			140.00	
and 17C-43 Tap and clean threads	1000	.17			170.00	
1000 units			8.80	3		26.40
STATION D			8.80	4	18.00 × 3.333	35.20 × 3.333
Coat and finish 300 units (300 DONE)	300	.06				
$\frac{1000}{300}$ = 3.333 X labor X material					60.00	117.32
			TOTALS		1560.00	262.52

LABOR	262.52
MATERIAL	1560.00
OVERHEAD	262.52
FACTORY COST	2085.04

Company mark-up	10	%
Repayments	20	"
Employee benefits	03	"
Estimating errors	03	"
NSF checks	02	"
Depreciation	08	"
Total mark-up	46	%

MARK-UP FACTOR 1.46

$$\frac{2085.04}{1000} \times 1.46 = 3.04$$

SELLING PRICE = 3.04

REVISED SELLING PRICE $2.99

ESTIMATING CONTROL SHEET

To make sure that you do not forget any item of expense when setting the price for your product, you can use a control form like the one shown on page 39.

This form is used for determining the cost price per unit of a certain number of units. If you have been ordering material and parts for one thousand units at a time, then 1000 will be the best quantity to base your prices on. Suppose that you ordered 1000 of one kind of part but for another part you had 600 in stock, so you ordered 400 to make your stock tally up to 1000. To make the invoices and costs for the 400 units conform to the control sheet, multiply the 400 invoice amount by 2.5, then the cost will be the same as for 1000. After a final price is established, we will divide by a thousand to arrive at a unit price.

Going across the form from left to right there are seven columns. The first column is for writing in the name of the department or the number of the assembly line station. The second column is a narrow one marked with a "Q". The "Q" stands for quantity, or the word quantity can be printed in. The next column is for price per part. The next column, mark "rate", is for the hourly wage rate paid in that department or at that assembly line station. The column after that is for the number of hours worked, usually broken into quarter hours for part time workers. The next column is for the total cost of the material and parts added on in that department or assembly line station. The final column is for the total cost of the labor in that department. An example is shown on the form.

If your labor record shows only 350 units made and the total cost for 350, then multiply by 1000 divided by 350. That will give you the labor cost for 1000 units. An example is shown on the form.

On the sample control form, some arbitrary numbers have been written in so you can see how it is used. If your product is a small simple kitchen gadget or something like that, you will only have two or three assembly line stations, but if it is an electronic device or complicated invention, you may have ten or a dozen assembly line stations. These stations are often just referred to by a number or letter, such as "A" station. Often the name of the station is just the name of the machine or system used at that work station.

ADDING OVERHEAD AND MARK-UPS

The overhead costs of a factory operation are quite stable and do not change very much. Therefore it a simple matter to calculate the overhead per hour just by dividing the monthly overhead by the number of hours usually worked in a month. Since the number of hours worked on 1000 units is determined by the labor total, it turns out that the overhead is always a simple fraction of the labor costs.

For very small operations, or just starting up, the overhead tends to be quite reasonable. It can be set at 60 to 70 percent of labor costs for 1000 units. However, in larger established firms with lots of employees (say, twenty or more employees), the overhead costs and the labor costs are just about equal. That is, overhead can be set equal to the labor costs. When you get up into big business with over a hundred employees, the overhead will be at least 110 percent of labor costs, and it can be as high a 120 percent of labor costs (per 1000 units). If the overhead is higher than 120 percent of labor, a more efficient accounting and office system is needed.

On the form a figure of 100 percent of labor has been arbitrarily used for overhead costs. Now the totals for material, for labor and for overhead are added together to give the factory cost for 1000 units. See the bottom of the control form.

The next step is add on the mark-ups. The best way is to set the factory cost at 100 percent. Because the mark-ups are based on percentages of factory cost, this makes it possible to merely add up the percentages for mark-ups to arrive at a total mark-up.

The list of mark-ups is:

(1) Profit for the company................10 percent.
(2) Repay backers............................20 percent.
(3) Employee pensions.................... 3 percent.
(4) Waste and estimating failures.... 3 percent.
(5) Non-collectable accounts........... 2 percent.
(6) Depreciation................................ 8 percent.
TOTAL MARK-UPS...........................46 percent.

To add on the 46 percent mark-up, the correct way is to multiply the factory cost number by 1.46. This is shown at the bottom of the control sheet.

SALES COMMISSION

If you have sales people on commission, there will be an additional mark-up to cover such commissions. Two other items that are often included in the mark-up list are "promotion" and "packaging". When your pricing structure for your product is analysed for competitiveness, it can then be decided whether to keep packaging and promotion in the overhead costs or add them on afterwards.

INVENTIONS FROM THE PAST

When it became well-known that an electric generator could be driven by a steam engine to supply electricity, the inventors of the day dreamed up many ways of applying the new source of power. One inventor vizualized an electric helicopter with a parachute attached as a safety precaution. This drawing was prepared for the popular press over a hundred years ago.

UNIQUE INVENTIONS AND IDEAS

The inventions and ideas illustrated and described in this section were originated by Peter Horsley for the entertainment of the reader. Any similarity between any example shown and any other patent or invention is purely coincidental. The initials P.A.F. mean "Patent Applied For".

ROTARY AND RADIAL ENGINES

Most internal conbustion engines burning gasoline or diesel fuel operate on a four stroke cycle. There is an intake stroke when the piston sucks in fuel and air, then a compression stroke when the fuel and air mixture are compressed, then a work stroke as the gases are exploded to drive the piston down, and finally the exhaust stroke during which the burnt gases are ejected. These factors account for the complexity of the engine in your car. An array of valves and other moving parts are needed to make the engine work.

An internal combustion engine can be greatly simplified by combining all four strokes into one cycle and then rotating the engine. The result is that no valves are needed. Only one spark plug is needed because the firing piston always comes around to the firing position.

THE RADIAL SWASH-PLATE ENGINE

An example of this kind of engine design is shown in the drawings of Figure Q-1. It is customary to tip the swash plate and rotate it. Since this engine is going to rotate, it is simpler to tip the engine so that the drive shaft from the engine remains level.

The pistons make two strokes while the engine rotates once. The rotary cycle of the engine is then made up of four quadrants representing the four strokes of a regular engine. The intake port occupies one quadrant and the exhaust port occupies the other quadrant. The spark plug is located at the beginning of the firing stroke.

The valve openings are designed in the shape of a cone. One part of the cone is polished steel. The other half is sintered copper or alloy. The snug fit of the cones keeps the intake and exhaust ports closed except when they are supposed to be open during the engine cycle.

All of the mechanical parts of this engine are exactly symetrical, opposite and identical. Therefore, the causes of vibration are also eliminated. Fan-shaped fins and flow through spaces make air-cooling possible, and this eliminates the need for a radiator and cooling fluids. A good electrical system will provide for a separate heat pump to take care of a comfortable climate within the car.

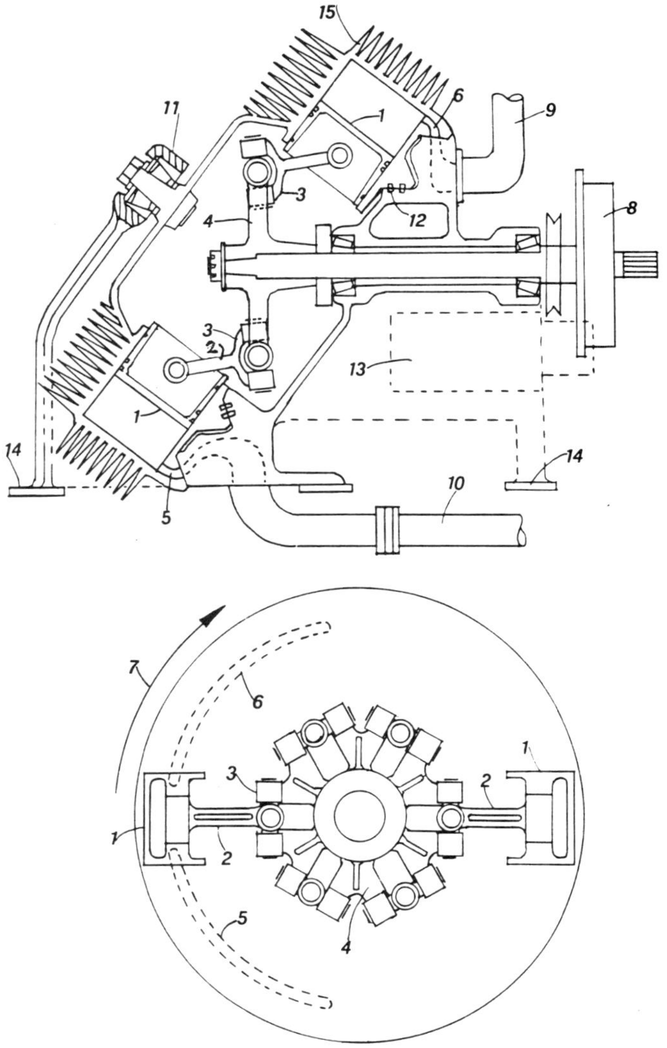

Figure Q-1: In these two views of the tipped radial engine the same number illustrates the same part in both drawings. The pistons are numbered (1), the connecting rods are numbered (2), the universal joints are numbered (3), the swash plate is numbered (4), the exhaust port and cones are numbered (5), the intake slot and cones are numbered

(6), the direction of rotation is numbered (7), the flywheel is numbered (8), the intake duct is numbered (9), the exhaust pipe is numbered (10), the main bearing is numbered (11), the cones for the oil seal are numbered (12), the starter is numbered (13), the motor-mounts are numbered (14) and the cooling fins are numbered (15).

OPPOSED PISTON ROTARY ENGINE

This engine is based on the mechanical motion shown in drawing (a) of Figure Q-2. Diagram (a) illustrates a method for making a rotating shaft turn a corner. As a side effect of the method, the angle rods also slide in and out of the openings in the ends of the shafts. The sliding motion is the basis for the engine.

Diagram (b) shows what the pistons look like. The angle can be much less than a right angle, as illustrated.

A side view of a typical engine adapted to work on this principle is shown in drawing (c). There are two sets of pistons, arranged in opposition. One set is adapted to act as the intake and compression system for the engine. The other set is the firing stroke and the exhaust stroke.

This system permits the engine to complete one cycle during one revolution. The result is that only one spark plug is needed. The valve openings are accomplished with cones and quadrant slots as described in the previous engine.

A unique feature of this engine is the method used to transfer the compressed air-gas mixture from one set of pistons to the other. In the center of each piston set there is a high-pressure duct, numbered (3) in drawing (c). It has a check valve on each end of the duct. The first check valve opens when there is any difference in pressure and allows the compressed gases to enter the duct. The second check valve opens only when the pressure in the duct exceeds a certain high pressure, which can be preset. The result is that there is a high pressure charge of the air-gas mixture injected into the firing stroke just previous to the spark plug firing. The timing of these phases of the cycle would be subject to adjustment for optimum operation.

This engine would be small and would operate at high speed. To avoid excessive gryro effect on the car from the rotating mass, the engine could be arranged in sets of two with a set of spiral gears connecting the pair. A top view of the two engines is

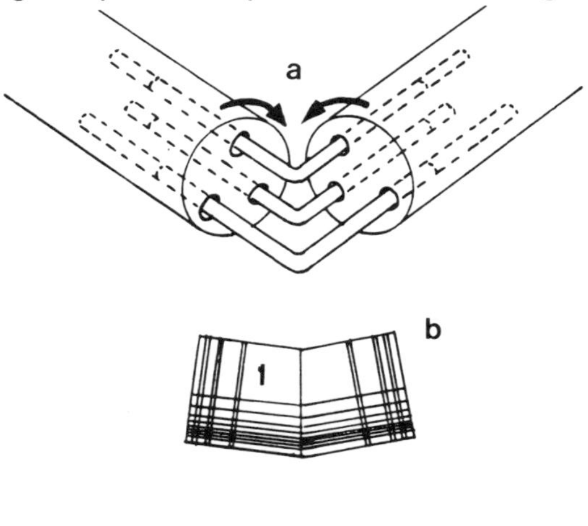

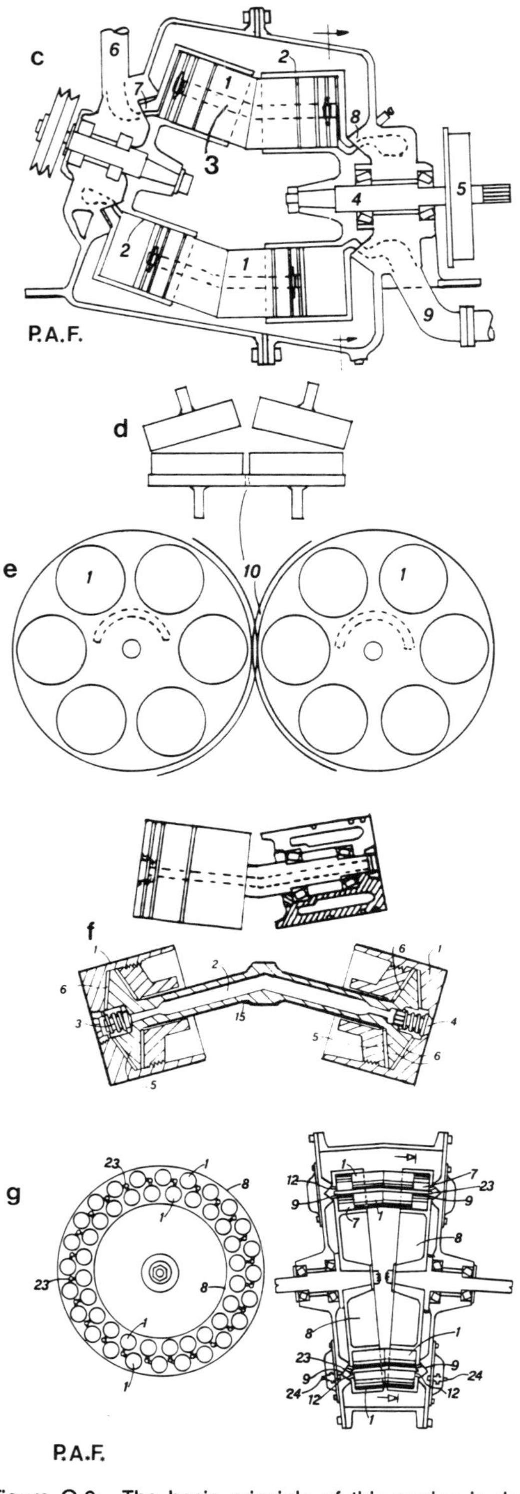

Figure Q-2: The basic principle of this engine is derived from the linkage shown in diagram (a) which is a method of turning a rotating shaft through a sharp angle. A typical pair of opposed pistons, marked (1), is shown in drawing (b). Drawing (c) is a side view in section of a simplified version of the engine. Only the main elements are shown. The pistons are marked (1), the cylinders (2), the forward shaft (3), the main shaft (4), the flywheel (5), the intake (6), the intake port and cones (7), the exhaust port and cones (8), and the exhaust (9).

The drawing shown in the diagram (d) illustrates a top view of an arrangement for two engines of the type mounted side by side and intermeshed with spiral gears (10). A rear view of the arrangement is shown in diagram (e).

The diagrams in drawing (f) illustrate two ways of eliminating the rotation of the pistons. In the upper example, tapered roller bearings are used. In the lower example a large area of bearing face (6) is provided, and designed to absorb the repeated impact of a working piston.

In drawing (g) two views are shown of a similar type of engine adapted for high speed rotation. Two rows of pistons are used, one row acting as intake and compression, and the other row acting as work stroke and exhaust. In this way the pressure on both ends of any one piston set would be balanced.

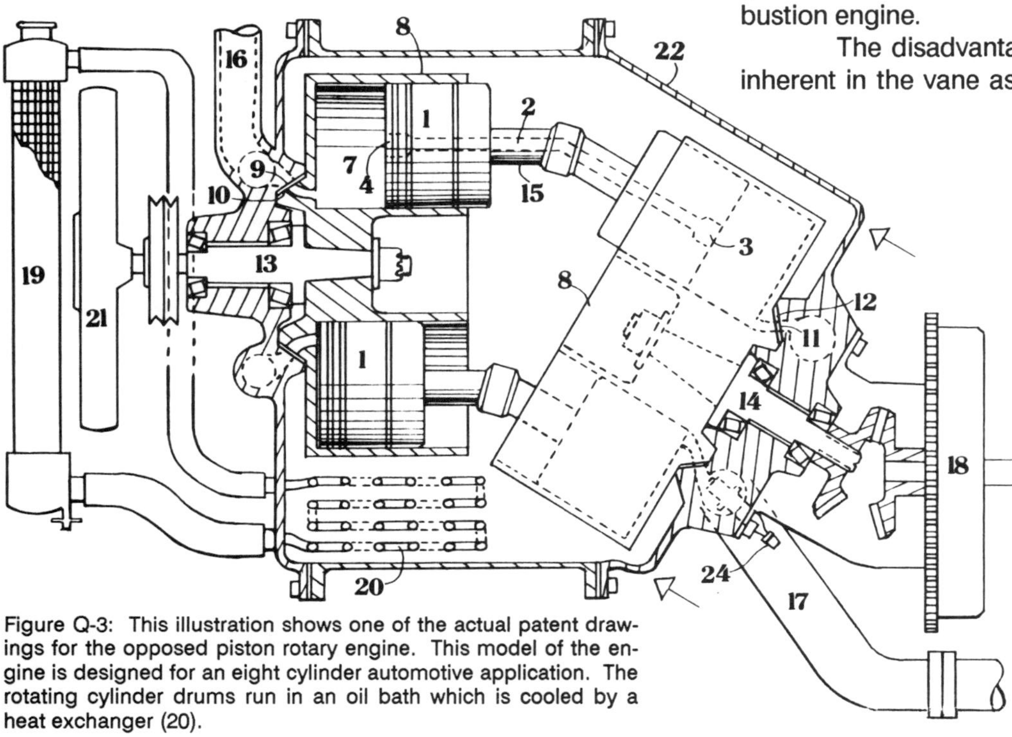

Figure Q-3: This illustration shows one of the actual patent drawings for the opposed piston rotary engine. This model of the engine is designed for an eight cylinder automotive application. The rotating cylinder drums run in an oil bath which is cooled by a heat exchanger (20).

shown in diagram (d) and a partial rear view in diagram (e).

In this engine, the pistons rotate as well as slide. The friction of this motion would be reduced by having a rotating insert inside each pair of pistons. This is shown in diagram (f).

The opposed piston rotary engine is also suitable for a positive displacement turbine, designed to operate at very high revs. Two rows of pistons are used so that each cylinder drum can act independently of the other. Two views are shown at (g). In this case, one row of pistons performs the intake and compression functions while the second row of pistons is adapted for the power stroke and the exhaust.

PARALLEL VANE ROTARY ENGINE

The only advantages this engine has is its compact shape and its few parts. Like the other rotary engines shown, one complete rotation of the engine includes the four stroke cycle of an internal combustion engine.

The disadvantages of this type of engine are inherent in the vane as a mechanical means for converting gas pressure to rotary energy. When a piston and cylinder are used, the combination comes with a century of experience. A modern piston with rings does not leak oil to any serious extent, whereas a lobe or vane displacement system is sure to give problems sooner or later.

The basic principle of this engine is derived from the parallel linkage on the drive wheels of the old steam locomotive. When the links are short and the "wheels" large, a number of links can be used. The dimension of the links is chosen so that they will not interfere with each other. The compression of the air-fuel mixture occurs when the narrow head of one vane follows the narrow base of the preceding vane through the narrow side channel. The valves are operated as slots in the intake and exhaust quadrants of the cones.

Figure Q-4 illustrates the parallel vane rotary engine. Any of the engine examples shown can also be adapted to function as a pump or compressor.

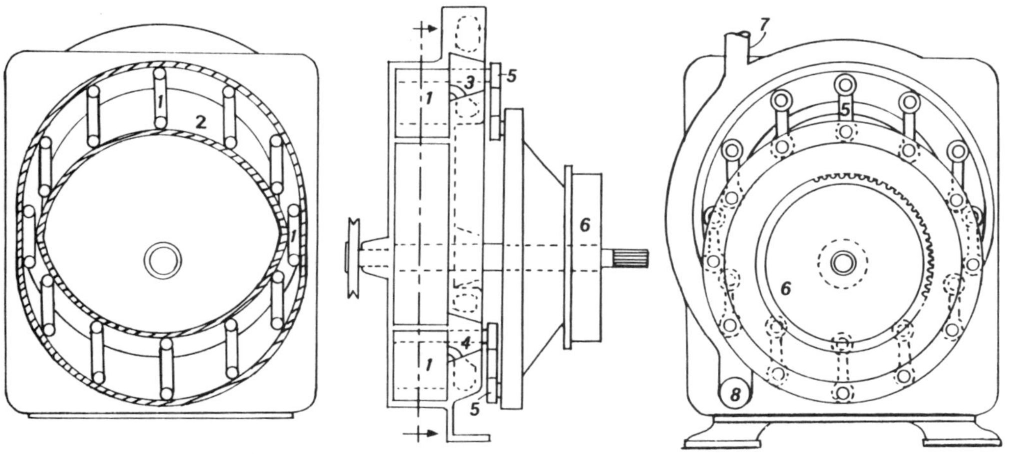

Figure Q-4: Three views are illustrated of the parallel vane rotary engine. The sectional drawing of the vanes reveals that the vane cavity is exactly shaped so that the ends of the vanes are touching the walls at all times. The vanes are numbered (1), the cavity (2), the bearing ring and intake cone slot (3), the exhaust valve quadrant (4), the parallel links (5), the flywheel (6), the intake (7) and the exhaust (8).

WAVE LOBE ROTARY ENGINE

This engine is a take-off from a gear pump. The important difference is that the wave shapes are computer generated so that when the internal lobes are inserted into the exterior lobes, every internal wave has a positive contact with an external lobe. The four sets of lobes interplay or mesh like gear lobes.

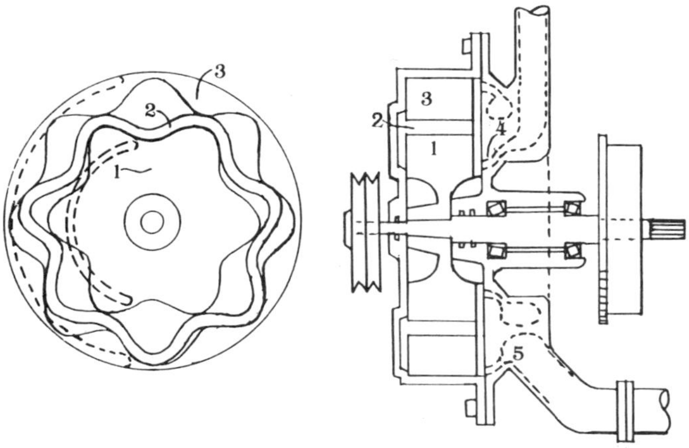

Figure Q-5: Computer generated lobe shapes make this wave-lobe engine function smoothly and efficiently. Because the displacement lobe type of engine is not as efficient as a piston engine, the fuel and lubricant are sometimes mixed together. This engine is designed to operate at very high revs, like a turbine. The high speed limits the amount of "blow-by" and oil drift that is characteristic of lobe displacement turbines. The central driving gear is numbered (1), the intermediate driven gear (2), the outer driven gear (3) the intake (4) and exhaust(5). The center lobe gear (2) has separate off-center bearings.

This means that the center gear is driving the outer sets. The intermediate gear rotates slower than the driving gear. The exterior set of wave lobes rotates slower yet. The intention is to operate this engine at high speed to limit blow-by and oil-drift.

In operation, when the inner lobes are opening up, that is the intake. The intake slot is semi-circular. When the spaces between the lobes are closing up, that is when compression takes place. The compressed gases go through a high pressure duct to be injected into the spaces between the outer set of lobes just as they begin to open up. The pressure of the injected mixture is controlled by a valve which only opens when the pressure exceeds a preset amount. When the spaces between the outer lobes start to close, that is the exhaust. The exhaust port is a semi-circle, as shown in the drawing.

Because of the difficulty of controlling the loss of oil in engines of this type, there is always the option of mixing the lubricant and the fuel together as they do in small marine engines and some motor cycles.

CIRCUMVENTING THE SECOND LAW OF THERMODYNAMICS

Before long mankind will be in space in a substantial way. Man will be in space to stay, including working and playing, solving problems and doing all the things that people do. One of the things that will be available in vast quantities will be free energy. To use energy you have to run it downhill. The sun represents the top of the hill. The blackness of space is a perfect cold sump. The blackness of space is the bottom of an endless hill.

To be blessed with free energy in space all that is necessary is to place something in the sunshine that will absorb energy. The really simple way to do it is to place an array of solar cells in the sunshine and get free electricity. Of course, you can use solar cells right here on earth, but the sun is always shining in space.

When man is in space, the idea of perpetual motion or any proposal to cheat the second law of thermodynamics will be meaningless. Why would anyone bother when energy is already free for the taking? However, here on earth, energy is still expensive. You have to buy gasoline for your car and you have to pay your power and heating bills every month.

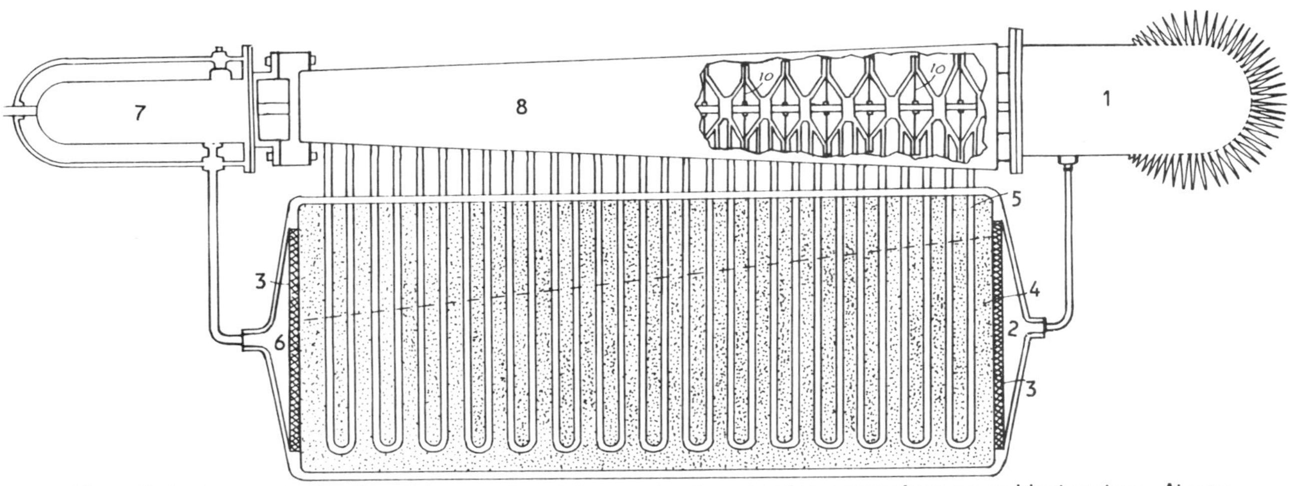

Figure Q-6: Schematic drawing of the experimental apparatus for extracting energy from an ambient system. At one end of the cycle the piston and cylinder (1) operates isothermally and exchanges energy with the air. The other end (7) operates in a cold vacuum with suitable insulation. In the diagram, the sand is numbered (4) and the heat exchanger is numbered (5). Numbers (2) and (6) are funnel-shaped compartments free of sand. The sand barriers (3) are fine mesh of metal or plastic. The expansion chamber is numbered (8).

Any method that might be suggested for reducing the costs of energy would be welcome.

Theoretically there are several ways of circumventing the second law of thermodynamics to derive free energy. Only two ways will be illustrated, because they are the easiest to explain. The second law states that there is no way that energy can be extracted from a system that is already at an even (ambient) temperature. Suppose that the air temperature is presently at 60 degrees F. The project is to take a sample of that air, extract useful energy from it and then replace the sample at a lower temperature than it was when taken. For these experiments it is permissible to take any path whatever, so long as the end result is as stated. During the process of the cycle, it is permissible to borrow energy or discard energy. The only rule is that borrowed energy must be paid back in kind. Energy borrowed at a certain temperature must be replaced at the same place at the same temperature.

THE COLD SAND SUMP

For this method the first step is to fill a large gas-tight tank with sand. The sand has to be of a specific average grain size. The size of the sand grains for this experiment will be such that the resulting chambers (open spaces) remaining between the grains will meet the following criteria. The molecules of air in the sand can only move around in the small spaces remaining between the grains (called interstices). The mean free path of the molecules will be thereby shortened. The mean free path is the average distance that a molecule of air will travel before it strikes another molecule. The mean free path that the molecules will have in the sand will be approximately one tenth of the normal mean free path at the given temperature and pressure. When this condition is met, the pressure of the air within the sand no longer follows the rules of ordinary thermodynamics. The new law that the air pressure follows in the sand is

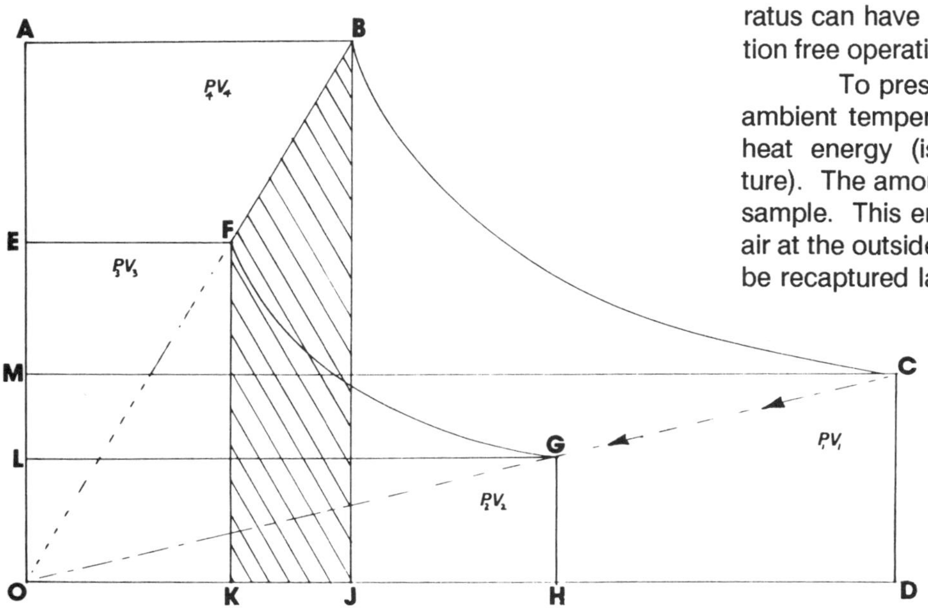

$$\frac{T}{P^2} = \text{a constant}$$

This relationship is derived from the fact that the number of molecules which may pass through a very small opening is dependent on the speed and number of molecules which enter the opening. In the case of a gas, speed means temperature, because the temperature is proportional to the velocity of the molecules squared. And the number of molecules near any one opening is another way of expressing pressure. The interesting result is that when the air in the sand is cooled, the amount of heat energy released is exactly the same as when the same amount of gas is cooled at a constant volume. However, within the sand, when the air is cooled it compresses itself into a smaller volume without generating any heat of compression.

The sand tank must have a large cross section to permit a measurable flow of air. A small sample of air at the ambient temperature is taken and the pressure, volume and temperature are measured. Pressure and volume together are called the PV of the gas. The PV is always proportional to the absolute temperature. This means that if the PV drops to .75 of the original PV, then the temperature also is .75 of the original temperature.

The first step is to get the air sample to enter the sand tank. A piston and cylinder are used for this. To reduce friction to a minimum, the pistons are made like the examples shown in Figure Q-7. In the diagrams, (a) and (b) are two views of a rubber diaphram (10) with imbedded metal or plastic ribs (12) and one-way check valves (11). The heat exchanger is numbered (5) with a flow-control valve (9). Diagram (c) is a cross section of a piston with rubber walls (C) which roll at (A) into the space (B) as the piston moves to the right. The heat exchanger fins (D) on the piston face are matched by recesses in the cylinder head. These are diagrams only. The actual apparatus can have many other refinements to assure friction free operation.

To press the sample of air isothermally at the ambient temperature into the sand tank costs some heat energy (isothermal means constant temperature). The amount of energy is equal to the PV of the sample. This energy will be stored temporarily in the air at the outside temperature. This is a loss which will be recaptured later. The PV diagram for the process

Figure Q-8 : To demonstrate gas cycles a PV diagram is used, as illustrated. The pressure gradient through the sand is shown by arrows. In actual practice, the gradient has to be slightly greater so that the air will flow through the sand. A large cross section of sand is used.

is shown in Figure Q-8. The sample is the box P_1V_1 with the corners lettered CDOM. As the sample travels through the sand tank it is cooled. There must be a pressure gradient in order for the air to flow. The pressure gradient is established by the cooling of the air by the heat exchanger (5). The path followed by the air in the sand is along the line from C to G, indicated by arrows.

At the cold end of the sand tank the sample is extracted. The amount of energy gained when the sample is drawn out of the sand tank is equal to the cold PV, shown by the box P_2V_2 , with the corners marked GHOL.

In a thermodynamic process a gas such as air can be compressed isothermally if it is done slowly to let the heat energy flow. When the sample is compressed isothermally, the pressure increase can be 2.71828 times. For this experiment, 2.7 will be near enough. When a gas is compressed by this much, the amount of heat energy developed is equal to the PV of the sample. The energy gained is used in a trade-off to compress the air. The compression curve in the diagram is the box with the sloped top, marked GHKF. The new PV is marked P_3V_3. The sample has been delivered to the first diaphram compartment of the heating and expansion process.

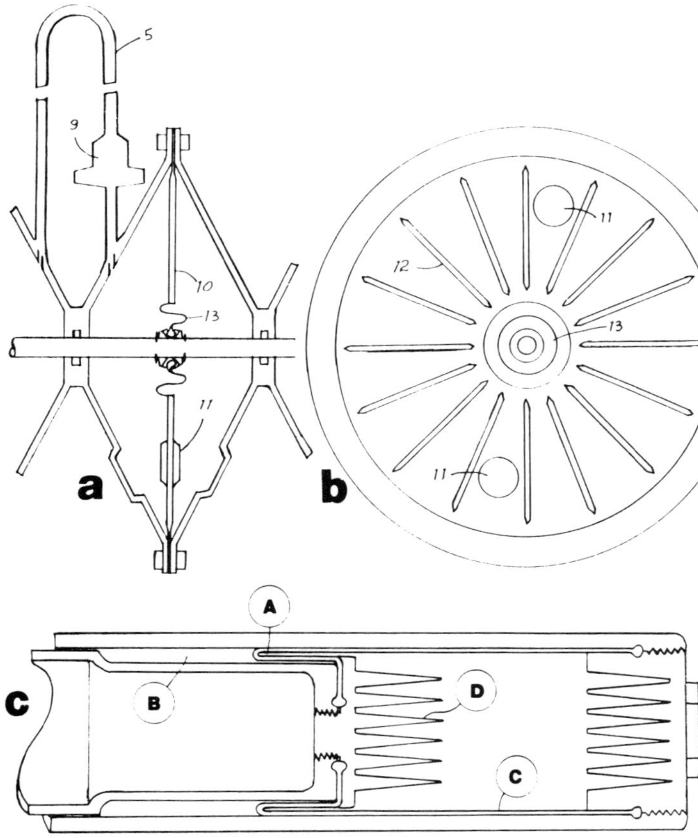

Figure Q-7: Special methods are used for transporting and compressing the air sample to reduce friction to a minimum. The sample is moved slowly through many identical stages like the flexible diaphram in diagram (a) and (b). Diagram (c) is a cylinder with flexible rubber walls and rolling seal. Isothermal processes are accomplished slowly with plenty of cooling fins and heat conductors.

The sample is heated by extracting heat from the sand. As the pressure starts to increase due to the expansion from the heat, the sample is allowed to expand along the line FB until the sample has attained the original ambient temperature. The box occupied at this stage is marked P_4V_4.

The expansion of the air converted some of the heat energy into mechanical energy. The amount of energy gained is equal to the cross-hatched area BJKF. A part of this energy can be used to operate a small refrigeration cycle with the expansion coil in the cold sump of the sand tank.

To get back the original PV that was stored in the outside air, the sample is permitted to expand isothermally along the curve BC. This is accomplished at the ambient temperature. The PV area BCDJ is equal to the original PV box CDOM.

It is true that in order for heat to flow you must have a temperature gradient. However, small the gradient may be, it represents a loss from the ideal. However, there is an amount of energy BJKF available for controlling losses.

CONSTANT PRESSURE METHOD

The words "constant pressure process" and also "isothermal process" are representing an ideal situation that is impossible to attain in the real world. Because it is true that heat will flow only if there is a thermal gradient and a gas will flow only if there is a pressure gradient, an isothermal process is not feasible and a constant pressure process is unattainable. However, the terms are useful in that the ideal defines the perfection that must be the target. In this way, the best possible alternative to perfection is created.

The method about to be described is based on a simple fact discovered many years ago. It is usually referred to as "free expansion". The main feature of the method is illustrated in Figure Q-9. Diagram (a) is an insulated gas-tight box with two compartments. In the partition wall there is a valve and some heat conducting fins of copper or aluminum. In diagram (a), both compartments contain air at the same temperature. However, the pressure in the first comparment is greater. When the valve is opened, the air from the higher pressure side rushes into the compartment with the lower pressure until the pressure in both compartments is equalized. This is shown in diagram (b).

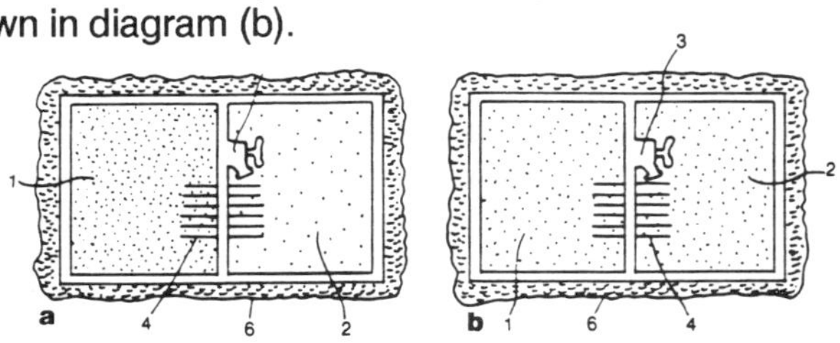

Figure Q-9 : When a gas is permitted to expand freely from a high pressure compartment into a low pressure compartment, as in diagram (a) there is no loss or gain of energy. When the temperature and pressure are equalized, as in dic

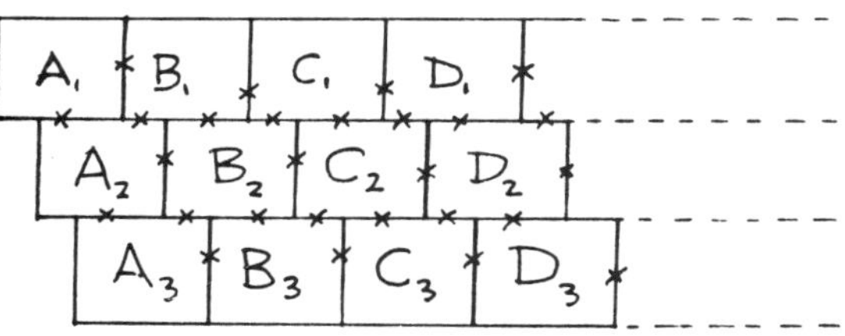

agram (b), the gas remains at the same temperature that it was prior to the expansion. In diagram (c), any two compartments that have a partition wall with a valve (marked X) can take part in a local "free expansion" of the type described. By cooling the gas at constant volume, pairs of compartments with interconnecting valves gradually permit the gas to flow in the direction of the cooling process. The opening and closing of pairs of valves assures that the pressure in all of the compartments remains nearly constant.

The important point is that when the air settles down in the new circumstances, the temperature is still the same. Whatever differences in temperature that may have occurred temporarily due to expansion or compression have been equalized. The amount of energy within the system remains the same.

At first consideration this looks like a loss of energy. A piston could have been used to extract a small amount of energy from the system during the process. However, the factor of interest is that the method provides a way to remove energy from a system at constant volume while the system remains substantially at constant pressure. The way this is done is to use a large number of small compartments separated by valves. Every pair of two compartments can then represent the free expansion method, by opening and closing every second valve.

The way in which the pressure is reduced in any given compartment is to cool the air in it. The result is that the air will be of a slightly higher temperature when it rushes into the next chamber, but that will not interfere with the method. If one compartment becomes slightly hotter than it should, then the one losing pressure will become slightly colder than it should. It will all even out. Diagram (c) shows how the compartments can be staggered in rows to reduce the temperature difference between adjacent compartments.

The small portion of the sample of air is placed into the first compartment by opening it to the ambient outside temperature and pressure. If a high pressure is desired to start with, a reservoir (tank) at the ambient temperature can be used. Every time an increment of the air sample is allowed to squeeze into the first compartment, a small isothermal compression will occur at the ambient temperature. This happens whether or not a piston is used. Measuring this energy shows that the total amount of energy released is equal to the PV of the sample. The extra energy that is released in this way can be temporarily stored outside in the air, which is at the same temperature.

The PV diagram for this process is shown in Figure Q-11. The sample of fluid gas is enclosed within the box with the corners marked CDOK. During the free expansion process within the many small compartments the gas travels along the wavy line from C to F. The average pressure remains nearly constant with only a slight pressure gradient to make the gas flow. Whenever one compartment loses pressure caused by cooling of the gas at constant volume, new increments of gas are permitted to flow in, and this keeps the pressure nearly constant.

When the sample is sufficiently cold, it has lost quite a lot of volume even although the pressure is the same. The cold sample fits into the box with the corners marked FGOK. To remove the sample from the free expansion system, it has to be released out in

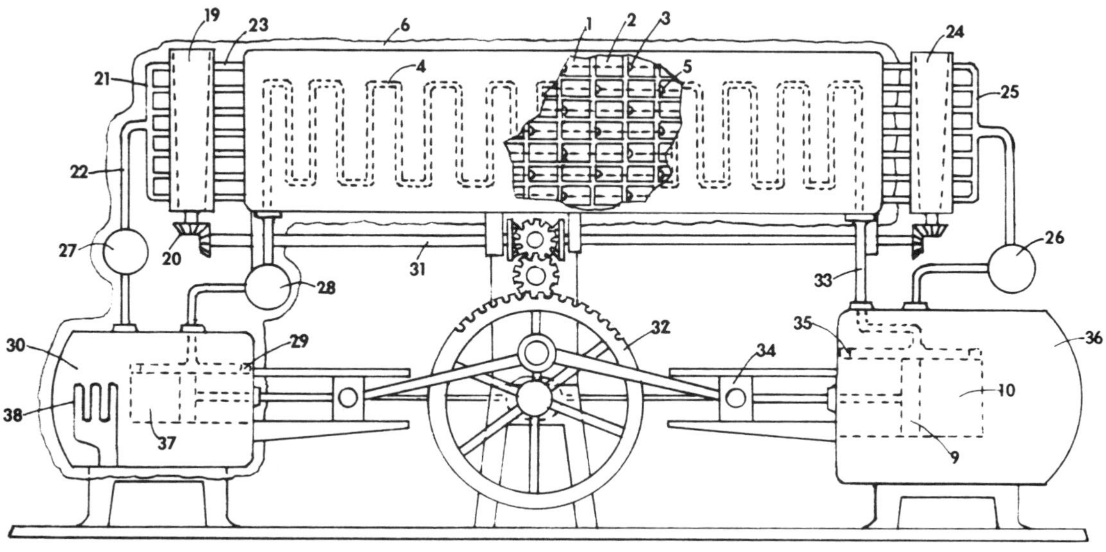

Figure Q-10: Schematic drawing showing the relationship of the parts in the constant pressure process. The alternate free expansion valves are numbered (3) and (5). The cone valves numbered (19) and (24) have a row of staggered ports so that the fluid gas is permitted to move through them in increments. In the actual device, processes are accomplished with rotary mechanisms as far as is possible.

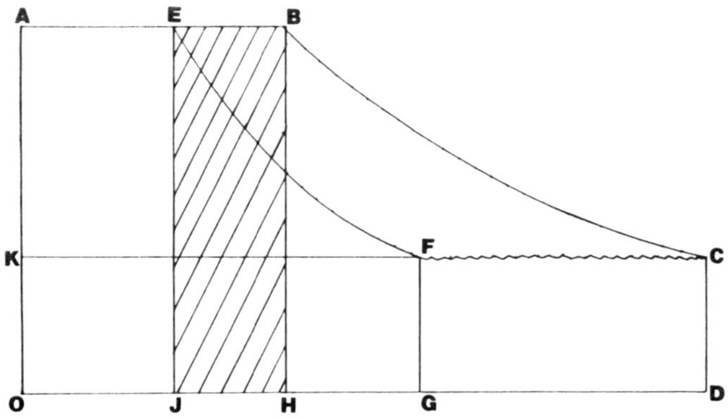

Figure Q-11: The diagram shows the PV changes for the constant pressure process. The gas is cooled at constant volume, but the "free expansion" process allows it to remain at nearly constant pressure. No heat of compression is generated except at the ambient temperature.

increments, using a piston and cylinder. As the piston is opened bit by bit, each increment of the air withdrawn from the last compartment goes through an isothermal expansion. The piston moves only when the valve "X" is open, as shown in Figure Q-12. The total amount of the energy extracted in this way is equal to the PV of the cold sample.

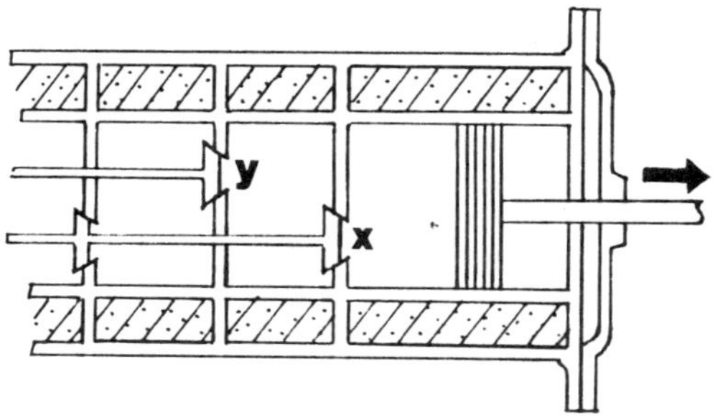

Figure Q-12: The sample has to be taken out of the free expansion process in increments. The piston moves to the right a small amount only when the valve "X" is open and the valve "Y" is closed. During this process an amount of energy equal to the PV of the gas is delivered to the piston.

The sample is now compressed isothermally along the curve from F to E. The compression ratio is 2.7 times. The amount of energy released isothermally is equal to the amount extracted from the free expansion system. A fair trade off is made, both in the amount of energy and in the amount of work received and accomplished.

The next step is to permit the sample to expand at constant pressure along the line EB. The amount of work and energy gained is equal to the cross-hatched portion EBHJ. At point B the sample has returned to the original ambient temperature. To get back the energy that was stored in the reservoir, the sample is allowed to expand isothermally. In this way the original work done on the sample is returned.

The schematic diagram in Figure Q-13 shows a cross section through some example compartments with the heat exchangers running through them. When the valves (3) are open, the valves (5) are closed, and when the valves (5) are open, the valves

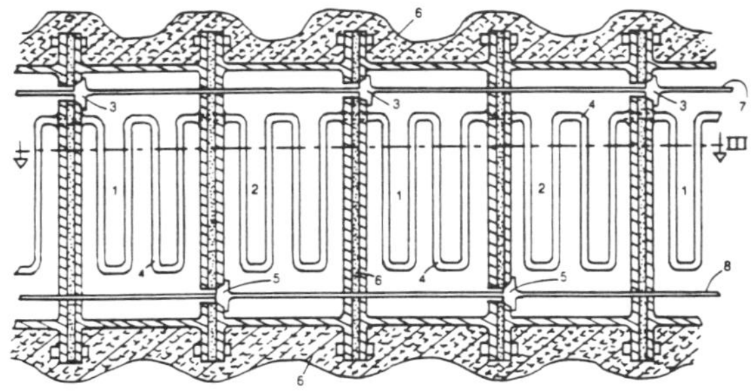

Figure Q-13: Diagram in section of the compartments, valves and heat exchangers. Only one row of compartment are shown. In a commercial installation, hundreds of rows of compartments would be used because the the output volume of gas would have to be sufficient to drive a large turbine installation which in turn would be driving an electric power generator.

(3) are closed. This permits the free expansion process to take place between the compartments (1) and (2), and alternately, between compartments (2) and (1).

WATCHING 3-D TV ON ORDINARY TV

The fact that we see in three-dimensions is mainly due to our binocular vision. Two images viewed from slightly different positions by our two eyes has the biodynamic effect of creating a three-dimensional world. If you close one eye and look around the room wherever you are, you will discover that you can see in three-dimensions with only one eye. This result has never been explained. A fact that is more useful is that if two flat images of the same subject are presented to the eyes, and if the two images were taken from slightly different positions, the eyes will accept the result as real three-dimensions. This is the basis of all three-dimensional products. The difference between the two images for 3-D effect is demonstrated in the diagram of Figure Q-14.

When you watch television, you are watching moving pictures. If you could see the moving picture image the way it looked a fraction of a second earlier, the differences you would see are almost exactly the same as the differences between a pair of stereoscopic images. It is only necessary to delay the moving picture for a fraction of a second and the result is a 3-D partner for the real-time image. The rea-

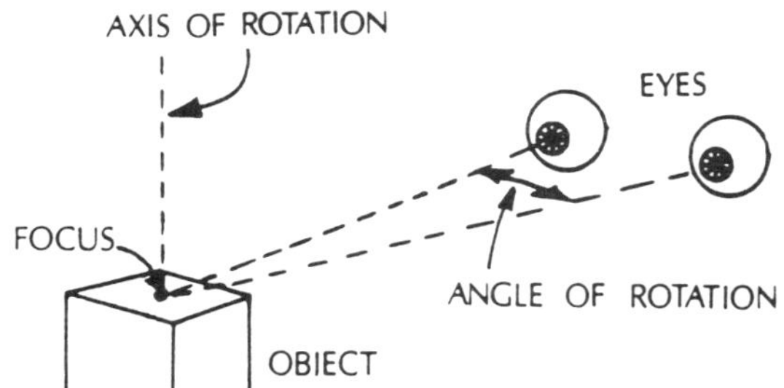

Figure Q-14: Moving pictures like television have a component of rotation in the image around a vertical axis. The rotation of the image provides a continuous pair of images for three dimensional viewing. The two views are separated by a small interval of time.

son that this happens is because in almost all types of motion, there is a component of vertical rotation. It only takes a small component of vertical rotation of the image to produce satisfactory 3-D pairs in moving pictures.

METHODS OF VIEWING 3-D TV

When you want to watch television in three dimensions, there are two basic problems to solve. One is how to make a time-lapse in a television signal. The other problem is how to get one eye to look at one image and vice versa. If you happen to have two television sets with the same size screen, and if you happen to have two video-recorders, you can solve both problems without special equipment.

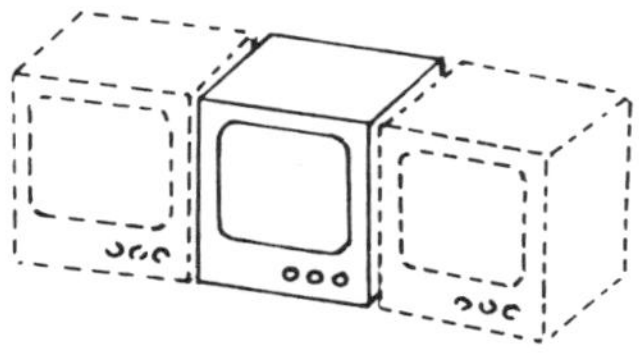

Figure Q-15: Place two TV sets side by side. Then hold your hand out in front of your eyes and gaze at your fingertip. The two TV's will merge into three. The center one will be distinct. Take your finger away from in front of your eyes but keep the TV screens merged together.

The way to get the time-lapsed images is to record the show on both of your VCR's. When you play back the show on your two TV sets, try ana get the timing as close as possible. You will find that you will have a slight time-lapse without much effort. To view the two screens as one, it is necessary to sit back as far as possible and then hold your finger up in front of your eyes, as shown in Figure Q-15. Staring at your finger tip will cause the two images to merge.

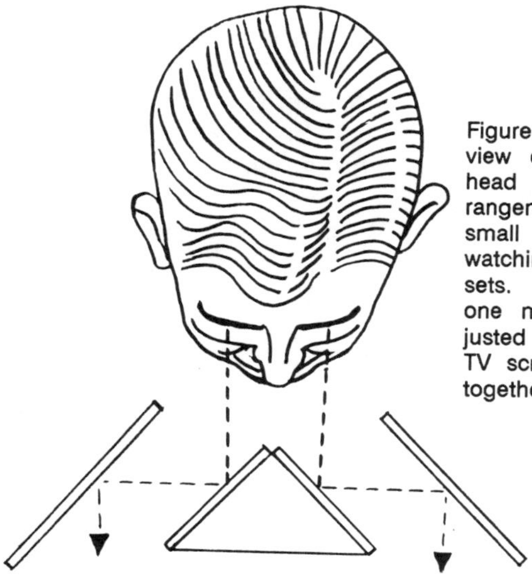

Figure Q-16: Top view of observer's head with an arrangement of four small mirrors for watching two TV sets. The angle of one mirror is adjusted until the two TV screens merge together.

Another method is to use four small mirrors, as shown in Figure Q-16, and position them so that when you look at the TV in the mirrors, the screens merge into one. Another method is to make a pair of viewing glasses like the examples shown in Figure Q-17. A prism acts to turn the line of sight to the center of the second TV set. A small opague panel on a stem effectively blocks out the distraction of the second TV, so that each eye is looking at one of the TV sets.

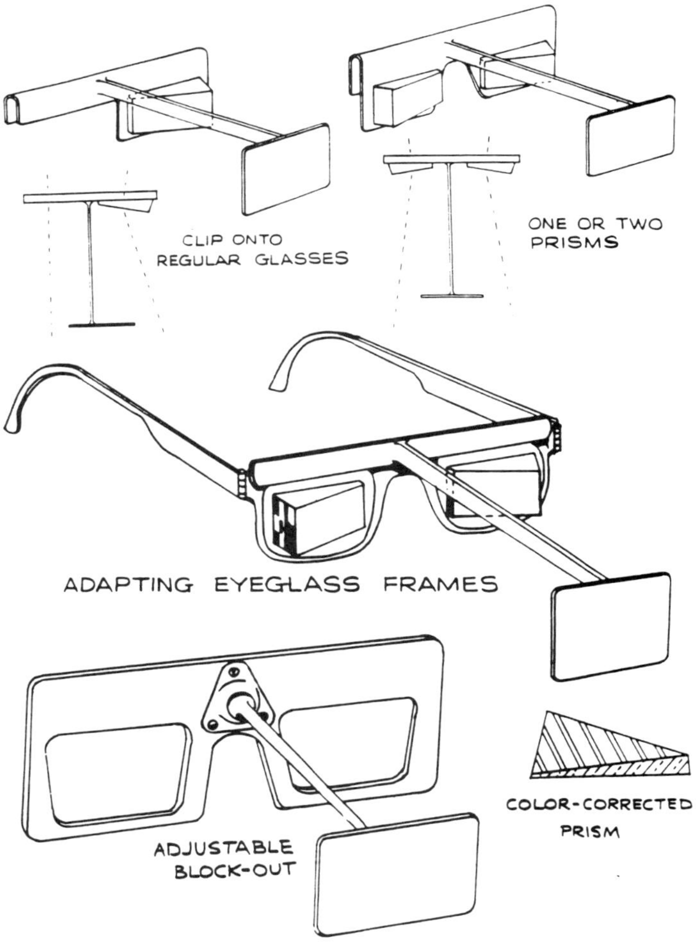

Figure Q-17: Methods of adapting eye-glasses as an aid in watching 3-D TV. One or two prisms are used to turn the line of sight, as illustrated.

SPACE PROPULSION FROM RELATIVITY

Perhaps the greatest concept to come out of relativity was Einstein's formula for the equivalence of mass and energy. During the early days of relativity there was naturally a lot of skepticism, especially when one of the results showed that a material object with a certain mass would become heavier when it was accelerated. The differences in mass due to kinetic energy are almost non-existent at normal velocities, but at velocities approaching the velocity of light, the effects become very noticeable. Near the speed of light the mass of an object increases towards infinity (according to the theory).

Although particle scientists now have to deal with mass variations in their daily calculations, during the nineteen twenties there was no way to prove that

Einstein's theory was correct, or even near to the truth. Einstein then drew attention to the orbit of the planet Mercury. Mercury travels on an orbit that is a long ellipse with the sun at one of the focii of the ellipse. Astromoners had long wondered why there was a discrepancy in the orbit of Mercury. The planet acted as if it became heavier as it passed close to the sun at perihelion. The increase in apparent gravity caused an unexpected change in the track of the orbit. Einstein produced some calculations based on his mass equivalence theory and showed that Mercury did become heavier, but the reason was because the velocity of the planet increased as it approached close to the sun.

When an object becomes heavier due to its increase in energy, it does not affect the conservation of energy. If one object increases in mass due to an increase in velocity, then some other object lost an equal amount of energy. The increase in energy exhibited by Mercury is at the expense of the sun's energy. The relationship becomes clear if the physical principles involved are considered from a different viewpoint.

SPACE PROPULSION USING FREE ENERGY

Figure Q-18 is a schematic diagram of a space ship. The space drive comprises a number of tanks full of compressed air, marked (a), (b), (c), etc. The number (2) indicates a bowling-ball or a canon-ball, and (3) is a track shaped like an elongated race track. The canon-ball fits into the track the same way that a shell fits into an artillery piece, that is gas-tight. The little door numbered (6) can be slid shut across the track.

The canon-ball starts out near the sliding door (6), which is shut. As the cannon-ball is rolled past the first tank of compressed air, the valve (1) on the tank is opened. The air rushes into the track and accelerates the canon-ball along. Every time the canon-ball moves past a valve, the valve is opened, releasing more compressed air into the track and accelerating the canon-ball faster and faster. By the time the canon-ball has reached the end of the track at the point numbered (4) it is going at maximum velocity. At the end (4) the canon-ball is forced to change direction. As the ball races back, the air in the track from the previous expansion is compressed and returned to the tanks. When the canon-ball returns to the starting point, all of its kinetic energy has been converted to compressed air in the tanks. The door (6) is opened momentarily to let the spent ball just trickle through, ready to start over again.

How does the system work? As the canon-ball is accelerated to a high velocity, it gains a certain amount of mass due to its kinetic energy. The track system loses a certain amount of mass because the energy imparted to the canon-ball came out of the track system. At the end of the track where the canon-ball is the heaviest and the track is the lightest, the canon-ball is forced to change direction. This is the same as an impulse. The force exerted on the track by the canon-ball is equal to the force exerted on the canon-ball by the track. The time factor is the same. The important point is that the ball was heavier than normal and the track was lighter than normal. This means that the track was susceptible to a push through space. And that is what happened. At the other end of the track, the canon-ball changed direction slowly, so that there is no relativistic force involved. The push that the track received in one direction is therefore greater than the push that it received in the opposite direction.

This is a propulsion system for propelling a star-ship through deep space. Of course, a canon-ball would not be used in a real life propulsion system. What would be used is electric fields and charged particles. However, the basic principle would be the same. A continuous stream of charged particles is accelerated to an extremely high velocity. While they have extra weight from their excess energy, these small masses are made to change direction. After that, the stream of particles is decelerated back down to slow motion. Then they are made to change direction again before the process is repeated. Whatever energy is used for accelerating the particles, the design of the track must include the means for decelerating the particles and converting the energy back to its original condition so that it can be used over and over again.

HOW DO FLYING SAUCERS WORK

Considering the shape that flying saucers are said to have, like two soup-bowls placed together with one of them inverted, perhaps the propulsion unit is

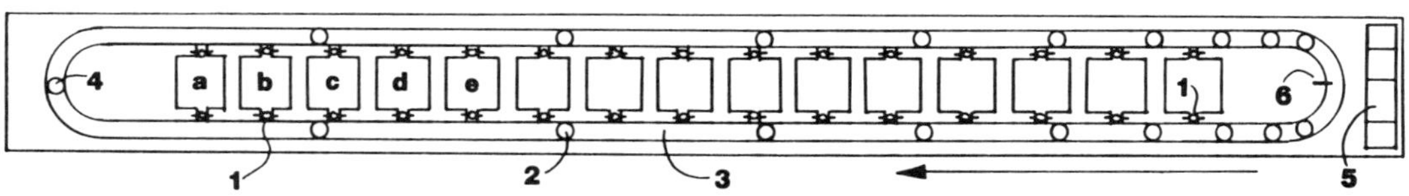

Figure Q-18: Schematic representation of a propulsion system for a star-ship. The principle of exchanging relativistic mass is used. A series of small masses are accelerated to a high velocity in sequence, and then made to change direction by 180 degrees. At the moment that the high velocity mass changes direction, it is at its heaviest due to increased mass from its kinetic energy, but at the same moment the star-ship (which is supplying the energy) is at its lightest. The result is an unbalanced impulse in one direction only. Whatever system is used, the small mass is made to give back the energy while it is being decelerated back down to non-relativistic velocity. In this way, the same energy is used over and over again.

something like the space-drive shown in the diagram of Figure Q-19. In this drawing the ball (1) is a hollow metallic sphere adapted to hold an immense negative charge. The charge might be built up in the same way that a charge is generated on a Van de Graff voltage accumulator. The plate (3) is a controllable positive-negative bias as a means to warp the orbit (2) of proton clusters. Only one orbit is shown. The orbits would criss-cross in every direction. Helium nuclei (or positrons) might be used instead of protons.

The proton clusters are formed when protons are accelerated to high velocities. Charges moving at relativistic velocities tend to lose their repulsion for each other and then cluster together in small clouds.. The objective of the orbit shape is to have the positively charged protons (2) moving near the speed of light when they go over the top of the negative ball (1), but slow down as they approach the warping plate (3). The effect would be like the orbit of Mercury around the Sun. The fast moving charges are changing direction when they are heaviest due to mass-energy equivalence.

The other parts of the drawing are numbered (4), the living quarters and control rooms, (5) an arrow showing the direction of lift or unbalanced force, and (6) is the outer shell.

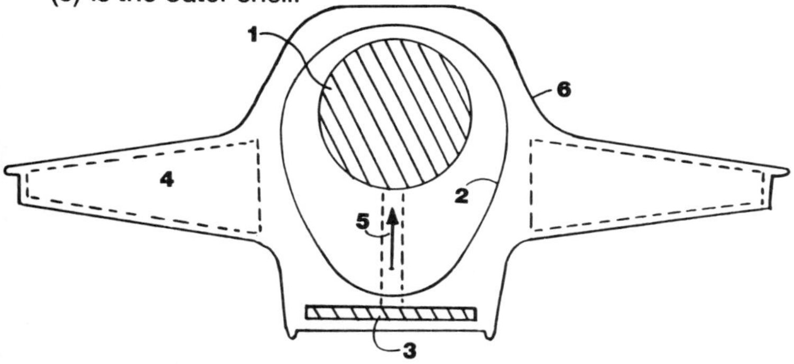

Figure Q-19: If flying saucers exist, the propulsion system they have might be like the method shown in the diagram (or it might be like the hot-plate method shown in diagram Q-20). In Figure Q-19, a sphere with a powerful negative charge has a large number of positrons orbitting around it. The orbit is warped by the lower plate, which acts to slow down the velocity of the positrons momentarily. Because the positrons are moving at relativistic velocities when they fly over the top, there is an unbalanced force in the direction of the arrow.

HOT PLATE PROPULSION

This method is based on a warm or hot area which attracts small fast moving particles towards it from one direction. A good example is the surface of the earth itself. The surface of the earth is warm. The small particles that are attracted to the earth are molecules of air. At the surface, the molecules are fast moving because of the warmth and because of another factor related to energy conversion.

Suppose there was only one molecule of air. If it happened to be high in the sky it would be cold and slow moving because of its distance from the earth. But as it fell towards the earth it would gain in speed (heat). At the moment that it strikes the earth, the molecule is moving at its maximum velocity. It has a slightly increased mass due to kinetic energy. The earth has a slightly smaller mass due to transfering energy to the molecule. The molecule strikes the earth with an unbalanced force.

Now in fact the atmosphere consists of millions of molecules doing exactly the same thing as the one molecule. When the molecules rise up, they lose speed and energy. When they move down, they gain speed and energy. You don't have to worry about the earth being pushed around by the atmosphere, however. The effect is very slight and also it is nicely balanced by the fact that the earth is a sphere.

The way these factors are applied to make a space drive is shown in the diagram of Figure Q-20. The device is aimed so that the Sun keeps the plate (1) hot. The gas inside the chamber (2) is mainly composed of positive ions. The plate (1) has a negative charge. When the positive ions are accelerated to the hot plate (1) they bounce off the hot surface and retreat. As the particles fly away they decelerate because they are tugged back towards the negative charge. The cycle repeats over and over. There is a continuous unbalanced force on the surface of the plate in the direction of the arrow.

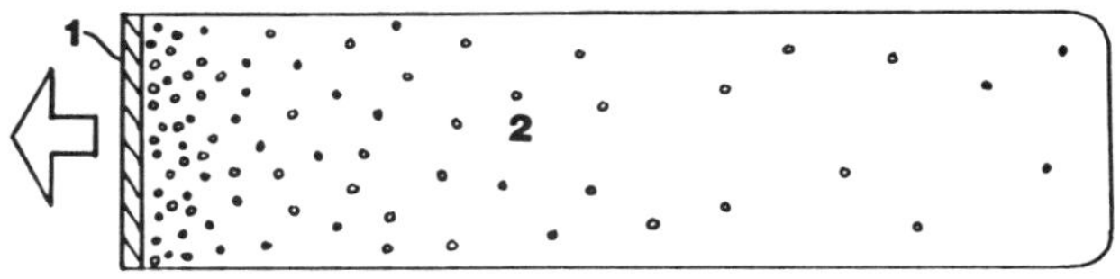

Figure Q-20: A simple propulsion system based on a sun-warmed plate (1) and a gas comprised of positive ions in the chamber (2). The plate has a negative charge. The ions are continuously attracted to the hot plate but bounce off the warm surface. Each such impulse represents an unbalanced force in the direction of the arrow.

INFORMATION MACHINES

When it is tied into a networking system, the personal computer is the ultimate communicator. In one handy device it combines a filing system, a reference library, a teaching machine, a typewriter and word processor, a calculator, an entertainment center, an imaging device and image interface. Besides that it can be hooked up to provide instant communication with the world.

The computer possesses many skills that are beyond human ability. On the other hand, the computer lags far behind the ability of the human brain to store and process a large variety of information. The human brain apparently has a nearly infinite capacity to store information. But the computer has three important features that the human brain lacks. Thes are (1) correct recall, (2) instant recall and (3) total recall.

The human brain remembers but it is often sketchy and doubtful about details. That is incorrect recall. The human brain often needs time to "think it over". It lacks instant recall. The human brain often forgets details or cannot remember the proper time sequence. It lacks total recall.

COMPOUNDING INFORMATION STORAGE

The human brain evolved slowly by processing natural information. The signals available to the evolving brain were produced by natural resonators. Therefore, the receptors of the human brain are designed to intercept natural frequencies. The ear is adapted perfectly to process sound waves that arrive through the air. The eye is adapted to perfectly process the frequencies of light waves.

In the brain information is processed and stored according to a system of patterns called network resonance. Each bit of information goes into a certain number of locations in which the attributes or characteristics of the information are impressed in or on special receptor cells. When an event occurs that again simultaneously energizes the same spots on the same set of cells, the original sound and image are restored and played back. Often an event that is similar to another will energize a certain percentage of a given pattern but not all of it. That is when you say, oh hey, that reminds me of another time.....etc.

Network resonance is one method of compounding a limited number of memory units in such a way that an infinite amount of memory is possible. The hologram must be included in available methods for compounding the recording and storage of information. The way that a hologram functions, the result is almost identical to network resonance. For this reason, there is probably a way to make the hologram work as an infinite memory method.

FREQUENCY SELECTION MEMORY

The airwaves, or broadcast spectrum, can be viewed as an immense memory unit. All of the the broadcast stations in the world, including radio, TV, microwave, ham radio broadcasters, etc., introduce information into this vast memory unit. When it is desired to select out information from the memory, the code that was used for storing the information must be known. In most cases the needed information is intercepted by using a device that selects only certain assigned frequencies. The frequency of the information then becomes the code for selecting it.

It is not necessary to have the whole broadcast spectrum for a memory unit. Exactly the same system will work on magnetic tape or other recording means. The information is stored on the tape with a code based on its frequency. By knowing the code, the information can be retrieved. The information can be piled up on the tape with new information on top of old. Since it is the frequency that selects out the information, overlapping of the signals on the tape is no problem. The only problem is that of running out of frequencies to use for codes.

ME, MYSELF AND I

The sense of identity that dominates the human brain is truly amazing. How did it evolve? What is the evolutionary advantage to the identity of self? These questions are important because it would be a great advantage if it could be possible to instill a sense of identity into a computer brain.

Apparently, in evolutionary processes, it was necessary to have a "self" to preserve so that the powerful sense of self-preservation could exist. Because that powerful need for self preservation is common to all living things, it must be assumed that the self identity feature became a part of living organisms very early in the process of the evolution of life. If insects have it, if snails have it, if the tiniest forms of animal life have it, then it is quite likely that the first evolution of the "self" was rudimentary. The question arises, what are the rudimentary steps necessary to derive a sense of "self"? What is it that imbues a group of cells with an awareness of self?

Suppose that a group of cells are organized to represent a living entity. Each cell of the group performs a certain function and contributes to the well-being and health of the group. What happens if there is not enough food to maintain the health of the group? What if there is not enough moisture for the cells to survive? If there is not enough moisture, the cell group begins to dry up. Now, dry cells are not alive, they are dust.

So it can be assumed that there is an evolutionary imperative to respond in a suitable way to a lack of sufficient water. In order to respond, the drying cells must feel the lack of water. And that feeling is awareness. That is what awareness is all about. Awareness is the imperative to control the surrounding food and water supply so that the cell group can survive. Now you can imagine tiny atennae or hairs attached to some cells of the group. Evolution has placed the antennae there. The antennae can sense the lack of water or lack of nourishment and initiate a suitable response, such as sliding or crawling towards moisture and food.

From the foregoing rudimentary response it is possible to expand the idea that awareness is built into each cell by evolutionary pressures. If there are tiny circuits or feelers which tell the cell when things are okay or not okay, the signals from those feelers and the built-in responses affect the cell at the "gut" level, and that is awareness of self.

NETWORK RESONANCE

The human brain is very complex, yet there has to be an underlying simplicity. The brain and the way it functions came out of a rudimentary application of evolutionary processes. What this means is that the brain was created by the information. The light and heat signals, the sound signals, the touch and chemical signals, were always there. The organs that intercept, sense and interpret such information signals were made step by step by the energy and force of the information itself.

If the nature of the information is understood, the organs that it created to process information will be understood. In the beginning there was light waves and heat waves. There were sound waves everywhere. To focus such information it was only necessary that there was lense shaped gobs of cells, protoplasm, scum, bubbles, water.....even electrolytes were common in the primitive earth. A sound wave in an electrolyte has an electric component, which would represent an important clue for a budding receptor.

Within a rudimentary living group of cells, the energy from the information is intercepted. The information bounces around, reflecting here and there and following a pathway here and there, and it cuts a wide swath of touching cells and caressing cells all through the glob of cells. The important concept is that every time energy and information arrives from a given direction and enters into the group of living cells, the information bounces around, reflecting here and there and following a pathway here and there, and cuts a wide swath of touching cells and caressing cells, exactly the same way every time. Every performance is a repeat performance. Everything is keyed for developing rudimentary receptors and sensing organs.

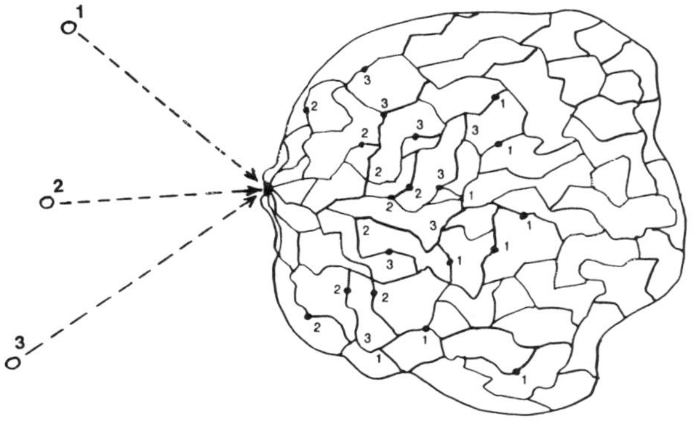

Figure Q-21: Rudimentary diagram of how network resonance works shows three sources of energy or information. A receptor on a network blob permits the energy to enter the network of cells and pathways. By a combination of distance of travel and sensitivity to the particular frequency, the energy flowing through the network automatically selects a certain group of cells or sensitive points to activate, thus recording the information in the form of a specific pattern. The groups of points are marked 1, 2 or 3 according to which energy source activated that point. Because the same source of energy reaches the same pattern every time, the recording points become sensitized by evolutionary processes. Any one cell can be used in a large number of patterns, resulting in a nearly infinite memory capacity.

Because of the nature of information, because of the wide swath of touching, caressing that occurs when energy moves through groups of cells, there was only one way that evolution could go. That was not a one-on-one system. It was thirty or forty energized pathways guiding a caressing thirty or forty energy flows through a pattern of receptors that respond only to that information. When that information resonates with 30 or 40 sensitized spots which are waiting for that particular frequency and strength of wave front, that is a a whole network of receptor cells responding simultaneously to the input signal. That is network resonance.

The more complex the pathways are in the developing brain, the greater is the number of sensitized spots in the responding patterns. In the diagram of Figure Q-21, a pathway network is shown. The numbers indicate the distance that an energy wave has to travel from the original receptor to the sensitive spot. The result is that energy and information arriving from different directions automatically sorts itself out into different receptor patterns. That is the first rudimentary step to discrimination and determining different types of information.

When the same patterns are energized again, the same response and interpretation is derived. Perhaps as many as a hundred different cells are programmed to respond to one kind of energy pattern. However, any one cell is available to function as many times as needed, as long as each pattern is different. The result is a nearly infinite capacity to record and store information. A region that would only be able to store a thousand bits of information on a one-on-one basis, can store millions of different patterns in the same region. Statistically, it is like a lottery in reverse. In a lottery your chances of winning depends on your choosing the correct pattern. In network resonance, the correct pattern is chosen every time by the information itself. In the 6/49 lottery, you have to get six numbers right in a group of 49 numbers. The chances of hitting the right combination is over one in 15 million. In network resonance, the opposite is true. There are millions of patterns for storing information in a group of 49 receptors.

APPLYING NETWORK RESONANCE

When the problem arises, how to apply network resonance to a manmade recording and storage system, there are a great many answers. There might be a thousand ways of doing it with magnetic belts and sensitive travelling recording heads. But there is an obvious solution. What is the fastest and most direct way known to science for the recording and retrieval of information. The answer to that question is the secret of how to apply network resonance.

The other problem is how to duplicate nature's way of interpreting information. Remember that the location of the receptor must be sensitive to the information, and to that information only. And thirty to forty recording spots must respond simultaneously to each bit of information, and only to that bit of information.

EXAMPLE RESONANCE NETWORK

Figure Q-22 shows how a resonance network memory unit can be designed using a magnetic belt of recording material. The belt is wide and travels in the direction of the arrow. One or more double backs, shown in the side view, increase the length of the belt to whatever is needed. The belt has about one hundred starting points for each roving recording head. The recording heads swivel back and forth across the magnetic belt as they record. The result is that each recording path is sinuous and irregular. The recording paths cross each other and crossover and overlap hundreds of times. This is the way it is supposed to function.

The signals that cause the recording arms to swing and swivel are selected according to the file number that is punched into the control monitor keyboard. There must be at least ten such signals, and there can be as many as desired. The objective is to have enough different selections in the signals and starting positions so that any file number used will provide a different pattern.

When information is recorded, a file number is selected and keyed-in. The recording is then made. All ten (or more) of the swivelling recording heads record the same sound track. Whatever starting positions were used for the ten heads, they were set by the file number. To play back the same recording, the same file number is keyed-in. The output starts when the belt and the heads have moved into the correct pattern of the file number. Then the output is delivered, identical to the original that was recorded.

The reason it works is because ten recordings are brought back simultaneously. That means that the output signal is ten times stronger than any spurious signal. All the crossovers from hundreds of other recordings tend to nullify each other by the fact that they have a short blip-time, and the blips are both positive and negative so they tend to cancel each other out. If there is an undesirable noise remaining, it is analysed for the dominant frequencies in the noise, and then a suppressor or filter for those frequencies is included in the output signal.

MACHINE INTELLIGENCE

Every definition of intelligence has to presume that any reasonable response to stimulus input is intelligent. This includes the simplest kind of instinctive responses by living creatures, providing that they are reasonable. If information input is to be processed in a reasonable way, then it is evident that a system for recording, storing, categorizing, filing and sorting must be a functioning part of intelligence. The experiences of "Goldilocks" exactly represents the methods of trial and error that are characteristic of thinking or intelligence. That is try it and if it doesn't fit try again. Have a sufficient awareness and memory to know

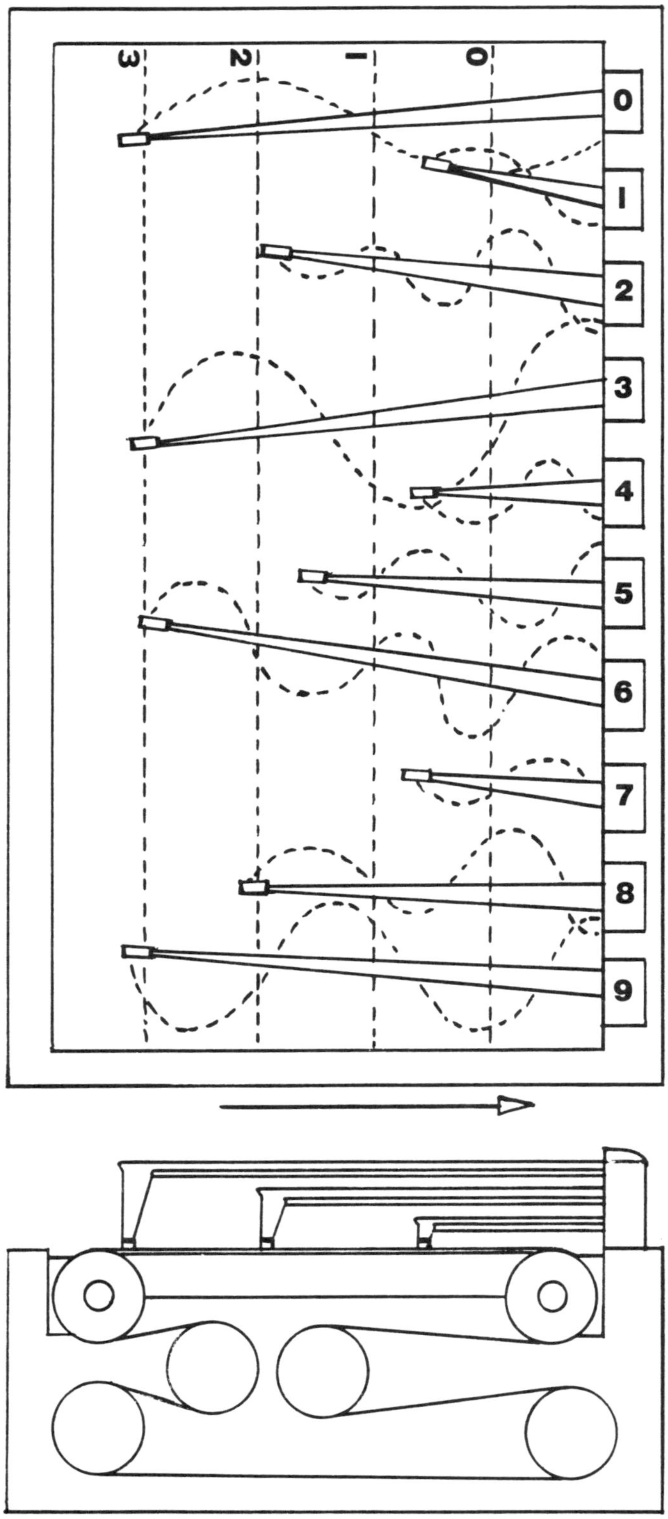

Figure Q-22: An endless belt of magnetic recording material and a group of swivelling recording heads are the operating elements of this example of network resonance. The positions of the heads are controlled by variable signals to provide a large number of repeatable recording patterns.

when something is too sweet (motherbear's porridge), too big (fatherbear's chair), too hard (fatherbear's bed), etc.

The process is simplistic in the beginning because intelligence developed out of rudimentary responses. When the abilities of the processor have evolved into a super processor (the human brain), the "Goldilocks" factor remains intact and in control. But now the equipment for recording, storing, etc., have evolved into super components. The recording method is multi-track audio-visual receptors, thermal sensors, chemical sensors and tactile sensors. The storage unit contains billions of cells and interconnections. The memory is a matrix of concurrent flow of many co-operating channels. The result is that the sight of an apple brings forth many memories of taste, smell, texture, color, halloween, bobbing, New York, PC computers, Johny Appleseed, orchard, don't sit under the appletree, and etc. There is no way that a one-on-one recording and storage process is ever going to mimic the reality of an intelligent response to multi-stimulii.

On the other hand, the concept and operating principle of network resonance is not difficult to apply, and it has the multi-channels necessary and the storage capacity required. When a small start has been made along the right path, the first very modest success will be the "open sesame" to real progress in multi-channel machine intelligence. But perhaps intelligence is too strong a word at this stage. The innovators and experimentors don't care what you call it as long as it is an improvement and advancement of the state of the art.

HOW MANY CHANNELS

The main advantage of network resonance is that the sorting and discriminating process is part of the information flow. This means that when the input data has been recorded, it is not necessary to run the data through a seive to sort and compare. The sorting is done by the network and the determination of information content is made as the flow touches the recording medium. In actual fact, the determination is made while the information is still flowing from the input receptor to the recording medium. The reason for this is because the information creates the system and the response.

For network resonance to function it has to be multi-channel. Forty or fifty (at least ten) parallel channels process the same information at the same time. The natural differences between channels give the matrix flow its special characteristics or imprint. What this means is that some channels are longer than others, so the time factor might be different. Some channels may have less resistance than others and some may impede certain types of input. The result is that the input information and the flow channels control the shape of the recording pattern. And the final shape of the pattern is the result when the sorting, sifting and comparing process are completed.

Applying the same technique to machine intelligence, the flow channels are in the control of the designer. The variations which create the recording patterns can be designed into the matrix of channels. This means that the designer can guide the process along a path of cause and effect.

In natural systems the information is mainly sound waves, light waves and other frequencies of radiant energy. The wide frequency range in natural information input provides an immense matrix of coded sine waves from which natural systems evolved. Although sine waves and natural frequencies are probably the most efficient type of self-sorting, self-identifying input, the principle of network resonance does not need natural information in order to function. It may have needed natural information in the first place in order to evolve into a reasoning brain. However, the method is basically structured and mathematical. There is good reason to expect that the main types of machine language will operate in a "network resonance" environment. The only criteria are those which make the process automatic in the sorting and discriminating attributes.

THE AWARENESS FACTOR

When you try to do something that your computer knows is an error and can't be done, it gives a little beep. That little beep lets the operator know about the error. It also lets the operator know that the computer has a certain quality of reasoning power. It is the self-awareness factor at work.

In natural systems of cells, the cells of the brain which actually account for the recording of memory, have a variety of input and output connections. These are:

(1) The input and output connections for the arrival and output of information signals.

(2) The input and output channels for power supply (nourishment, blood supply).

(3) The input and output channels for the self-awareness factor.

(4) The input and output channels for the lymph system (enveloping fluid or supporting medium).

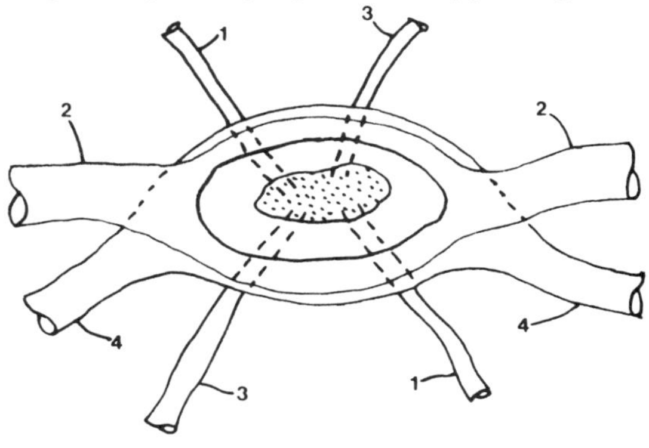

Figure Q 22.1: The minimum number of connections to one recording site are shown in this diagram. The most important function of the cell (or receptor spot) is to record the input information and reproduce it on command. Including both input and output channels, the minimum is eight channels. The enveloping lymph fluid is a dynamic system with definite flow characteristics, and like the blood stream, has an input and an output.

The most important input-output channels are the self-awareness monitors. This is equivalent to a separate test circuit which continuously measures all of the things that are happening in the cell. If something does not measure in the normal range, the self-awareness response becomes evident and there is an alarm signal, which can be a mild reminder or an urgent response or it can be a panic situation.

When all of the many cells are registering normal, the central core of the self-awareness factor provides a feeling of well-being. Everything is okay. But when some or many of the cells are registering not okay, that is when the self-awareness factor provides an appropriate directive. You may feel like taking a nap for half an hour, or you may feel a sense of unease or foreboding. Perhaps you feel as if a headache or a cold is going to arrive soon. The panic situation occurs if you get dizzy and fall down. Then you know you need help. The telephone is within reach.

Perhaps these same kind of features, adapted to function as testing monitors in computer circuits, will become the rudimentary self-awareness core of the thinking machine. The process is one of refinement and sophistication. But the first step is to provide parallel circuits to all information circuits. The job of the parallel circuit is to immediately convey what is happening to every component and connection in the computer memory.

BINARY AND DIGITAL PULSE COMPOUNDING

The use of analog devices was well established before the computer became a household word. Very accurate meters and monitors are still finding a useful application in the measurement and control of electricity and fluids such as water and natural gas. Binary numbers provide unparalleled accuracy because there is only a one and a zero. But digital pulse technology has also made wide inroads into the music and communications industries. Errors are mainly eliminated by the number concept.

Suppose that a very accurate current meter is used to measure the amplitude of a certain pulse of electric current in a circuit. The measurement yields a number like 8.77264093875. What happened is that the meter is sufficiently accurate to produce a string of numbers. It would be a straightforward engineering problem to make an output device for the meter so that the string of numbers would come out as a string of pulses with the same sizes as the string of numbers. This example is indicating that the single pulse of current was in fact representing ten or twelve digital pulses.

An information storing and transmitting method would then be possible in which a single pulse would represent several pulses with digitized values. This method works best with binary numbers because there are only two types of numbers to distinguish, a zero and a one. A zero is quite different than a one, which means that the method does not have to be perfectly accurate to realize useful results.

Figure Q-24 shows a type of electronic gate with one bar. When a pulse is directed through the

Figure Q-24: Shown at drawing (a), three things are necessary for decoding compound binary bits. One is a gate with a bar which cuts off a pulse at the height of a "one". The second is a feed-back loop. The third is an amplifier in the feedback circuit that exactly doubles the value of the remainder of the pulse. Shown in drawing (b), the part of the cut-off pulse that is left over is fed back through the amplifier and then goes through the gate again.

gate, if the pulse hits the bar, the gate says that a "one" went through. If the pulse goes under the bar, the gate says that a "zero" went through. If a "one" went through, the portion that was higher than the bar then goes to an amplifier. If the pulse went under the bar, it goes to the amplifier. Whatever part of the pulse remains, it is amplified exactly by two. Then it is directed through the gate again. The process is repeated until the pulse yields only zeros. This gate and amplifier permits 8 or 10 binary numbers to be recorded in one pulse.

THE TEN-HIGH GATE

The ten-high gate works exactly like the binary gate. The only difference is that the amplifier must accurately amplify the leftover signal by a factor of ten. As shown in the illustration, Figure Q-25, the ten-high gate has nine bars placed like a ladder. When a pulse is directed through the gate, if it goes under the number one bar, the gate says that a zero went through. If the pulse hits the number 7 bar, the gate says that a seven went through. Whatever part of the pulse remains after the seven units have been cut off of it, that part is amplified by ten times. The process is repeated until the pulse yields a string of zeros. This gate and amplifier permits 10 to 15 digital pulses to be recorded in one pulse.

The process for coding bits of information to form compound pulses is simple. For binary bits, the string of bits to be compounded are run through the processor in reverse order. Each bit is reduced to half of its value as it goes through the processor. The reduced pulse is added to the next pulse in the string just before it too is reduced in size. Since the pulses are all ones or zeros, the reduction is always the same amount.

For digital bits, the compound pulse is formed by selecting the string of bits and running them through the processor in reverse order. Each bit is reduced to one tenth of its size as it passes through the procesor. Then the reduced bit is added to the next bit in the string just before it is also reduced in size.

Figure Q-25: The process for decoding digital compound pulses is the same as for the binary pulse except that the gate has nine bars and the amplifier increases the leftover pulse by a factor of ten. The code for compounding bits of information is direct. Each digital pulse is reduced in size to one tenth of its value and then added to the next pulse in the sequence just before it is reduced in size.

HOW A LASER WORKS

The way that a laser operates is by generating a specific frequency within a lasing type of material. By flashing an intense burst of energy through the lasing material, many of the electrons in the material absorb energy and then represent potential radiators. Any small trigger at the correct frequency will cause the electrons to release their energy.

The release of energy starts with one electron. The radiation from that one electron is the correct frequency to make all the electrons respond and radiate. At that moment over ninety percent of the potential radiators release their energy simultaneously, and in phase with the "trigger" wave.

The released energy is coherent, which means that it is all in phase at the same frequency. The energy radiates away in every direction at the velocity of light. To capture and guide the radiation, the lasing material is made in the shape of an elongated tube or strip. Some of the new lasers are in the shape of a stick of gum. The transparent lasers have flat mirrors at the ends of the tube to make the radiation bounce back and forth inside.

The radiation mainly leaks away, but the small amount of radiation captured by the tube performs the lasing function over and over again, causing more and more electrons to release energy along the linear direction of the tube. At one end of the tube the mirror is only half silvered, so that bursts of the laser beam are released.

Coherent radiation does not spread out but remains in a tight beam or ray. A ruby laser that the writer used for night demonstrations would place a red spot of light the size of a dime on a white post several blocks away.

SELF-FOCUSSING LASER

The self-focussing laser operates exactly the same as a normal laser except that the lasing substance or lasing substance container is in the shape of an "egg". As shown in the diagram of Figure Q-26, an ellipse has two focal points. The distance between one focal point and the other, including one reflection off the wall of the "egg", is a constant. That is, **A** plus **B** is always the same, regardless of the direction of the line through the focal points. This constant dimension then is chosen to be the base-line for the laser.

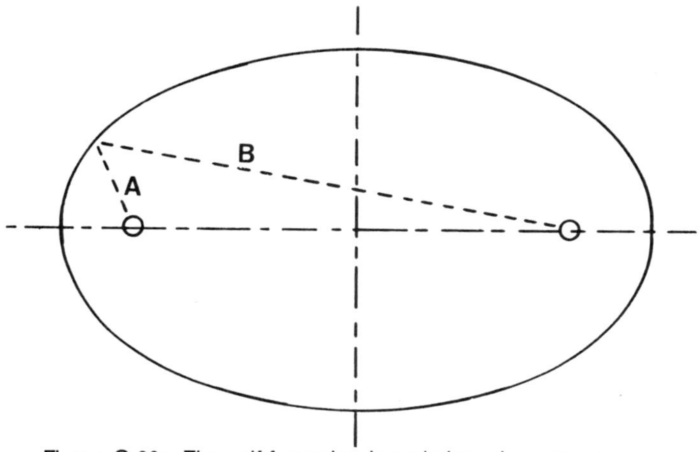

Figure Q-26: The self-focussing laser is based on an egg-shaped container of the lasing substance. As shown by the diagram of an ellipse, the two focal points are always an equal distance from each other including one reflection from the interior surface of the "egg". That constant distance is chosen as the base-line for the laser.

The interior of the "egg" (or exterior in transparent solids) is mirrored so that light rays passing through the focal point will reflect back to the other focal point, thus establishing the base line.

Another physical fact about an oval is that any line passing through one of the focal points will change its direction during each reflection until it passes through both focal points. This self-aligning feature causes any light rays passing through one of the focal points to become aligned exactly with all such other light rays. The more often the light rays reflect from the inside of the "egg", the more focussed the rays become.

The desired degree of focussing is controlled by the size of the opening that is made in the end of the mirrored "egg". The opening is made exactly in line with both focal points. The smaller the opening, the more tightly focussed the laser beam will be when it releases from the laser.

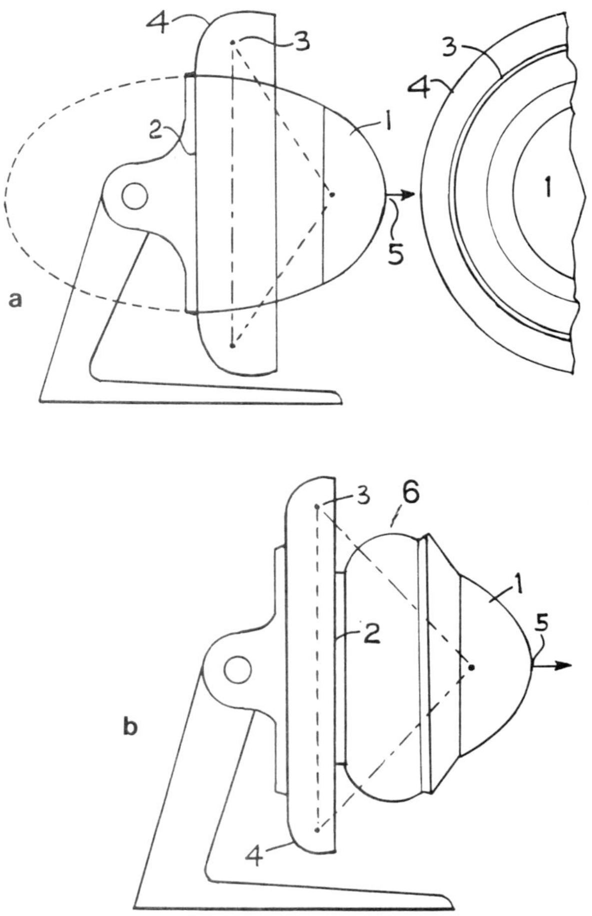

Figure Q-27: Side view of "egg" laser showing the portion of the inside of the "egg" that is silvered (1), the plane mirror bisecting the "egg" (2), the flashing filament (3) with a reflector (4) to converge the rays to the focal point. The focussed rays escape through a small opening (5). In diagram (b), the transparent solid of the lasing substance is shaped as a lense (6) to converge the rays onto the focal point of the laser.

As shown in the diagrams, Figure Q-27, the oval part of the laser (1) is cut in half and the plane cut is made into a plane mirror (2). The whole "egg" then still exists in terms of the optics of the laser, as in drawing (a). Half of the egg is a reflection.

The flash of energy used for giving the laser its starting power is provided by a circular filament (3) with a parabolic reflector (4) or other converging lense (6), as shown in drawing (b). After the light rays have packaged themselves into a parallel coherent bundle, they escape from the small opening (5) in the nose of laser. The laser beam does not escape but continues to reflect back and forth inside the ellipse until the focussing is automatically completed. The size of the opening controls the degree of focussing.

THE PHOTON GENERATOR

The photon is the name given to the particle that is believed to represent light waves. A photon is created when an electron gives up energy. The charge in the electron vibrates in a radial mode and so the resulting waves of radiant energy are toroid-shaped or doughnut shaped. The next following doughnut in the photon is of opposite polarity, and the polarity changes at the frequency of the radiation. This series of toroidal fields is extremely durable and persistent. It is for this reason it is often called a particle.

Since it is known that the toroidal-shaped fields are generated by charges vibrating in a radial pattern, then the next step is to make large man-made photons by designing devices that will cause electrons to vibrate rapidly in a radial way. This places the size and the power of the photon under the control of the designer. They can be made sufficiently powerful so as to have a significant effect.

The uses to which man-made photons can be applied are (1) communication systems, (2) magnetic projectiles, (3) traction at a distance, and (4) for launching space vehicles. The magnetic projectile requires a sender only, but the other three items need both a sender and a receiver. The photon is aimed at the receiver in the same way that a laser is aimed. The receiver then responds to whatever type of control pulses there may be in the transmitted signal.

Besides being used as a means for launching space vehicles, the photon generator can also be used for building bridges and skyscrapers. Someday when photon technology is well advanced, it may be used for levitating and transporting people across a city or from town to town.

HOW THE PHOTON GENERATOR WORKS

The methods about to be described are "PAF", which means Patent Applied For. In mentioning this the intention is not to scare off inventors from working on the idea. Improvements in any field can be patented, including an improvement on a newly patented invention.

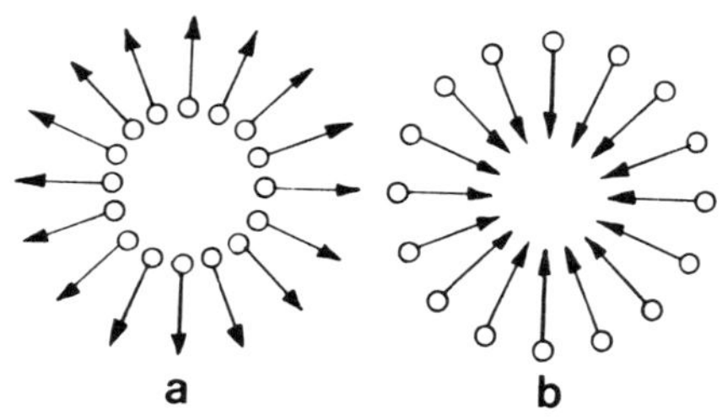

Figure Q-28: When charges are vibrated in a radial pattern, the resulting fields that are generated take the shape of a toroid or doughnut. Each change of direction of the vibration also changes the polarity of the doughnut. The projected ray is like a stack of doughnuts in a line along the axis of the toroid shapes.

The main concept that makes a photon generator function is a method for making charges vibrate rapidly in a radial pattern. The diagrams (a) and (b) in Figure Q-28 show the radial pattern. Diagram (a) shows the charges when they are at the inside of the vibration and diagram (b) shows the charges at the outside extent of the vibration. The charges vibrate in unison between the two positions shown.

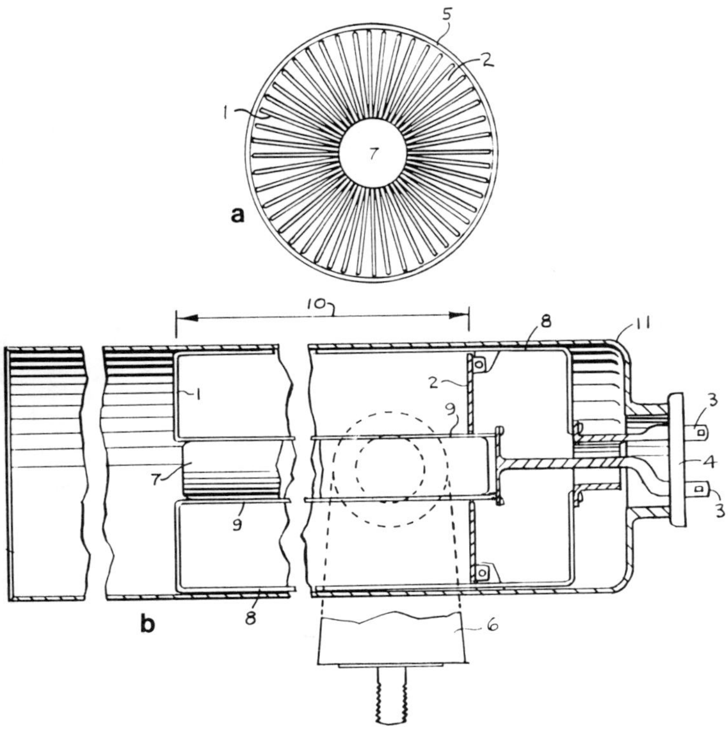

Figure Q-29: Drawing (a) is a front view or axial view of the radial pattern of conductors in the photon generator. When an alternating current goes through the radial conductors, the effect is the same as when charges are vibrated in a radial pattern. Drawing (b) is a sectional side view of a photon generator. The radial conductors are indicated by the number (1) and the reflector is numbered (2).

What happens next is amazing. Each of the individual charges generates an acceleration field, which is a circular field of magnetism at right angles to each axis of vibration. The circular fields combine by vector addition to form doughnut-shaped fields. The doughnuts leave in a straight line along the axis of the doughnut at the speed of light, and every doughnut is of opposite polarity to one next to it. Each doughnut-shape is boxed in by opposing fields. This makes the projected beam of radiation coherent and persistent, similar to the photons in a ray of light.

The projected ray leaves the photon generator in both directions. The ray that is going in the reverse direction is reflected back by a plane reflector. The plane reflector is positioned so that the reflected ray re-inforces the ray that is going in the right direction.

THE PHOTON GENERATOR

An example of a photon generator is shown in the drawings of Figure Q-29. Drawing (a) shows how conductors are arranged in a radial pattern. When an alternating current of the correct frequency is introduced through the radial conductors, the magnetic doughnuts are produced. Drawing (b) shows a side view. The radial conductors are indicated by the number (1) and the reflector (2) is also shown in section. Figure Q-30 shows a space vehicle in the process of being launched. The cross-section shows a pattern of five photon generators in the base of the rocket. There is a duplicate set of photon generators in the ground launcher.

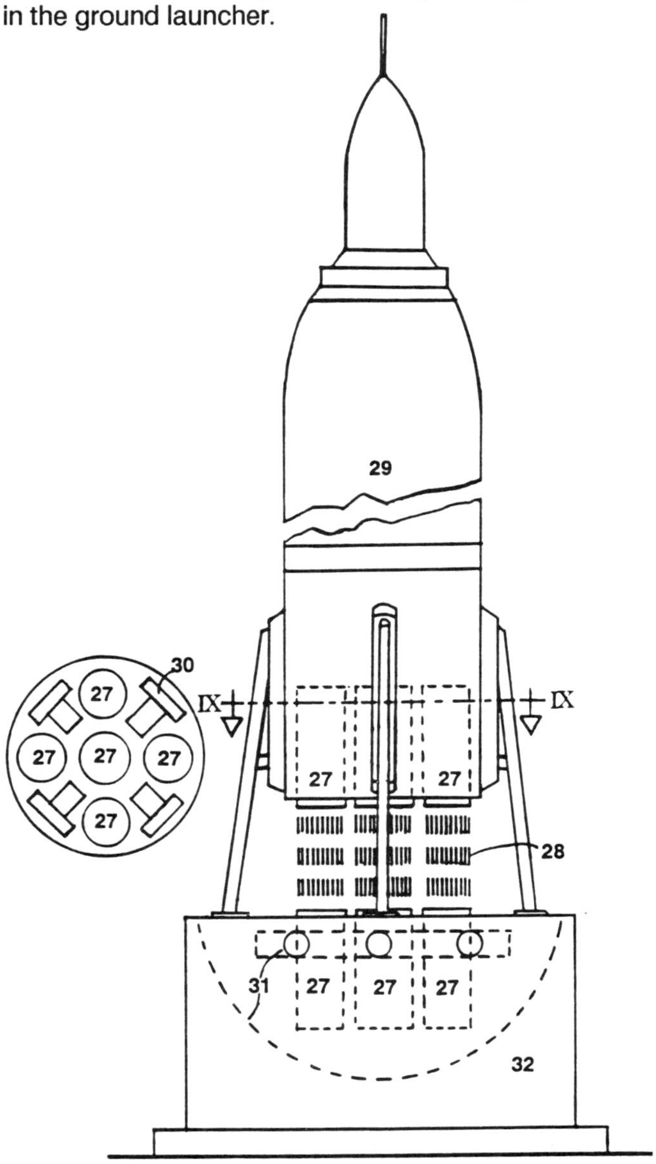

Figure Q-30 is a schematic drawing of a space vehicle during launch. A cross section near the base of the rocket shows a pattern of 5 photon generators. There are 5 matching photon generators in the ground laucher. The set of photon generators in the ground launcher exchanges energy with the set in the rocket.

The repulsion that lifts the vehicle off the ground is supplied by powerful magnetic fields. These fields are in the shape of doughnuts and they are of opposite polarity. The photon generators in the rocket are designed to get their main source of energy by absorbing the fields projected by the photon generators in the ground launcher. The same energy is then projected downwards from the photon generators in the rocket. Since there is very little in the way of chemical fuel on board, the rocket weight is mainly payload.

The control of the vehicle, to keep it upright and on track, is accomplished by inertial guidance and compensating motions in the photon generators. In other words, the guidance is the same as for a chemical combustion rocket.

The circuits of the photon generators in the vehicle are designed to intercept and absorb the energy from the ground launcher. In this way 90 percent of the energy used is supplied by the ground launcher, not the vehicle. The disadvantage is that the space vehicle has to be tracked and controlled from the ground right up until the time that it is in orbit. The technical know-how to make such a launching system work is still in its infancy. This means that the whole field is wide open for innovation and new inventions.

THE MECHANICAL MAN

Power steering makes it possible for a man to easily steer and control huge tractor trailer units and other large machinery. The power steering idea permits a man's strength to be greatly multiplied.

The mechanical man is actually a man-like structure of articulated joints. Each joint, such as knees, wrists, elbows, etc., are powered by hydraulic, pneumatic or electric servo-motors. The power source can be remote or back-packed by the machine.

An ordinary man, trained in its use, is placed inside the device. He is protected by steel armour-like plates and internal padding for comfort. What happens is that when the man moves, the device moves, but with strength 1000 times as strong as a man. When there is an earthquake or a trainwreck, the mechanical man goes to work rescuing trapped people. He is strong enough to gently remove slabs of steel and concrete from trapped victims. He can lift up the end of a railroad coach by himself to free the helpless and injured.

A man's natural balance that prevents him from falling over is duplicated by the machine because it follows every move that the man makes. In effect, the mechanical man can do anything that an ordinary man can do except that the mechanical man is 1000 times stronger. Augmenting regular disaster relief, this machine could save thousands of lives around the world in the course of a year.

CONCENTRATING RADIANT ENERGY

Solar panels that generate electricity from sunlight and solar panels that are suitable for heating homes are an important source of energy in these times. Once the solar panel is installed, it is virtually trouble free so that when the initial costs have been amortized by the resulting energy supply, the energy is nearly free. One reason for the great interest in solar panels is because of the rapidly diminishing world stocks of coal and oil. Also, when there is a threat of war or revolution in oil producing countries, the world oil supply is at risk.

In the past twenty-five years hundreds of designs for solar cells and solar heat collectors have been patented. Many of these designs are effective, and have the result of reducing energy costs for those who are using them.

PARABOLIC REFLECTOR PANELS

The most efficient concentrator of radiant energy is the parabolic reflector. One reason is because it concentrates rays of every type and frequency by reflection, whereas equivalent lenses may not pass all of the important energy frequencies. The parabolic reflector is made in strips, but the cross-section through a strip is not much different than the cross-section through the reflector in a flashlight or in the headlights of your car. This is shown in the drawing of Figure Q-31. The element at the focus of the parabolic concentrator is usually a copper pipe painted black to absorb the maximum energy.

The disadvantage of the parabolic reflector is that it has to move around with the sun, always facing the sun. If the sun goes behind a cloud, or at night time when the sun is gone, the power shuts down temporarily.

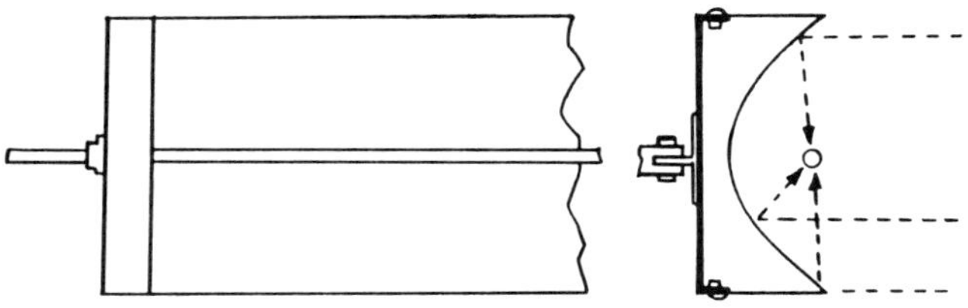

Figure Q-31: A side view section and a front view of a parabolic reflector strip. These reflectors and also lenses function by converging incoming rays to a point. They are most efficient when lined up facing the sun.

AMBIENT RADIATION CONCENTRATOR

Another kind of radiation concentrator operates by increasing the amount of radiant energy that is arriving at a thermal element. The reason that a particle of matter remains at an ambient temperature is because the character of the radiation and the amount of energy leaving the particle is equal to the amount arriving. If the radiant energy arriving at the point is increased, the point will become hotter. It will increase in temperature until the radiation it is emitting is of the same wave lengths and equal to the radiation being received. Then the particle will hold that higher temperature even although it may be hotter than the surrounding air. The sun does not have to be shining for this to work. However, the sun is a great booster to any kind of radiation collection system.

The ambient radiation concentrator works by diverging the angle at which radiation is arriving at any particular spot. This is the opposite to what a lense or parabolic reflector does. They converge radiation.

The parabolic reflector takes parallel rays and converges them to a point. The objective of the ambient radiation concentrator is to take radially incoming rays and converge them to a point. This effect can be seen if you have a fish tank. The diverging rays are shown in the diagram of Figure Q-32.

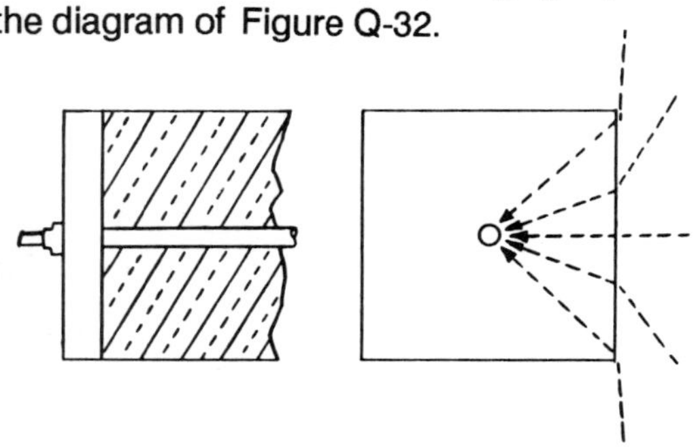

Figure Q-32: Example of radiant energy concentrator that uses diverging rays instead of converging rays. The result is that the radiant energy that was approaching a square glass rod from an enclosed angle of 170 degrees is made to arrive at the thermal element enclosed within a angle of 90 degrees. Since the square glass rod has four sides, all sides accept radiation coming from any direction.

There are many ways of achieving an elevated temperature by using diverging techniques. A special Fresnel lense can be designed to accomplish a divergence in radiant energy. A concave glass lense is another example. An example of a reflector-type diverging system is a parabolic reflector strip in reverse. The back of the parabolic surface is mirrored rather than the front. This is shown in Figure Q-33.

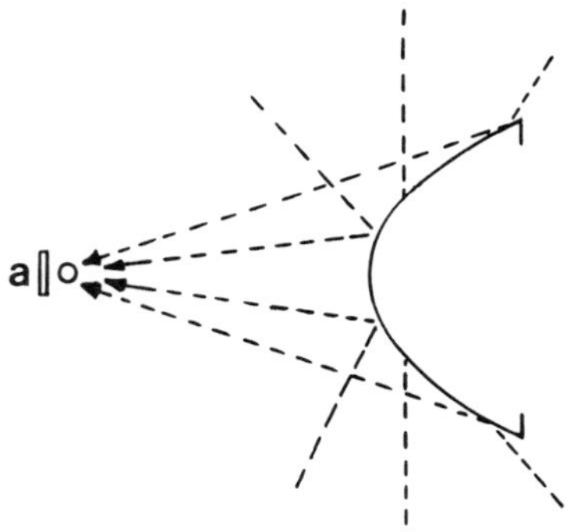

Figure Q-33: The back of a parabolic reflector functions as a diverging lense for concentrating ambient energy. The effect is increased by placing a small plane mirror, (a), behind the thermal element.

The reversed parabolic mirror easily diverges a full 270 degrees of enclosed angle down to 30 degrees, a concentration increase by a factor of 9 times.

ELLIPTICAL MIRROR CONCENTRATOR

Another type of radiant energy concentrator uses the attributes of an ellipse to concentrate converging rays of energy. As shown in the diagram, Figure 33.1, the two attributes of an ellipse are (1) that any ray passing through a focus point and reflecting from the inside surface of the ellipse will then pass through the second focus point, such as the ray (a) and (b), and (2), any ray passing through one of the focus points will, in subsequent reflections, become parallel and co-incident with the line (g), which is the main axis of the ellipse.

The ellipse is cut-off on the line (h) and only the small portion (e) is used. A plane mirror placed on the vertical axis at (d) will then perform as the optical image of the part of the ellipse that is cut off. The size of the mirror (d) is determined by the line (c).

Some cross sections of the concentrator are shown in Figure 33.2. A small opening or aperture in one of the mirrors on the axis permits the concentrated radiation to enter and impinge on a hollow thermal element (4) with a blackened surface (5). A cylindrical jacket with a mirrored inner surface (7) reflects back escaping energy.

Figure Q-33.1.

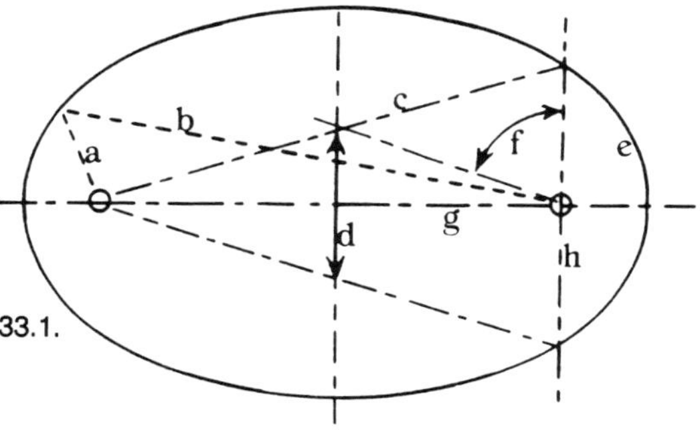

P.A.F.

FIG. 1.

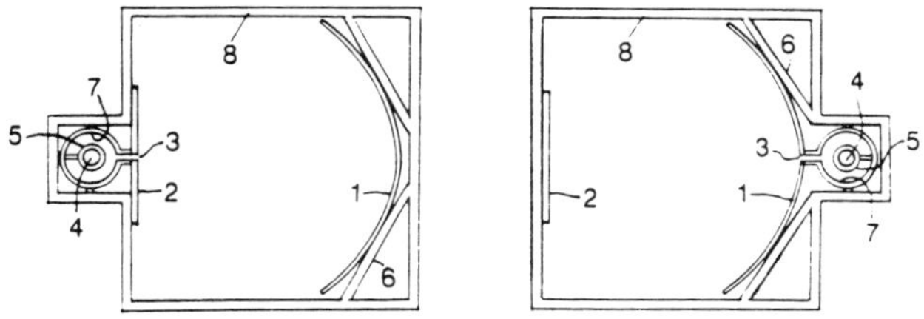

FIG. 2.

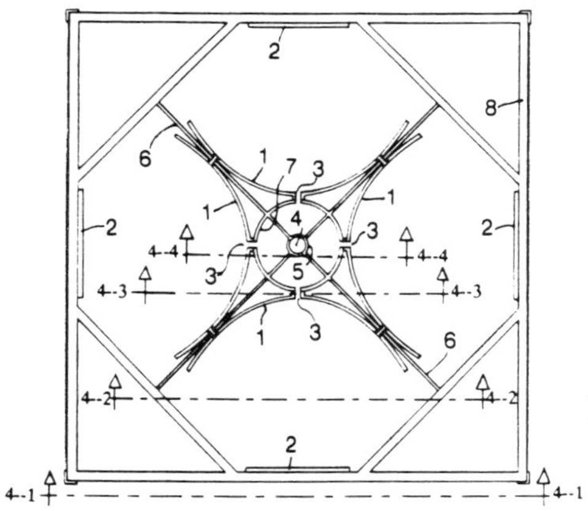

FIG. 3.

Figure q-33.2: These drawings are cross sections of the radiation concentrator reproduced from the patent drawings. The lower drawing is a cross section of a tower with the elements placed vertically. The radiation concentrator faces in all directions so that if the sun is shining, the rays are utilized.

INVENTIONS FROM THE PAST

Alexander Graham Bell was the pioneer in the technology of communicating with light rays and parabolic reflectors. This drawing depicts Bell and an assistant sending audio signals by varying the intensity of a ray of light. The signals were sent to a receiver several miles away. This experiment forecast the use of micro-wave dishes for transmitting communications.

This drawing was made for the popular press over a hundred years ago. The artist is unknown (unless somebody out there can remember).

PHYSICS : UNDERSTANDING HOW IT WORKS

THE MAGDEBURG HEMISPHERES

In l642 Otto von Guericke was the Mayor of a small town in Germany called Magdeburg. His office allowed him time to indulge in his hobby. He was an amateur scientist. When he talked to people about his ideas, they thought he was wasting his time. People laughed when he claimed that air could exert a powerful force. Nothing in the world is so fluffy and light as air!

Burgomaster Guericke thought that we were situated at the bottom of a huge ocean of air. He didn't know how much pressure that would develop but he was sure it would be considerable. So he designed and made a vacuum pump. He hired the local blacksmith shop to help him to make two hemispheres out of metal about two feet in diameter. He fitted the two hemispheres together and made a leather gasket soaked in oil to make them airtight. Then he used his vacuum pump to create a vacuum inside the spherical cavity.

The Mayor called a holiday and got all the townspeople and the local farmers to watch his demonstration. A team of horses was hooked up to each hemisphere. Then the horses were whipped into a tug-o-war to try and pull the two hemispheres apart. When they finally let go it was like a bomb going off. The air rushed into the vacuum. Then Otto von Guericke showed everyone that only a leather gasket was separating the two hemispheres. Air pressure on the outside of the hemispheres held them together like glue. Everyone was impressed. Otto von Guericke became known throughout the scientific circles of Europe.

GIANT STEPS FORWARD

About the time that Otto von Guericke was demonstrating his ideas about the ocean of air around us, Isaac Newton was born in England. He was to have a profound influence on the lives and the knowledge of mankind. He was a keen observer. He was also very curious. He watched the way that objects moved and how they bounced when tossed about. He noted that the rougher the surface was, the quicker a rolling apple slowed down. What would happen if the surface was perfectly smooth? Finally he realized that if there was no friction at all, the apple would keep going at speed forever. Nobody had ever seen such a thing happen because there is friction everywhere. But Newton looked up at the moon and thought to himself that it was a perfect example of what he was thinking about. Apparently there was no friction where the moon moved.

Newton wrote notes about his ideas and eventually resolved all motions into simple laws. These are:

(1) That a mass at rest will continue at rest and a mass in motion will continue at the same speed in a straight line unless acted on by a force.

(2) That a mass will accelerate if acted on by a force. The acceleration will be in the exact direction of the force and will be proportional to the force.

(3) That to the action of every force, there is an equal and opposite reaction along a straight line.

These laws may seem obvious to us now, but in Newton's day, nobody would have believed that a moving object would move forever. Everyday experience showed that everything slowed down and stopped.

IN THE GRIP OF THE GALAXY

Early scientists were very religious. Many of them were seeking a confirmation from the heavens that they were walking on a path ordained by God. One of the amazing things about the universe that really impressed them was the fact that everything in the solar system moved according to a space allowed by the stars. This fact alone seemed to reveal the hand of God. How could the distant stars control our destiny without divine guidance?

This profound knowledge was available only to the scientists of the day, and then only to those who understood the motions of the planets. In 1851 an astromoner named J. B. L. Foucault got an idea that would demonstrate to everyone how the stars controlled our lives. He made a heavy iron ball and attached it to a strong wire 200 feet long. He suspended the ball from the top of the center dome of the pantheon, which is a Roman temple dedicated to all Gods. Two hundred feet below fine sand was smoothed over a concave floor made exactly to Foucault's specifications. A pointer attached to the bottom of the ball was adjusted so that it would draw a line in the sand when the suspended ball was moved. To start this huge pendulum in motion, it was carefully held to one side of the room and tied with a thread. Then the thread was burned off to release the ball. This prevented any unwanted motions from affecting the pendulum. Duplicates of Foucault's pendulum can be seen in many museums around the world. There is one in the museum in the park at San Francisco.

What Foucault's pendulum does is to swing back and forth on a line dictated by the stars of the Galaxy. This makes the pendulum seem to change direction and turn right around every twenty four hours. Of course what is happening is that the earth is turning while the pendulum remains fixed in space.

How does this affect our daily lives? Well, when you go around a corner in a car or on your bike, you feel centrifugal force pulling you to one side. You may think that is an effect of the earth, but it is not. The power of the stars is more than a million times stronger that any slight effect that earth exerts on us. Remember that we are talking about intertia, not gravity. The earth is a tiny speck within an immense ocean of matter we call the Galaxy. The forces that give rise to inertia are contributed to by all matter. So you can imagine how much more powerful the whole Galaxy is when compared to the earth alone.

WHAT NEWTON'S LAWS TELL US

In the same way that the Magdeburg hemispheres demonstrated the enormous weight and pressure of the atmosphere around us, Newton's laws and Foucault's pendulum demonstrate the power and pressure of space. In the same way that the air seemed to be too light and fluffy to have a real effect on anything, the space around us also seems to be without substance and of no consequence. However, by examining Newton's laws, and applying them to the events around us, space also reveals its power. Whereas the pressure of the atmosphere turned out to be 14.7 pounds per square inch, the pressure of space turned out to be almost infinite in strength.

Newton's laws in the form of formulas, as well as simple laws derived from them, are as follows:

$F = ma$ (The force necessary to accelerate an object is equal to the mass of the object multiplied by the acceleration attained).

$ma = ma$ (The mass multiplied by the acceleration of two reacting objects will be equal and opposite).

$mV = mV$ (The momentum of reacting masses is equal and opposite. The momentum of the individual masses is the mass multiplied by the individual velocity).

$mV + mV = mV + mV$ (The conservation of momentum states that the total of the masses multiplied by the velocities of colliding objects will remain constant, before and after collision).

$Ft = mV - mV$ (The difference in momentum in a pushed object is equal to the force applied multiplied by the length of time the force is applied).

$$Ek = \frac{mV^2}{2}$$

(The kinetic enery of a moving mass is equal to the mass multiplied by the velocity squared, divided by 2).

$$\frac{mV^2}{2} + \frac{mV^2}{2} = \frac{mV^2}{2} + \frac{mV^2}{2}$$

(The conservation of energy states that the kinetic energy of colliding masses will total the same before and after collision).

$$Fd = \frac{mV^2}{2} - \frac{mV^2}{2}$$

(The difference in kinetic energy in a pushed object is equal to the force applied multiplied by the distance over which the force is applied).

The most significant factor revealed by Newton's laws is that space is not light and fluffy at all but rather extremely tough. Space will resist whatever amount of force you can apply with an equal and opposite force. The resulting acceleration of the pushed object is a measure of the energy required to overcome the resistance of space.

There was no intention on Newton's part to separate inertia from matter. Newton assumed that inertia was a characteristic of mass. When Newton's detractors wanted to know which frame of reference Newton's space coincided with, Newton pointed to the vault of the heavens. The bulge around the equator of the earth convinced Newton that the earth was spinning in absolute motionless space.

Ernst Mach, the Austrian physicist, pointed out that the earth might not be rotating at all, and that the bulge around the equator could as well be caused by the rotation of the star fields. This was the first indication that inertia might not be an attribute of local mass at all, but an attribute of all mass throughout space. It was also a clue to the development of relativity which was just around the corner.

However, many scientists agreed that the evidence of Foucault's pendulum indicates the almost total separation of local mass from inertial forces. Dennis Sciama, a British university professor and author of note, described inertia in an article in Scientific American magazine. Sciama left no doubts as he put the finishing touches to the theory of inertia as a property of all of the matter in the universe.

THE DEATH OF THE ETHER

The ether was invented by early scientists in order to explain how gravitation and light rays could act across great distances of space. If a light ray was a wave, there had to be some medium which would permit that wave to travel. The ether was supposed to be a fluid that filled all space.

Two American physicists, Michelson and Morley, teamed up to build a device that would measure the flow of the ether as the earth travelled through it. The device is called an interferometer. Their idea was that if the earth is travelling through the ether, a ray of light would go faster in the direction of ether flow than it would in the opposite direction. They proposed to send a ray of light upstream and back while at the same time a ray of light was sent across stream and back. The mechanics of the experimemt are shown in Figure p-1.

The interferometer is extremely sensitive. It would have detected a difference in the velocity of light thousands of times smaller than the difference that was expected. So it was a surprise when there was no indication of ether flow. The results of the experiment were negative. Many later re-runs confirmed the first result. Apparently the ether did not exist.

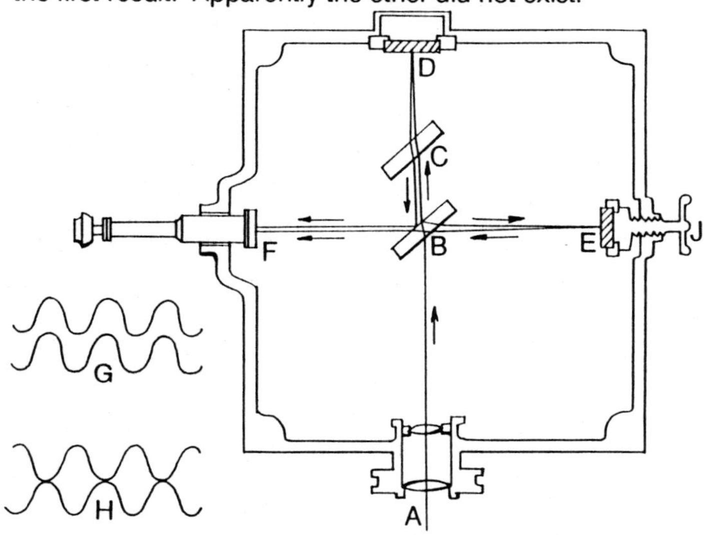

Figure p-1: Schematic diagram of Michelson's interferometer. A ray of light is introduced into the lense system at (A). The angle mirror (B) is dichroic (partly silvered) so that half of the ray is reflected to the plane mirror (E). The other half of the ray goes to the plane mirror (D). Since the ray going to (E) passes through the thickness of the glass (B) three times, the ray going to (D) is made to pass through a piece of glass twice at (C) to even out the distance of travel. Both parts of the split ray are recombined in an optical system with a wave comparator (F). If one ray is out of phase with the other half, the difference is measured. The diagram (G) shows waves that are in phase and diagram (H) shows wave trains that are out of phase by the maximum amount.

The ether was a basic part of the universe in those days so it took a while for scientists to get over the shock. Some suggested that perhaps the ether did not exist. But of course it had to exist. Every kind of wave or propagation of energy has a medium through which to travel. Something has to wave. Of that there is no doubt. There had to be some simple reason why the ether did not show up.

At first, tongue-in-cheek, the suggestion was made that perhaps material objects shrink a little bit in the direction that they are moving. Both FitzGerald and Lorentz came up with mathematical ways to show the degree to which things would have to shrink in order not to show up. Their combined effort was called the Lorentz-FitzGerald contraction. But somebody soon pointed out that mathematics was not enough. They showed that an accurate clock would still reveal the ether flow if it existed. Well, why not say that clocks in motion slow down a little bit, also? (This idea may seem funny, but they were really serious scientists looking for a way to explain the nature of the ether). They were willing to consider any kind of a suggestion that would save the ether.

Meanwhile, Einstein didn't concern himself as to whether or not the ether existed. What interested him was the evidence that a material object must shrink in the direction of its motion. And of great consequence also, the evidence that time slows down within an object that is accelerated into motion. Einstein got busy and applied the new evidence to the theory he was devising. That theory did not need the ether; it was Einstein's theory of relativity.

THE SCIENTIFIC IMPERATIVE

The basic tenet of science, the foundation on which science rests, is the imperative for physical cause and effect. The evidence of an event sends the scientist along a path of discovery at the end of which he will reveal the physical nature of the event. Included in the scientists imperative is the need to measure and quantize the contributing mechanical interactions which caused the event to occur.

The scientific method was devised in an original form by the ancient Greeks. In more recent times, Gallileo is given the honor of reviving the method and applying it to natural events. The scientific method comprises five steps:

(1) The evidence of the event is observed and recorded.

(2) An explanation is advanced. The explanation may be a hypothesis or a visualized model. It may include deductions that are based on the evidence.

(3) A mathematical model, called quantization, may be devised to fit the evidence.

(4) Experimentation proceeds to make the event repeat under observation or controlled conditions, permitting a comparison of new evidence with the quantized model.

(5) Repeat all steps, adjusting the model until it matches the evidence. Examine and illuminate until the event is thoroughly understood.

IGNORING THE EVIDENCE

The scientists who were deeply concerned about the loss of the ether had plenty of evidence and the model. They knew the ether was an elastic medium filling all space. But they lacked a mathematical model that fitted the evidence in the way that they wanted it to fit. They did not only want a theory and a mathematical model, they also wanted it to behave in a certain specified way.

Meanwhile, there were no convincing explanations as to what was happening. As time slipped away, the FitzGerald-Lorentz contraction gradually became part of physics. Einstein accepted it without a shrug. All during this time the evidence of what the ether was and how it worked was all around them. The scientists could not see it because that is not what they were looking for.

Something began to happen to science that had never happened before. A mathematical model was accepted into the body of scientific knowledge without a physical model to back it up. That set a precedent that was to become the norm in the following decades. The general feeling among scientists was that when they entered the micrometer world of the electron, they had entered a realm where physical models were no longer possible. They began to depend on mathematics alone, regardless of weird results that were often generated.

As this new method gained favor, it was supported by a new concept called the uncertainty principle. This axiom states that when you try to observe something that is very small, you disturb its true condition. Therefore, you will never be certain that you have learned everything about it.

The uncertainty principle was never a hindrance to Einstein or to other theorists. Their mentally visualized experiments could not be disturbed by a principle of the physical world. And so new knowledge about the atomic nucleus continued to accumulate. Nuclear fission became a reality. The atomic bomb was built. The technicians were building incredible things......the hologram, the laser, color television, etc. The uncertainty principle was no hindrance to progress in any serious way.

THE LINES OF FORCE

What was the evidence for the existence of the ether that the scientists ignored? The first hint came from Faraday who conceptualized the idea of electric lines of force and magnetic lines of force.

* Foucault's pendulum proved that the whole solar system was floating in a field-space generated by the star fields of the Galaxy.

* Maxwell used Faraday's lines of force to develop a mathematical description of magnetism and electricity.

* Michelson and Morley proved that the ether was attached to matter rather than being a fluid throughout space.

* Hertz proved that man-made waves of electromagnetic energy could be made to propogate through empty space.

* A Faraday experiment showed that magnetism co-incides with the inertial field.

* Experimenters became aware of the fact that every negative charge is attached to evergy positive charge by elastic lines of force.

* Scientists proved that a light ray is an electromagnetic wave. Therefore, electromagnetic fields must extend all the way to the distant stars from which the light rays emanate.

All of the above facts were known. It was only necessary to put them all together to realize that space is filled with electric lines of force. Electric field lines criss-cross through space, running in every direction. This model of the ether explains all of the evidence, including an explanation of why the interferomerter behaved the way it did. The reasoning is as follows: The ether is an electric field attached to matter. The interferometer and its optical system are composed of matter. The ether field is attached to the optical system. Hence, there will be no measurable ether flow in any experiment performed with the interferometer.

THE DOPPLER EFFECT

If the ether is an elastic medium of the type expected from the evidence, it should be possible to detect some effects of the ether. This is true. Light rays reflected from a moving mirror surface will show a doppler effect if the source of the light ray is motionless or moving at a different speed than the mirror.

In the early 1800's a scientist named Christian Doppler was standing by the railroad tracks one day. When the train roared by the engineer sounded the whistle. Doppler was surprised to hear the whistle suddenly drop a tone in frequency as it passed the point where he was standing. He studied the phenomenon. The frequency of the sound is increased when the source is moving towards you, and decreased when it is moving away from you. Doppler's name became associated with the effect of frequency changes caused by moving sound production or reflection. The same kind of phenomenon show up in any energy propogation where frequency is a factor.

How does the Doppler effect show up in the ether? Since the ether is an elastic medium, and since it is attached to matter, then whatever effects can be measured in a piece of elastic will also show up in the ether. A sound wave travelling in an elastic band while the elastic is being stretched will have its frequency lowered. A light ray travelling in the ether while the ether is being stretched will also show a reduced frequency, called a red shift.

THE ELECTRIC LINE OF FORCE

The electromagnetic structure of the ether can be determined by examining the structure of one electric line of force. This is demonstrated in the diagrams of Figure p-2 . Diagram (a) shows a segment of a live conductor. The concentric magnetic fields are shown to be active around the conductor. However, the electric lines of force which provide the potential gradient in the conductor are imbedded within the conductor.

The electric lines of force can be detected by using three conductors spaced out in the form of a triangular space, as shown in diagram (b). When an alternating current is sent simultaneously through the three conductors, the magnetic field enters the triangular space. It cannot be detected, however, because as the magnetic fields move into the triangular space, their vector sum is zero. The magnetic fields remain on the outside of the triangle as shown in the diagram. Within the triangular space there is an alternating electric field, generated by the magnetic fields which cannot be detected. The electric lines of force are parallel to the conductors.

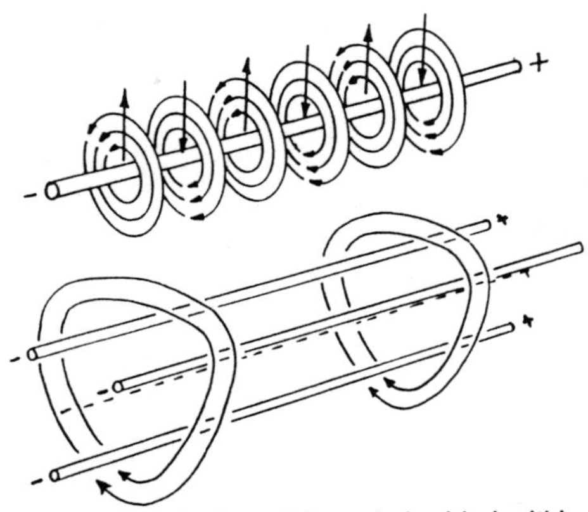

Figure p-2: The electric line of force imbedded within the conductor, as in diagram (a), can be detected by arranging three conductors within a triangular space as in diagram (b). When a current is introduced simultaneously into the three conductors, the magnetic fields entering the triangular space have a vector sum of zero. Only an electric field can be detected in the triangular space. The magnetic fields remain outside the triangular space as shown in the diagram.

Analysing the result of this experiment, the structure of an electric line of force must look like the example shown in Figure p-3. The circles represent moving magnetic fields. Some are expanding and some are collapsing, the same as in the experiment. The vector sum of the magnetic circles is zero, so that they cannot be detected. Only the electric line of force can be detected.

The conservation of energy assures that magnetic fields which are collapsing towards the center line do not stop there, but continue on through and become expanding circles of magnetic potential.

The conservation of energy is also effective when electric fields are combined. When two equal electric fields are opposed, both fields disappear. They cannot be detected. But the conservation of energy assures us that the energy of the electric fields is still there. That energy is detected when you try to move a mass through the region. That is to say, the electric fields have become inertial fields.

THE INERTIAL FIELD DEFINED

The structure of the inertial field is described when the following results are combined:

(1) The magnetic field: Standing waves of expanding and contracting magnetic potential within which the vector sum is not zero, leaving a detectable field with a polarity bias.

(2) The electric field: The vector sum of magnetic fields moving through a plane is zero.

(3) The inertial field: The vector sum of the electric fields through any plane is zero.

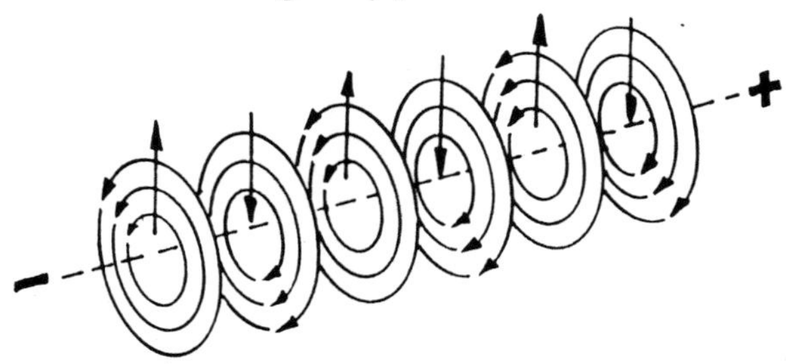

Figure p-3: Expanding and collapsing magnetic fields result in lines of electric force. A magnetic field that collapses to the center line passes through the center and then must expand out again instantly to comply with the conservation of energy. The vector sum of these magnetic fields is zero.

The relationship between electromagnetic fields and the inertial field is supported by evidence from many sources. The following paragraphs draw on these sources to define the causal mechanisms of inertia and gravity.

THE ULTRAGAMMA SPECTRUM

The existence of a viable spectrum of electromagnetic radiant energy in the ultragamma region is supported by evidence from several sources. In this region the wavelengths are so short that technicians have not been able as yet to detect them.

The sources from which the evidence arises includes the basic wave mechanics of the quantum theory. Other evidence arrives from (2) an analysis of the inertial mechanism, (3) the characteristics of the neutrino, (4) the evidence derived from conversions between mass and energy, (5) the ultragamma spectrum provides the medium for the causal mechanism of gravity, and (6), the successful application of the ultragamma spectrum to reveal the causal mechanisms of relativistic effects.

The evidence begins with Prince Louis de Broglie of France. In the early part of the century, the apparent dual nature of light was a problem of some significance. In 1925 de Broglie advanced a proposal to provide a causal mechanism to explain the dual nature of light. De Broglie surmised that any material particle in motion would generate an associated wavelength in accordance with his formula

$$\lambda = h/mv$$

showing that this wavelength would be equal to Planck's constant divided by the momentum of the mass in motion. This would explain why a light ray could exist as both a particle and a wave.

A successful demonstration in support of the de Broglie concept was carried out by Davisson and Germer. They used a crystal to defract electrons, which had been accelerated through a voltage difference of 100 volts. Subsequently, the de Broglie theory was rigorously tested not only for electrons, but also for atoms and light molecules. Throughout these experiemnts, the de Broglie formula was found to hold true. On the basis of this positive confirmation, both Schroedinger and Dirac adopted the de Broglie formulae. The Louis de Broglie concept became the fundamental tenet of the quantum theory. The successful application of the quantum theory to particle physics and other phenomenae have had far reaching consequences.

THE DE BROGLIE WAVELENGTH

The de Broglie formulae apply to matter in motion. The effect can be likened to a marble rolling across the ridges of a washboard. To climb over one ridge, the marble must convert some of its kinetic energy to potential energy and then convert potential energy back to kinetic energy. The amount of energy converted is the energy unit represented by Planck's constant. The frequency of the marble's energy conversions is then derived from the velocity of the marble as it overtakes the equally spaced ridges of the washboard.

The frequency of the de Broglie wave is

$$f = \frac{mvC}{h}$$

from which the relativistic forms are derived,

$$f = \frac{mvC}{h\sqrt{1 - \frac{v^2}{C^2}}}$$

and

$$f = \frac{mC^2}{h}$$

These formulae can be applied to show that the de Broglie wavelengths change proportionately in accordance with the amount of mass in motion. The individual de Broglie frequencies of the components of a moving mass are added together when the components are added together. The results indicate that the de Broglie frequencies are generated from a particular type of beat frequency. The de Broglie formula can be applied to show that when two masses with a common velocity are combined to form one mass, the de Broglie frequencies behave like beat frequencies (A beat frequency is the difference and the sum of two frequencies that are combined).

For example, the de Broglie frequency of a water molecule in motion is the sum of the frequencies of the oxygen atom and the sum of the frequencies of the two individual hydrogen atoms that combined to form the water. That is

$$f_H + f_H + f_O = 2\,\frac{m_H vC}{h} + \frac{m_O vC}{h}\,.$$

When masses larger than the very lightest molecules are combined, the range of the de Broglie frequencies very quickly moves into the region of the ultragamma spectrum, within which the frequencies are many times higher than the highest frequencies that can be detected by technologists.

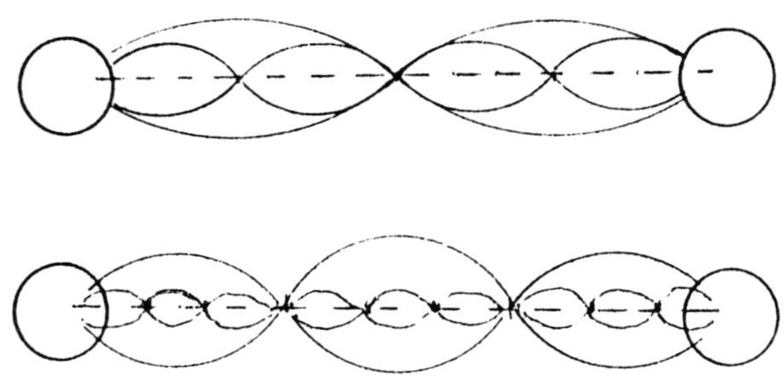

Figure p-4: Within the standing wave system between a pair of charges, harmonic frequencies form in accordance with the available energy. The equipartition of energy allots a proportion to each new harmonic as it is formed. There is no limit to the number of harmonic frequencies.

Ultragamma frequencies are also generated by other mechanisms. The most significant of these is the harmonic effect when an atomic oscillator is resonant with an interacting wavetrain of radiant energy. When the components of a mass come together to form a larger mass, the harmonic frequencies associated with bound electrons include integers of any half wavelengths which can be fitted into the distance between the electron and any other charge. Because they have resonant nodes, the dominant harmonics are related to the squares of the interacting frequencies. For example, if the interacting frequency has generated two half wavelengths within the new chemical bond, the harmonics generate four half wavelengths, sixteen half wavelengths, etc., producing a doubling and redoubling of the original interacting frequency. Similary, the dominant harmonics can be odd number integers, namely one, three, nine, and eighty-one, etc. This effect results in a tripling and retripling of the original interacting frequency when matter components are brought togther.

The harmonic effect, which is fundamental to the nature of energy vibrations in confined wave systems, is limited only by the velocity of light and the available energy. The doubling and tripling of harmonic frequencies continues to generate shorter and shorter wavelengths on into the ultragamma spectrum of frequencies, and towards infinity.

The spontaneous generation of harmonic frequencies is spawned by the beat-frequency mechanism. The causal mechanism of beat frequencies is the interference and reinforcement of wave crests as they combine. Any kind of disturbance in the energy flow within a confined wave system encourages the harmonic frequencies to form.

THE INERTIAL MECHANISM

The force of inertia is evident only when a mass is accelerated or decelerated. The causal mechanism generating this effect can be defined by imagining two masses set apart from each other a certain small distance. In this mechanical demonstration, a mass M and a mass P are separated by a distance R. These two masses have a similar composition and a common local temperature.

Suppose that wavetrains of radiant energy are travelling from the mass M to the mass P and from the mass P to the mass M. (Perhaps the two masses are exchanging radiant thermal energy, which will serve as an example spectrum of frequencies for this demonstration).

The dominant frequencies of thermal radiation are directly related to the temperature. This permits the atomic oscillators within the mass M and the mass P to be resonant when they are the same temperature. Radiant energy emanating from the mass M can be absorbed by the mass P and vice versa.

When the same frequency of radiant energy travels in opposite directions through the same space, standing waves of a certain kind are generated. These standing waves have a stationary node at the end of each half wavelength. The halfwave nodes are like the markers on a measuring rod. This condition is imposed on the mass M and the mass P because in order for the two masses to remain resonant with each other, the distance R has to be a multiple of the half wavelength of the dominant frequency. The half wavelength then sets the unit of action for the system. In a resonating system, this distance cannot change. If it does change, the resonanance will be thrown out of tune.

Regardless of this limitation, the resonanant condition interacting with masses M and P can sustain during a change in the distance R if the distance R changes at a steady rate. For example, if the mass M and the mass P are approaching each other at a constant speed, they can remain resonant by adjusting their internal atomic oscillator frequencies to match the doppler effect generated by the motion. This has the effect of increasing the temperature of the two masses as they approach each other.

Similarly, if the mass M and the mass P move away from each other at a constant speed, they can again sustain the resonant condition by adjusting their internal frequencies to match the doppler frequencies produced by the motion. This has the effect of reducing their temperature as they move away from each other.

During the time that the distance R is changing from one rate of change to another, there is a phase shift affecting the wavetrain frequencies where they engage the atomic oscillator frequencies. When there is a phase shift, the two masses are not quite resonant and the magnetic potentials of the interacting waves of energy are partly overlapped in relation to the oscillator vibrations. If the magnetic fields in the overlap portion have the same polarity,

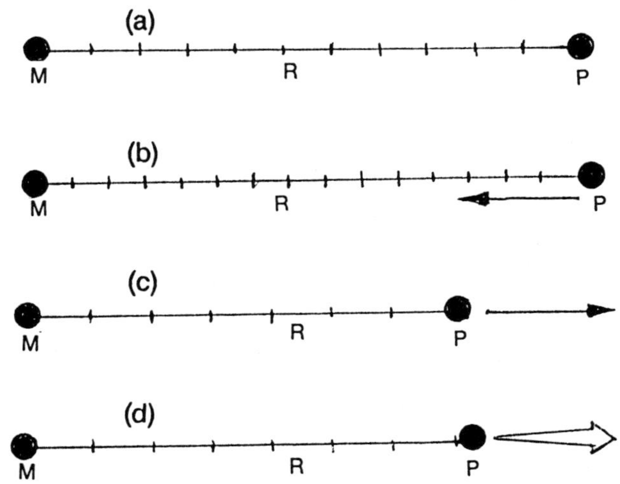

Figure p-5: Diagram (a) depicts a mass M and a mass P exchanging radiant energy along the connecting path R. When the internal oscillators of the mass M and the mass P are resonant, the distance R is a multiple of the half wavelength of the dominant interacting frequency, as shown.

In diagram (b) the mass P is moving towards mass M at a constant speed. A doppler effect in the energy exchange now impresses on the vibrating oscillators within the masses, forcing them to speed up to match the doppler frequency caused by the motion. The resonant condition is restored although the mass M and the mass P had to increase their temperature in order to accomodate the change in frequency.

In diagram (c), the mass P is moving away from the mass M at a constant speed. The doppler frequency of the motion forces the internal oscillators within the masses to reduce their frequency so that the resonant conditioon is sustained. The pair of masses become cooler due to the separating motion as the distance R increases.

In diagram (d), the mass P is accelerating away from the mass M. The doppler frequencies try to force the internal oscillators within both masses to reduce the frequency to maintain the resonant condition, but the distance R is changing at an increasing rate so that a resonant condition is impossible to attain. The result is that there is a phase shift in the wavetrains of radiant energy as compared to the vibrations of the resonators. The phase shift generates an attractive force acting along the path R. The force is caused by the phase shift, which places the magnetic fields of the wave systems into overlap. As magnetic fields with opposite polarity try to merge, there is an attractive force that persists as long as the phase shift exists.

they repel each other. If they have opposite polarity they attract each other. The phase shift causes an interaction between the mass M and the mass P.

In proportion to the degree of overlap, a force is generated along the line connecting the mass M and the mass P. If the rate of change of the distance R is positive, the force is repulsive. If the rate of change of R is negative, the force is attractive.

From this mechanical demonstration it will be seen that without any other masses existing in the whole universe, the mass M and the mass P share a mechanism that controls their mutual temperature and can exert a force along the line connecting the two masses if there is a rate of change imposed on the intervening distance R. The force can be positive or negative depending on whether the two masses are accelerated away from each other or towards each other.

The purpose of the above demonstration is to show that the behaviour of the mass M and the mass P is exactly like the behaviour of two masses in an inertial field. When the internal oscillators within matter are resonant, there is no interacting inertial force. When there is a phase shift, the inertial force is evident. The phase shift occurs whenever the distance R is changed at an accelerating rate.

Foucault's pendulum responds to these same forces. The mass in the pendulum is responding to ultragamma frequencies which cannot be detected. Now look up into a clear night sky and you will see that the universe is jam-packed with matter that is in the business of exchanging radiant energy. The ultragamma frequencies to which Foucault's pendulum is attuned, those frequencies are the dominant frequencies of the great river of matter in the Galaxy and all of the matter in the universe beyond. The inertial forces affecting local matter and the temperature range of local matter is directly affected by the ultragamma frequencies generated by all the matter in the Galaxy and in the universe.

THE NEUTRINO

The neutrino has no discernible mass, but since it carries off a definite amount of energy from a radio-active source, it is thought to be a nearly massless particle moving at close to the velocity of light. During the radio-active disintegration process, there is a distribution of energy in the form of beta rays and alpha particles as well as gamma ray photons. Careful determinations measuring the total and balance of the energy distribution indicates that a portion of the energy in the transmuting particles cannot be accounted for.

As no other reason presented itself, the energy disparity was attributed to a massless particle that was invented to balance the energy account. As the new particle had no charge, it was called the neutrino (Italian for "little neutral one").

Suppose, at that time, when physicists were trying to account for the energy disparity, that the existence of a viable ultragamma spectrum of radiant energy was well known. Then, when energy was being carried off in an invisible way, the natural conclusion they would have reached would be that the lost energy was being carried off by one or more photons in the ultragamma range. That would explain why the energy could not be detected and why the neutrino appears to be a massless particle moving at the velocity of light.

When the neutrino photons are released, they behave exactly like photons in the visible spectrum. They fly away on straight paths until they encounter some matter with which they can interact. The ultragamma frequencies of these neutrino photons must be characteristic of many exchange bonds in atomic components. It is likely that they interact frequently with matter, but in ways that are difficult to detect because of the ultragamma frequencies. Perhaps the neutrino photon is the trigger that initiates the radioactive disintegration of the unstable atoms and isotopes, of which there are several hundred.

THE CREATION OF MATTER

There is an apparent conversion of radiant energy into mass, and the converse, when certain conditions prevail. It appears that the energy in radiation can convert to matter in the form of an electron and a positron pair when the frequency of the radiant energy attains the level

$$2.48 \times 10^{20} \text{ hertz,}$$

and if the radiation with this energy is directed into a source of nucleonic matter, so that the needed conservation of momentum may assist the process.

When the de Broglie relativity formula is applied to the electron or to the positron, the maximum frequency these charges can attain is

$$1.24 \times 10^{20} \text{ hertz,}$$

and it will be noted that this frequency is exactly one half of the matter creating frequency. Furthermore, when the electron and positron come together and annihilate each other, the frequency of the radiation released is double the frequency of the individual charges.

These relationships indicate that a beat frequency is operating. They also show that there is standing wave system in operation. The generation of such frequencies in double and half quantities is evidence of harmonics. Harmonics would only be generated when there is an exchange system of wave energy interacting between the pair of charges. Furthermore, these processes are reversible, as needed to match the evidence.

GRAVITATION

In a previous paragraph, the force named inertia was attributed to the ultragamma spectrum of frequencies throughout the universe. Inertia was shown to be caused when the distance from mass to mass is changed at an accelerating rate. When a rate of change in the velocity of one of the masses occurs, a phase shift is generated in the wavetrains of energy which interact from mass to mass. The force of inertia persists as long as the phase shift persists.

The force is attractive when the interacting masses are accelerating away from each other. While the attractive force exists, there is an increase in the wavelengths of the radiant energy. This wavelength increase is similar to a doppler shift. The attractive force between masses is accompanied by a lengthening of the wavelengths, similar to a red shift. The red shift is evidence of a force that exists between masses.

In this regard it is of interest to note that two masses may be accelerating away from each other even when the distance from mass to mass is decreasing. An example is when you toss a ball up towards the ceiling in a room. The ball begins to accelerate away from the ceiling the instant that it leaves the hand. However, the distance between the ball and the ceiling is decreasing because the ball is still in the process of moving upwards, although it is slowing down.

THE GRAVITIC MECHANISMS

An amazing fact about Euclidean space is that it is impossible for one mass to move in relation to another mass without generating a phase shift and an attractive force. Although there are several gravitic mechanisms that are operating, only two of the most active are described below.

The gravitic mechanisms are demonstrated by deploying a mass M and a mass P into a region of Euclidean space. In diagram (a) of Figure p-6, a mass M is moving with a constant velocity V along the tangential vector MN. The distance R is the line connecting the mass M with a second mass P, which is not moving. The curve E is the radius curve equidistant from the mass P.

The objective is to find the rate of change of the distance R. By inspection the actual change in the distance R is the short segment S, but the rate of change of R is the acceleration vector a. The equation shows that the numerical rate of change of the distance R is equal to the velocity squared divided by the distance R. The amount of force associated with the rate of change of R is proportional to

$$F = \frac{V^2}{R}\left[\frac{MP}{R^2}\right]$$

The example shown in diagram (a) is a special case in which the vector MN makes a right angle with the radius distance R. The general case is shown in diagram (b). Here the vector V may take

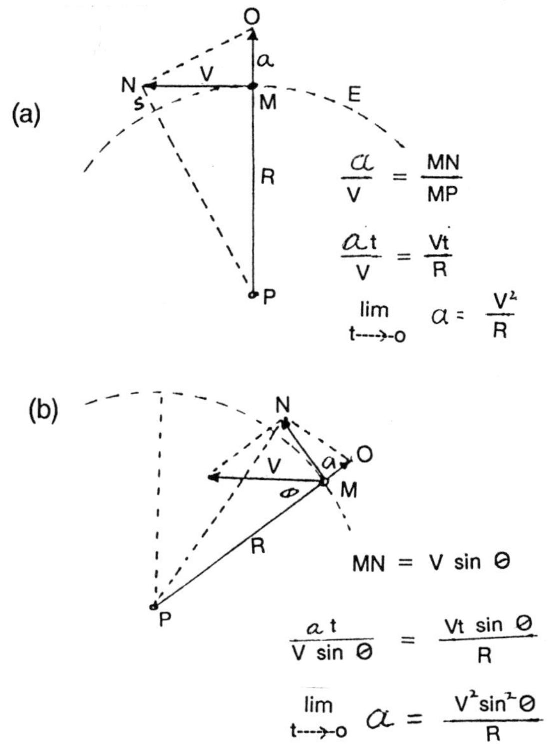

Figure p-6: When a mass M is moving in a straight line at a constant speed, there is a negative acceleration and an attractive force in relation to any mass P located in the surrounding space. The intervening distance R can be increasing or decreasing, but the rate of change of the rate of change of the distance R remains negative and the force attractive.

any angle with respect to the line R. The tangential component of the motion, from which the acceleration factor derives, is again the vector MN. The vector MN is equal to the sine of the angle Θ. The equation shows that the numerical rate of change of R is equal to the velocity squared multiplied by the sin of the angle squared, divided by the distance R. The force generated as a result of the motion of mass M is proportional to

$$F = \frac{V^2 \sin^2 \Theta}{R}\left[\frac{MP}{R^2}\right]$$

The other significant gravitic mechanism is shown in Figure p-7. In the diagram, the large circle is a mass M that is rotating or spinning. It could be a nucleon or a planet, as size is not a factor. The smaller circles, P1, P2, P3, etc., are other masses located any distance from mass M. Choose any small mass component N near the edge of mass M and it will be apparent that each small mass component N traces a harmonic vibration in relation to any of the surrounding masses Pn.

There will be a force acting along the line R connecting to mass P1, because the mass element N is alternately accelerating towards mass P1 and accelerating away from mass P1. This motion is equivalent to a harmonic oscillation, as shown by the hairpin arrows in mass M. The acceleration force of mass element N on mass P1 is strongest when the element N is at the upper and lower boundary regions E and F.

During the portion of the rotation when mass element N occupies the region from E to O, the acceleration of the mass element N is positive with respect to the mass P, and the mass P experiences a repulsive force. Conversely, when mass N occupies the region from F to O, the acceleration of N is negative with respect to mass P1 and the force is attractive. As these interactions are electromagnetic, the law of inverse squares applies. Since the region F is closer to mass P_1 than the region E, the attractive force is stronger than the repulsive force. That is

$$a = 4\pi^2 f^2 r \left[\frac{MP_1}{(R-r)^2}\right]$$

is greater than

$$-a = 4\pi^2 f^2 r \left[\frac{MP_1}{(R+r)^2}\right]$$

There are other gravitic mechanisms as well, and of interest is the mechanism that generates the main cosmological red shift from the distant galaxies. The effect is called the "lighthouse effect", because it is generated when masses are spinning slowly. This effect is described and quantized in a separate volume.

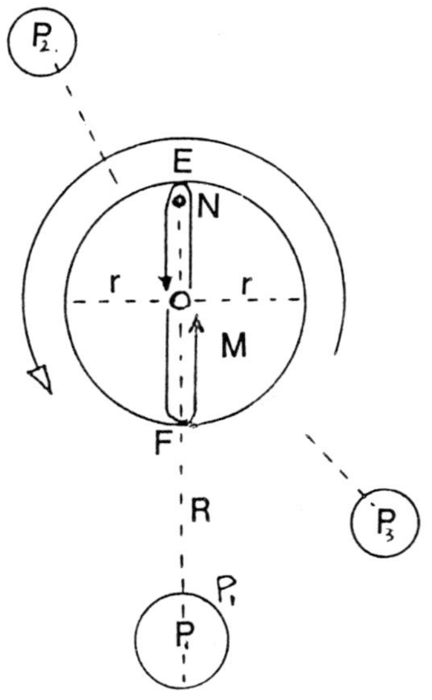

Figure p-7: Any mass that is spinning, like mass M is the diagram, generates an attractive force which interacts with any other masses P1, P2, P3, etc. The attraction is provided by the acceleration and deceleration of each small mass component N of mass M. During one revolution of the component element N the accelerating force on mass P_1 is repulsive for half of the cycle and attractive for half of the cycle. However, the attractive force occurs when the element N is nearest to mass P1. The attractive force is then stronger because of the inverse square law for distances.

EFFECTS OF RELATIVITY

To account for the evidence from experiments like the Michaelson--Morley experiments, it became apparent to Lorentz and FitzGerald that something was happening to the dimensions of matter and time. When a mass is accelerated into motion, there is a shortening of the mass in the direction of motion. At the same time, the rate of time flow within the object is reduced.

To better understand these transformations, both Lorentz and FitzGerald developed a mathematical expression to illustrate the effect. From these formulae, Einstein confirmed that the mass of the moving object must also change. There would be an increase in mass as the flow of time was reduced. The simple versions of these formulae are, for mass

$$m = \frac{m_o}{\sqrt{1 - \frac{v^2}{C^2}}} \quad \text{and for time,} \quad T = T_o\sqrt{1 - \frac{v^2}{C^2}}.$$

Einstein combined the Lorentz formulae with his evidence that the velocity of light is a constant for all observers. The results generated the important fact that energy and mass are interchangeable. Soon after, Einstein published his now famous formula $E = mC^2$. Incorporating the velocity of light as a constant of nature, these concepts formed the basic tenets of relativity.

A result of the Lorentz formulae is that the flow of time and the relativistic mass in a moving object form a constant when multiplied together. That is

$$mT = \text{a constant}.$$

THE MECHANICAL CAUSE OF TIME FLOW

In the year 1958, Rudolph L. Mossbauer inroduced a concept to phyics which made it possible to measure the flow of time in nanoseconds. The method measures the rates of vibration of the atomic oscillator systems within matter. This method (called the Mossbauer effect) is so sensitive that the rate of change in the flow of time can be detected when a mass is accelerated into motion. The internal frequencies of objects clamped to the edge of a spinning disc changed in accordance with the amount predicted by Einstein's theory.

The question then arises why time should slow down when the kinetic energy of a system is increasing. Logic says the rate of time flow should increase. The answer to this question is rooted in the doppler effect. Consider that when an object is moving at a high velocity through Galactic space, it is overtaking the wavetrains of Galactic energy in the forward direction. The doppler effect increases the frequencies. However, in the rearward direction, the wavelengths are stretched out longer.

Mass is based in kinetic energy, and the dominating factor in kinetic energy is the velocity of the mass squared. Velocity in matter includes the vibrational speed of the resonating oscillators. Suppose that the normal frequency of a resonator is 100 units. When the mass is accelerated to a velocity so that the basic frequency of the internal oscillators is increased by 10, the resulting doppler frequencies are 110 in the forward direction and 90 in the rearward direction. When these frequencies are squared and averaged, and the root taken, there is an increase in the mass/energy frequency to 100.5.

On the other hand, when the two doppler frequencies are multiplied together and the root taken, the time-flow frequency has dropped to 99.5. The comparison between the attributes that cause mass and the attributes that relate to time are;

$$\text{for mass} \sqrt{\frac{(c+v)^2+(c-v)^2}{2}} \qquad \text{for time} \sqrt{(c+v)(c-v)}$$

$$= \sqrt{\frac{110^2 \text{X} \, 90^2}{2}} \qquad = \sqrt{110 \text{ X } 90}$$

$$= 100.5 \qquad = 99.5$$

VARIATIONS OF MASS WITH VELOCITY

When a mass is accelerated, the applied force acts to increase the internal frequencies of the atomic oscillators within the mass. From the physics of vibrations it is known that a higher frequency takes more effort to change the frequency (as anyone who has tuned a stringed instrument is aware). The amount of applied force is the criteria that is used to measure mass. Therefore, the mass of an object is greater when it is in rapid motion compared to its mass when at rest. It takes a greater applied force to change the rates of vibration of the internal resonators within a speeding mass.

MATTER COMPRESSION

The forces that bind matter together are the standing waves of energy interacting from charge to charge, resonating with the vibrations of the charges. The dominant frequency sets the unit of action for the system, which is related to the distance between centers of charge, described in the next chapter.

When small masses like electrons and charged nucleon are very near each other as they are in the nucleus of the atom, the inertial forces interacting between vibrating charges are extremely powerful. The unit of action, like Planck's constant, is stable and unchanging.

Suppose that such a structure, comprising a tight system of vibrating charges, is accelerated to a high velocity. The kinetic energy of the structure is increased. The doppler frequencies generated by the wavetrains of galactic energy change the internal characteristics of the mass. As the dominant frequencies increase, the wavelengths in the direction of motion become shorter. When the distances between charges in the direction of motion are reduced, the whole structure becomes shorter in length in the direction of travel. The change of dimension, predicted by relativity, is shown in Figure p-8. The normal structure is shown in diagram (a) and the compressed structure is shown in diagram (b).

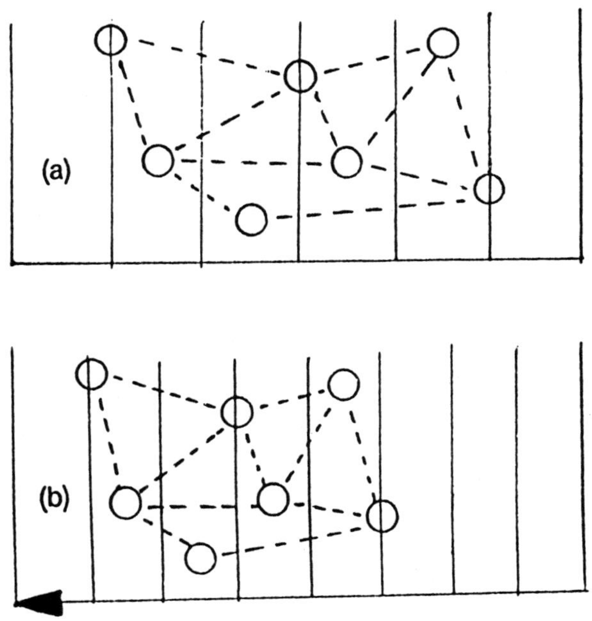

Figure p-8: When a mass is moving through the Galactic frame of reference, the doppler effect caused by the motion increases the frequency with which the moving mass overtakes the Galactic frequencies. All of the frequencies within the mass then resonate with the doppler frequencies of the motion. As frequencies rise, the distances between interacting charges are shortened in the direction of motion. The shorter wavelengths imposed by the doppler effect control the dimension of the mass, which becomes shorter in the direction of travel.

INVENTIONS FROM THE PAST

There is a saying that some people make things happen; some people watch things happen; and some people say, "What happened, man?" The reader is challenged to determine what is happening in this antique drawing of a water mill. Is it part of an irrigating system? Who knows?

THE FINE STRUCTURE OF MATTER

NATURE'S FORCES

At the center of the atom is the nucleus, the components of which are bound together by powerful forces. Scientists have been unable to define these forces except in abstract mathematical terms, derived from the quantum theory. There are no pictures available to illuminate what is taking place inside the atom.

The modern resonance particles are called bosons. The one that causes gravity is called a graviton. The one that generates electric and magnetic forces is called the photon. The one that provides nuclear force is called a gluon. Another kind that permeates all space is called the Higgs boson. The Higgs particles are the conveyors of mass.

To make use of such knowledge, inventors must be able to comprehend the underlying mechanical causes. Maxwell believed that all physical phenomenae, including forces, could ultimately be explained in simple mechanical terms. In the spirit of Maxwell's philosophy, the scientific imperative is here applied, with the objective to illuminate the mechanics of natural forces. To the degree that every physical phenomenon derives from waves of energy, the starting point of such an endeavor begins with the study of waves and the forces that are associated with waves.

WAVE DYNAMICS

If you have ever spent a summer by the lake or by the seashore, you probably have a good idea of how waves exert pressure on floating objects. As the waves come in towards the shore, a floating ball or a small piece of driftwood will bounce up and down with a rythm that matches the arrival of each wave crest, permitting the waves to pass by. The floating object remains where it is. However, if the floating object is sufficiently heavy, it is unable to bounce up and down fast enough to keep in step with the waves. The result is that the object spends a small interval of time on the forecrest of each wave, causing it to be gradually driven to shore.

The frequency of a wave is the number of wave crests that pass a given point in an interval of time. The velocity at which the waves move past a stationary point is equal to the frequency divided by the wave length measured from crest to crest.

Figure p-9 is a diagram of experimental wave motors. In diagram (a), two hollow floats are arranged to have motion up and down in opposition to each other. The springs give the floats a characteristic frequency in the up and down direction. Example (b) has an adjustable pendulum which controls the frequency at which the floats want to move up and down.

These experimental devices are placed in a flat bottom tank and the tank is filled with water until the floats are half submerged. A train of waves is pro-

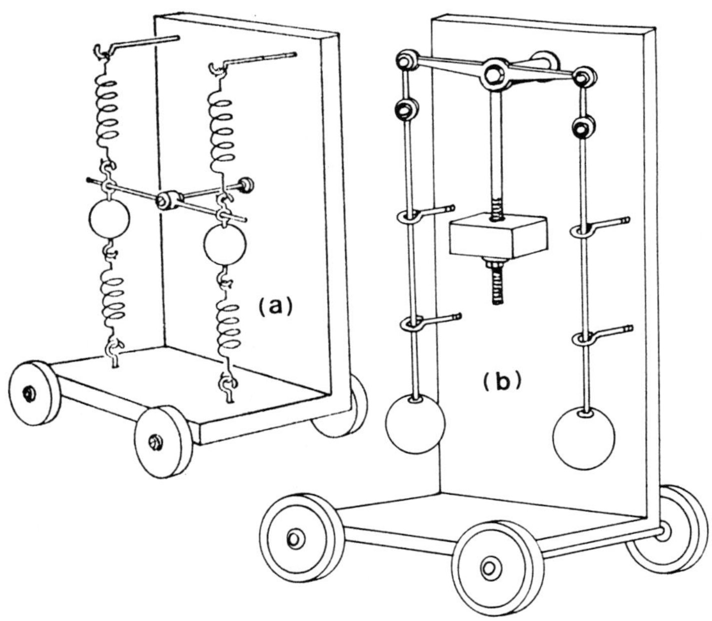

Figure p-9: Experimental wave motors demonstrate the dynamic effects of waves on objects. The floats move up and down in opposition, one half wave length apart. The natural frequency of the float motion is controlled by adjustable springs (a) or by an adjustable pendulum (b). By adjusting the frequency, the wave motors will proceed with the direction of the wave train or against the direction of the wave train. If the natural frequency is the same as the train of waves, the wave motor stays in one place.

duced at the right end of the tank. The horizontal distance between the two floats has been chosen to measure one half of a wavelength, as shown in the diagrams of Figure p-10.

As the experiment proceeds it is apparent that when the frequency at which the floats want to vibrate is slightly more than the frequency of the waves, the float apparatus will move to the right. In other words, the train of waves is exerting a pulling force on the float motor.

When the two floats are on the crest of a wave as at (a), the float on the right is forced deeper into the water because it is trying to go faster than the wave frequency. The float on the left is rising out of the water faster than the wave frequency. Similarly, when the two floats are in the trough of a wave as at (b), the float on the right is rising out of the water too fast and the float on the left is forced deeper into the wave because of the frequency difference. As shown by the force vectors, the result is a preponderance of force towards the direction from which the waves are proceeding. The float apparatus tends to move in a direction which eliminates the frequency difference.

When the train of waves is adjusted so that the wave motor has a frequency slightly lower than the wave frequency, the apparatus tends to move in the same direction as the train of waves. Again, the direction of motion is such that the difference in frequency is eliminated by the dopper effect created by the moving system.

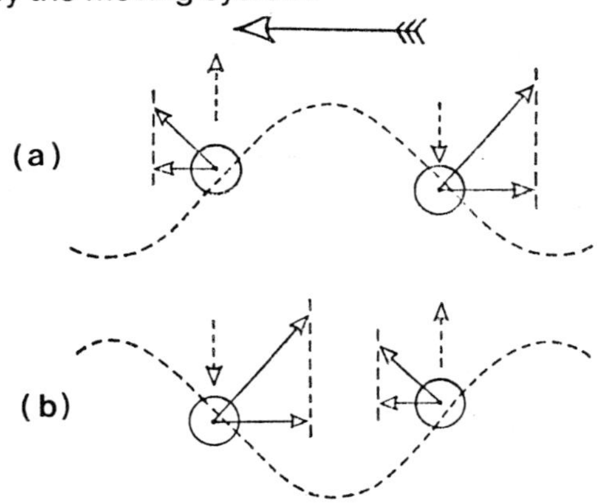

Figure p-10: The forces on the floats, shown by vectors, are always in the direction so that the resulting motion will eliminate the frequency difference.

DYNAMICS OF WAVE MECHANICS

Experiments similar to the water wave experiments, except using electromagnetic waves, have similar results. The devices that are suitable for experimentation are, for example, matching coils energized by alternating currents. Matching capacitors also have an electromagnetic effect on each other which can be harnessed to explore the forces of interaction. The magnetic effects within electric motors also yield clues about the nature of forces generated by a wavetrain of energy.

In each case a train of waves is caused to interact along the line connecting two of the resonating devices. When the train of waves is allowed to flow in both directions, standing waves are detected. Secondly, the phenonenon of quenching can be detected. Quenching occurs when components of the waves totally intefere with each other. And thirdly, the phenomenon of instantaneous action at a distance is demonstrated.

The results of such experiments show the following to be true:

(1) A train of waves interacting between resonating charges exerts a force on the charges.

(2) The force is evident only when there is a phase shift between the wave train frequency and the charge vibration frequency. A phase shift occurs when the arriving waves in a wavetrain are out of step with the waves generated by a vibrating charge.

(3) The phase shift can be the result of a doppler frequency caused by the motion of one or the other of the charges, or it can be the result of frequencies that are not quite resonant and are out of tune.

(4) The force is always in the direction that will eliminate the phase shift or eliminate the doppler frequency.

(5) The force occurs instantly when the phase shift develops, regardless of the distance between the interacting charges.

STRUCTURE OF THE EXCHANGE PARTICLE

Matter is built up out of charged particles. The vibrations of these particles generate wavetrains of energy which interact from charge to charge. The forces generated by the standing waves of resonating energy act to hold the charged particles apart, or bind them together. Or they perform both functions at once, binding the charges into a system of electromagnetic girders. These systems of standing waves are called "exchange particles".

The electromagnetic structure of an exchange particle can be visualized by considering the effects of Maxwell's formulae on a vibrating charge. The electric lines of force radiating out of charge point in every direction. When a charge is vibrated along an axis,

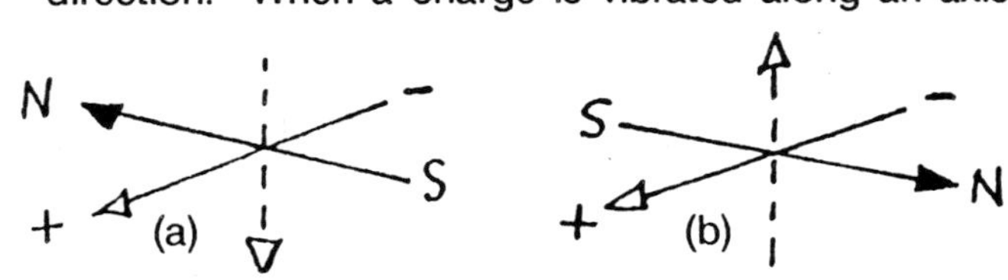

Figure p-11: Illustrating Maxwell's relationships, in diagram (a) the electric field segment + ---- - moving in the direction of the dotted arrow generates the magnetic field segment N----S. In diagram (b) the magnetic field segment N-----S moving in the direction of the dotted arrow generates the electric field segment + ---- - .

the axis of vibration turns until it is aligned with the nearest charge of opposite sign. Meanwhile, each vector component of the radial electric field (that is, components that are perpendicular to the axis of vibration) generates a segment of magnetic field. All of these segments together form concentric magnetic circles around the axis of vibration. This is illustrated in the drawing, Figure p-12. The circular magnetic fields alternate in polarity as the vibrating charge alternates, changing from deceleration to acceler-

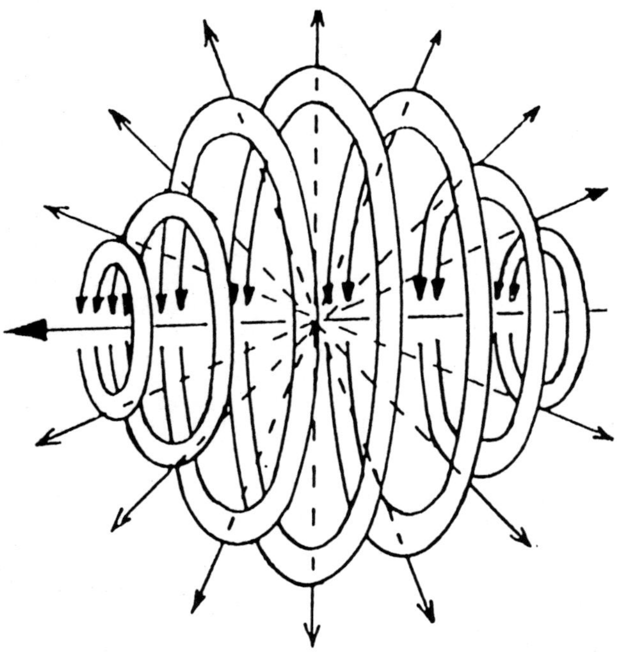

Figure p-12: When a charge is accelerated along an axis, the electric field vector components perpendicular to the axis of motion generate concentric circles of magnetic field, as shown. The polarity of the fields alternate at the same frequency as the frequency of the vibration of the charge.

ation.

The magnetic fields move away from the vibrating charge at the velocity of light. But they don't get very far. They attenuate as their expanding motion generates electric fields. The electric fields instantly fly away from each other by mutual repulsion. The expanding electric fields instantly generate new circles of magnetic potential. But in accordance with Maxwell's formula, these secondary magnetic circles are collapsing towards the axis of the charge vibration at the velocity of light.

Meanwhile, the circular vector components of the magnetic field which are moving parallel to the axis of vibration are not attenuating. They are not generating electric fields. What they are doing is following the electric field lines towards the nearest charge of the opposite sign. These doughnut shaped groups of magnetic circles alternate in polarity at the same frequency as the vibrating charge. When they reach the nearest unlike charge, they set it to vibrating at the same frequency as the original vibrating charge. The pair of charges begin to resonate, exchanging energy.

When the fields generated by the affected positive charge race back towards the electron, standing waves of energy are formed from charge to charge. With the exception of the standing wave system, all extraneous fields along the axis of vibration extending in both directions are extinguished and quenched by inteference.

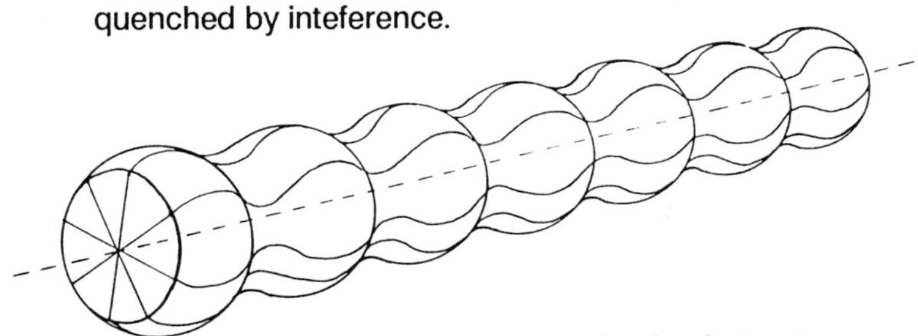

Figure p-13: Electric field lines connecting the electron to the positive charge in the proton are convoluted by alternations of polarity. The wave energy generated by the resonating charges acts to increase and diminish the electric field in time with the alternations of potential. The resulting electric field takes the shape shown in the diagram. These are standing waves in the photon-like exchange particle. The convolutions bounce because the wave energy is travelling in both directions.

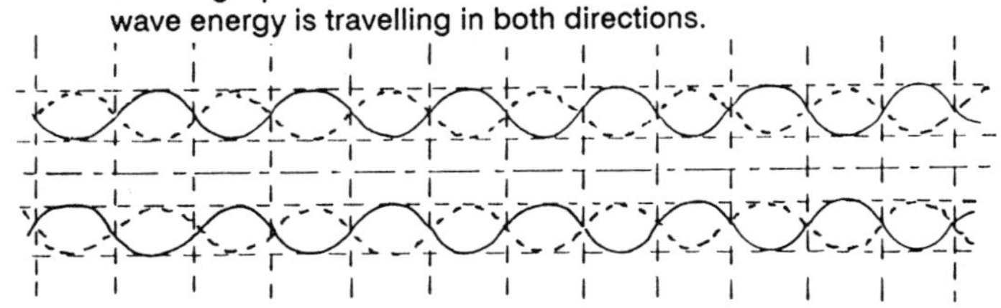

Figure p-14: Diagram of the shape of the electric field extending from the electron to the positive charge in the proton. The alternations are standing waves. The magnetic fields associated with this exchange particle are similar to those shown on page 68 (Figure p-3).

The shape of the standing wave system between resonating charges is shown in Figure p-13. As the waves are travelling in both directions, they merely bounce up and down in place. The diagram in Figure p-14 shows a cross section of the standing waves. The alternations are shown with dotted lines.

MECHANICS OF QUENCHING

The diagrams in Figure p-15 illustrate the mechanics of the quenching effect. In diagram (a) a negative charge is vibrating and generating trains of waves which travel away from the electron in both directions. The curved arrows show the polarity of the magnetic fields. In diagram (b) a train of waves is arriving from a vibrating positive charge. The polarity of the magnetic waves from the positive charge is opposite to the polarity of the negative charge, as shown. In diagram (c), the waves between the two charges have the same polarity and repel each. As they are going in opposite directions, they form standing waves. The extraneous portion of the wave system from the positive charge proceeding into the region of the dotted arrow near the letter "C" is going the same direction as the wave train from the negative charge.

But as the polarity is opposite, and the waves blend exactly together, total inteference occurs. Only the standing wave system, interacting between the charges, remains.

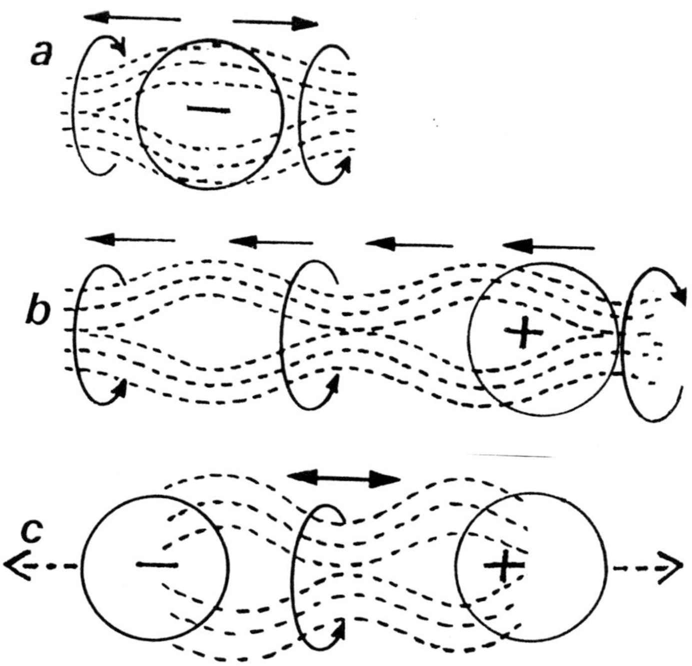

Figure p-15: The extraneous fields generated by the resonating electron and proton are extinguished by interference. In diagram (a), the fields generated by the electron proceed in both directions along the axis of the vibration. In Diagram (b), waves of potential arriving from the positive charge have opposite polarity when compared to the electron, as shown by the curved arrows. In diagram (c), when wavetrains of energy from both charges merge, the confined fields have the same polarity so they oppose and re-inforce each other. The extraneous fields proceeding beyond the resonating charges are quenched because they have opposite polarity and they are moving in unison. The regions of quenching are shown with dotted arrows.

THE BASIC STRUCTURE

The exchange particle described, represented by a system of standing waves, is the fundamental mechanism operating between vibrating charges in matter. Because the standing wave system sustains the distance between the interacting charges, the geometrical structure of matter is controlled by wave energy. Four basic factors affecting matter emerge from the exchange particle mechanism. These are:

(1) The mechanism accounts for the chemical bond between atoms.

(2) The structure of atoms and molecules are designed in accordance with the line-of-sight imperative of the standing wave system of the exchange particle.

(3) When the natural harmonics are introduced into the standing wave system, the electron levels of the hydrogen atom are revealed.

(4) The energy component of Planck's constant is revealed.

The relationship between the fundamental frequency and the harmonic frequencies is shown in Figure p-16. The charge associated with the proton is at level zero. The electron levels are located at the nodes where Planck's frequency and the fundamental frequencies act to reinforce each other. The electrton levels are located where the square of the quantum numbers generate the greatest amount of reinforcement of the wave systems.

Whereas only two frequencies are shown to demonstrate the electron levels, normally many more frequencies are generated. Any frequencies with odd numbers of half wavelenths that will fit into the distance from the center of the proton to the center of the electron, are likely to arise.

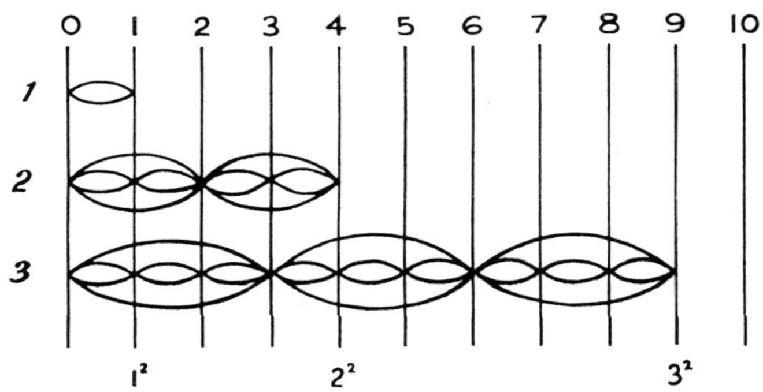

Figure p-16: Harmonic frequencies combine with fundamental frequencies of visible light to form the quantum levels of the hydrogen atom. The proton is at level zero, in the diagram. The electron levels are located where the squares of quantum numbers reinforce the fundamental waves. Planck's harmonic frequency controls the electron levels, as shown. Planck's constant is the energy content of one of the small half waves.

THE STABILITY OF CHARGE STRUCTURES

Every kind of structure that exists within matter does exist because it is stable. This is true of the atoms and it is probably true of the particles themselves. The most recent revelations about particles shows that many of them are built up from smaller particles called quarks. The quarks carry charges that are fractions of the charge on the electron. The probability is that all matter is made from smaller particles with fractional charges. The charges on the smaller particles give matter its inherent stability.

The kinds of charge structures that have inherent stability are shown in the diagrams of Figure p-17. The diagrams can reveal only a vague hint of reality. One reason is the matter of scale. The electron and the proton are the same size, but the proton is 1830 times heavier than the electron. The distances between particles of matter is immense in the sense that if you imagined that the electron was as big as the earth, the proton would be out there where the sun is located, over 93 million miles distance. However, there is a substantial possibility that our methods of measuring the sizes of atoms are not very refined. Our ability to measure the dimensions of charges is even less refined.

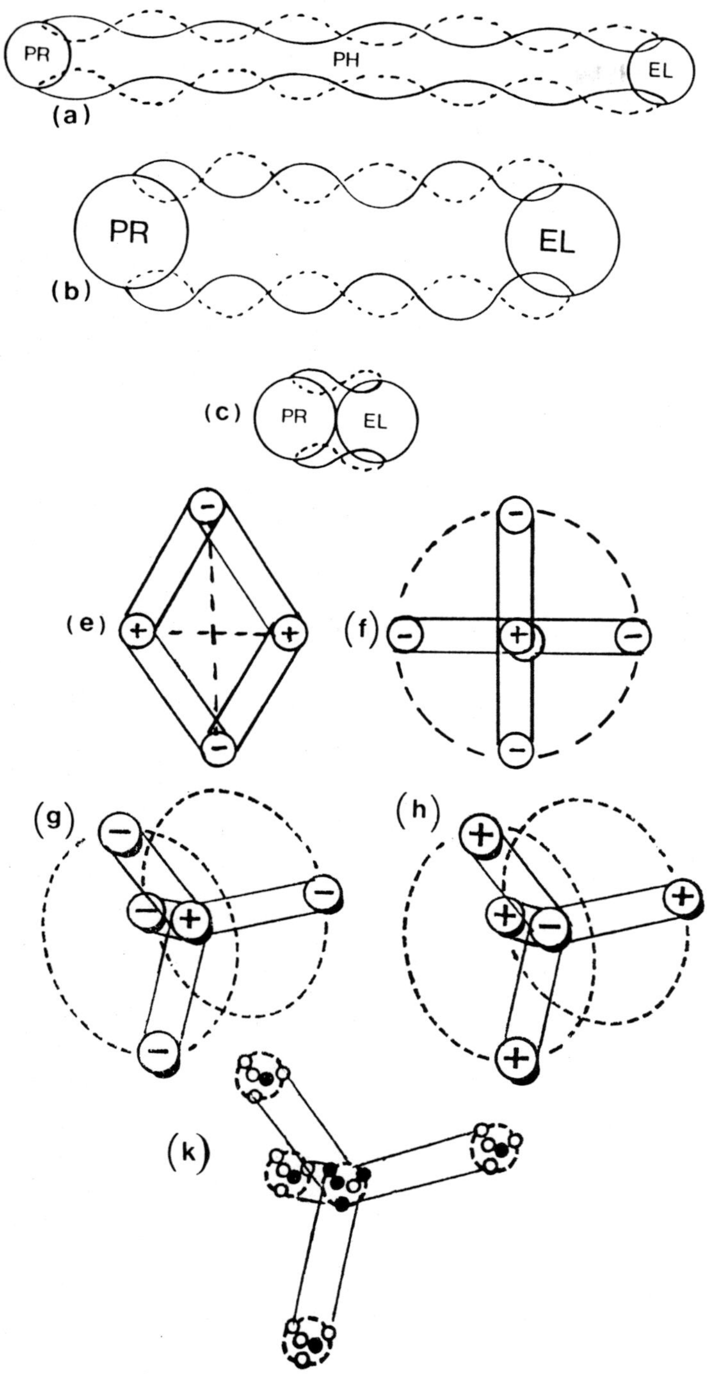

Figure p-17: Whatever structural combinations of charges that are stable will be found in nature. If a structure is unstable, it will not exist, or it may have a temporary existence. In the drawings, (a) represents an exchange particle connecting an electron and a proton (the hydrogen atom). One theory holds that the neutron is a closely bound hydrogen atom. The neutron as a stable particle may be represented by drawing (b) or by drawing (c).

Two views of more complex systems are shown at (e) and (f). This structure is shaped like a wheel with a short axle. There is a charge of one kind at each end of the axle. The rim of the wheel is comprised of charges of the other kind, with a minimum of three charges. The drawings at (g) and (c) are not much different. If you take drawing (e) apart and rearrange it the form of a tetrahedron, the structures (g) or (h) are the result.

The way in which more complex structures are adapted from smaller systems is shown by diagram (k). This system is made from four systems like (g) and one system like (h). Note that the (k) structure contains a surplus of nine chanrges of one kind.

Whatever structural combinations of vibrating charges are stable and persistent, such structures will be found in nature. Structures that are not stable will not be formed, or if they are formed they will not sustain. The radio-active decay of matter is related to this principle.

In Figure p-17, the drawing (a) represents the basic exchange particle, comprising a system of standing waves connecting a pair of resonating charges. Unlike charges generate fields with opposite polarity. The result is that the extraneous wave energy from one charge acts to quench the extraneous wave energy from the other charge.

One theory holds that the neutron is a tightly bound system comprising an electron and a proton. The neutron may actually have the proportions of drawing (b) or drawing (c).

Charge structures that would have natural stablility are shown in the diagrams (e) and (f). Such structures may or may not rotate. They would be stable either way. Photon-like attachments between unlike charges are shown as a double line connecting the pair of charges. Drawings (g) and (h) show structures in the form of tetrahedrons. All corner charges are of the same sign, and the central charge is of opposite sign. To illustrate how such structures can represent more complex systems, the structure at (k) illustrates a structure made out of four systems like (g) and one system like (h).

VARIABLE MASS

When the similar charges within a mass are spread out with large distances between them, and the unlike charges are closer together, the potential energy of the whole structure of the mass is greatly reduced. In accordance with Einstein's mass/energy equivalence formula, the reduced energy translates into reduced mass. It will be seen that in this way the same structure of charges could have a mass that would be quite different in special circumstances.

For example, when a spring is stretched, the internal charges are forced to take positions where the potential energy between the charges is greatly increased. The weight of the spring is slightly increased by the increased energy content.

The subtle changes in the arrangement of the charge structures also affects the inertial mass of an object. When a mass is accelerated, magnetic fields rise from the component charges. Charges that are different are assisted in their acceleration when each charge passes through the magnetic field of the other. Conversely, similar charges that are close together are hindered in their acceleration by the act of passing through the magnetic field of the other. In this way, the exact determination of weight, mass and inertia are all affected by the arrangement of the charges within matter.

THE STRUCTURE OF THE NUCLEUS

The imperative of line-of-sight bonding between electrons and the nucleons in the atom places a limit on the kinds of geometric structures that can represent the atom. There are clues to the character of nuclear structure in the specific peculiarities of the atom. Some of these are:

(a) The limits in the numbers of electrons in the quantum levels of the atom, that is 2, 8, 18, 32, 18, 8, 2. See the schematic in Figure p-18.

(b) The analysis of the products of disintegration of the atom.

(c) The nature of the radiation emanating from the nucleus of the atom.

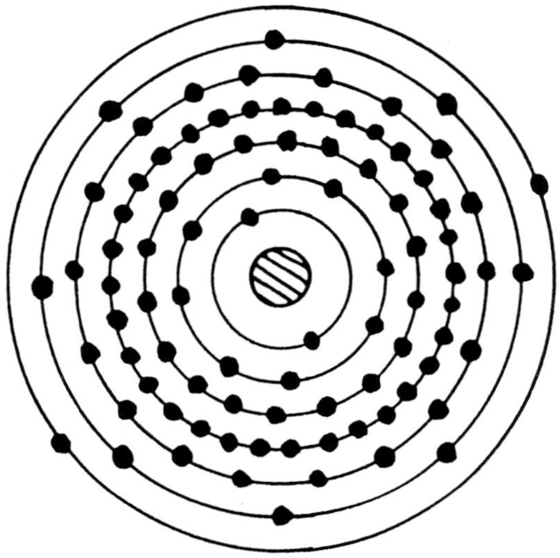

Figure p-18: An important clue to the nature of the atom is revealed by the numbers of electrons in the quantum levels. As shown by the diagram, there are only two electrons in the lowest level, 8 in the second level, 18 in the third level and 32 in the fourth level. The formula for these numbers is 2Nsquared, where N is 1, 2, 3 or 4.. Strangely, the numbers of electrons reduce in the reverse order as the levels go further away from the nucleus. The next level is 18, the next after that is 8 and finally the outer level is 2. Two possible reasons are advanced in the main text.

(d) The peculiarity of the nucleon bonds.

(e) The stability of the "magic" numbers of nucleons in the atom.

(f) The degree of success of existing models for nuclear structure, ie., the liquid drop model and the hot gas model.

As shown in the diagrams of Figure p-19, there are only two structures that can account for the existence of only two electrons in the lowest quantum level of the atom. Diagram (a) is a drum shape or barrel shape with positive charges grouped at the ends. The two electrons are then attracted to the positive ends of the barrel. Diagram (b) is a wheel shape with the positive charges forming the rim. The two electrons are then located along the axis of the wheel, opposite each other.

On the next level it is necessary to find places for 8 electrons. It must be assumed that the negative electrons will find locations which permit the largest possible spaces between them.

Whatever shape the nucleus of the atom takes, that shape is dictated by the shapes of the individual nucleons and the forces that exist between them. Every condition that is found in nature is there because it is inevitable. This rule holds because the bricks of matter will only fit together in key ways. Each nucleon adds to the number and position of the energy nodes and energy shadows which attract or repel particles.

One of the important factors is that each electron is held captive to a proton by an intervening photon (a web of standing waves of electro-magnetic

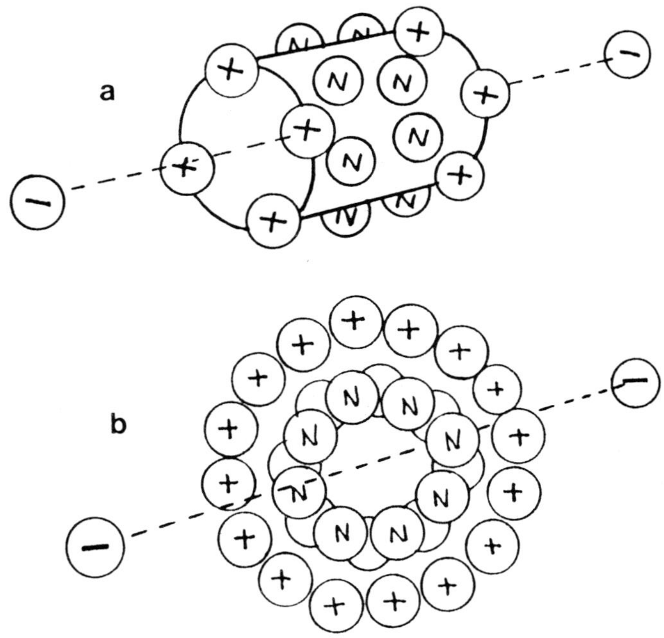

Figure p-19: To explain why there are only two electrons on the first quantum level of the atom, a definite structural configuration of the nucleus may be the reason. Two examples are shown. Diagram (a) is barrel shaped with the positive charges grouped towards the ends of the barrel, causing the nearest electrons to take up axial positions. In diagram (b) a wheel shaped nucleus tends towards the same result.

energy). Each electron then represents an energy shadow which eliminates the possibility of another electron being located in the same region. Whereever there is a space in which a proton, photon, and electron system can exist without interference, there will be such a system in that location.

The formula for the numbers of electrons in any quantum level is 2Nsquared. N is equal to 1, 2, 3 or 4. The region as described by the formula is two dimensional, whereas a nucleus occupying space is a three dimensional entity. There are two ways in which the three dimensional form can be represented by 2Nsquared. One way is described in the following.

It had been assumed in the past (because of

the orbital shell concept for the atom) that the quantum levels represented different distances from the nucleus of the atom. However, the wave-energy or photon concept does not relate to higher or lower levels, but only to higher or lower energy states. Any one energy state can be located anywhere, dependent only on the amount of wave-energy and the strength of the photon necessary to hold the electron in that position. This means that an electron can be positioned close to the nucleus, yet, because of the energy shadows that control it, will be loosely held. It will have the characteristics of an "outer shell" electron so far as the energy relationships are concerned.

From these considerations it is evident that the "outer shells" of electrons which are reducing in number, that is, 18, 8, and 2, may be actually energy states that are closer to the nucleus in the exact reverse order. The second shell of 18 electrons is in the same region as the first 18. The second 8 electrons are in the same region as the first 8, and the final 2 electrons are actually close to the nucleus along with the first 2. When the formula 2Nsquared is considered in this light, then the result is a three-dimensional nucleus rather than a two- dimensional one.

THE PRODUCTS OF DISINTEGRATION

The other way is revealed by the kinds of particles that fly out of an unstable nucleus. If you happened to be standing at the side of a road when a car drove by, you might recognize the make of the car by the parts that fell off of it. If you were an expert you might say, "Well, that distributor cap came off a Chevy (For example)." In the same way it should be possible to surmise something about the nucleus of the atom by studying the parts that shoot out of it. Including radio-active isotopes, there are a whole range of atoms that have parts falling off of them. Also enlightening is the nature of these atoms after a piece has been lost from them. The number of radio-active atoms, both natural and artificial, is over 500.

The kinds of particles that shoot out of disintegrating atoms are of five types:

(1) Neutrons: For example, when Boron is irradiated with alpha particles, some atoms transmute and become radio-active Nitrogen, which, after a few minutes ejects a neutron and another particle. The Boron has become an isotope of Carbon.

(2) Positrons: The other particle that is ejected in the above example is a positron, which is a positive electron.

(3) Alpha particles: The alpha particle is the nucleus of Helium, comprising two protons and two neutrons. Alpha particles are ejected with high velocity from many radio-active elements.

(4) Electrons: Electrons (called beta particles originally) are ejected from radio-active elements along with alpha particles.

(5) Neutrinos: High speed particles that are neutral and nearly weightless.

A type of photon with extremely hard radiation is also ejected from radio-active elements. These are called gamma rays. Gamma rays are released from decaying atoms along with alpha and beta particles, so that it is likely that the gamma ray is the photon-like web that was holding the particles together before they were ejected.

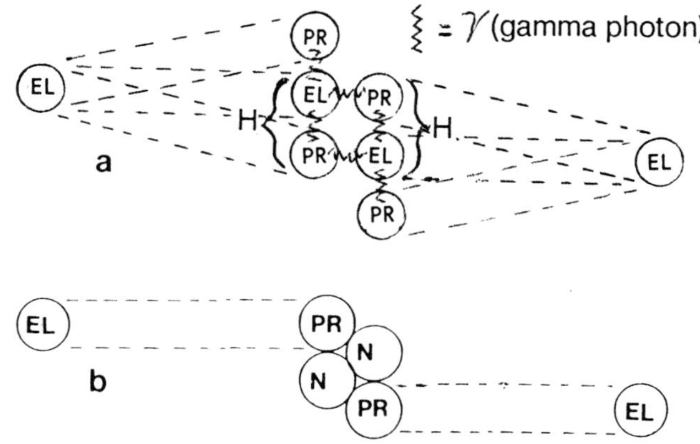

Figure p-20: The Helium atom may comprise 4 electron-proton sets as shown at (a). In this case the photon web may be complex. If the Helium atom looks like diagram (b), the photons are confined to act only between electrons and protons.

One factor to consider in the products of disintegration is that gamma rays can be produced in two ways. One way is when mass is converted to energy, the energy escapes as gamma rays. The other way is when closely bound particles, including electrons, become more closely bound, and hence, more stable. An increase in the stability of an atomic nucleus is accompanied by a release of energy in the form of gamma rays.

When positrons (positive electrons) are being ejected from a substance, the positive charge is the result of energy being converted to mass. The most likely reason for the creation of positrons is because electrons are being created inside the nucleus at the same time. In this case gamma rays that are sufficiently energetic can transmute directly into a pair of unlike charges. The missing electron from the pair is assumed to have joined with a proton to form a new neutron.

The other particles that are shot out of atoms, the alpha particles and the beta particles, are actually the components of Helium atoms. Helium is a very stable substance, so there is a good possibility indicated here that the heavier atoms are all made out of Helium and Hydrogen. One theory holds that all matter can be represented by electron-proton sets. The neutron is then just a closely bound electron and proton. In this case, the Helium atom comprises four electron-proton sets. This is shown in the diagram of

Figure p-20. (The Helium atom is actually two neutrons and two protons along with two electrons).

All atoms that have an even number for their atomic number can be built entirely out of Helium. Atoms that have an odd number atomic number have at least one hydrogen atom or isotope of hydrogen in

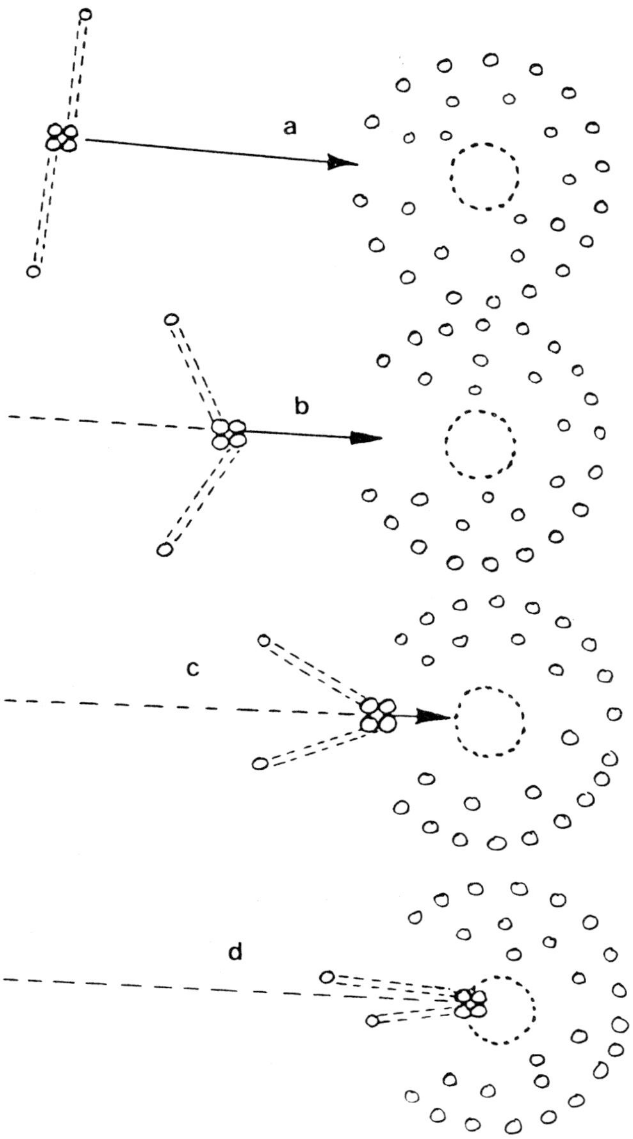

Figure p-21: Diagrams showing how the shape of the Helium atom changes when it is shot toward the nucleus of a larger atom. As the nucleus is transmuted by the intrusion of an additional Helium atom, the two new electrons share the photon spaces remaining in the outer shells.

their make-up. All the other mass can be represented by Helium.

Helium does not lose its identity when it is used as a building block for heavier nuclei. This can be shown in two ways. When larger atoms are transmuting, whole Helium atoms are ejected. This indicates that the Helium probably never lost its identity as a component of a heavier nucleus. The other way is revealed when alpha particles (Helium nuclei) are shot at other atoms. The Helium atom is often absorbed into the nucleus of the other atom, forming a new atom with a heavier nucleus.

Using a criteria of Helium components in heavier atoms, it becomes evident that the numbers of electrons in the quantum levels of the atom, that is, 2, 8, 18, 32, 18, 8, and 2, are representing pairs of electrons belonging to Helium atoms. Where there are two electrons in the inner shell and two in the outer shell, there are two Helium atoms of equal energy, each of which "owns" one of the lower electrons and one of the higher electrons. Similarly, the 18 electrons on the second level are related to the 18 electrons on the 4th level by being "owned" by 18 Helium atoms, each holding an equal share of the energy within the heavier atom. The 32 electrons of the middle level are "owned" by 16 Helium atoms, which when combined with the original 2 Helium atoms, become a full set of 18. The result is that the heavy atom (in this case Radium) is comprised of 44 Helium atoms.

THEORY OF NUCLEAR STRUCTURE

Many models for the nucleus have been proposed in the past. The two that are the most popular are Bohr's liquid drop model and the "hot gas model". These models of the nucleus were developed in order to predict what the nucleus would do under a given set of conditions.

The liquid drop nucleus is like a drop of water. The surface tension keeps the drop together and keeps molecules of water from escaping. However, the drop of water becomes quite unstable when more molecules of water are introduced. Finally the surface tension gives way and "fission" occurs. The similarity occurs when nucleons in the form of alpha particles are introduced into a nucleus. When the nucleus exceeds a certain size, it becomes unstable. Only a small amount of additional energy is necessary to make the nucleus "fission".

The hot gas model is quite similar. The difference is that molecules of a hot gas are confined within a small space. The equipartition of energy shares the energy among the affected nucleons. Additional energy causes evaporation or "radioactivity". More energy, from introducing new alpha particles, causes the contents of the nucleus to "boil" and that is, to fission.

The diagrams of Figure p-21 illustrate what might happen to a helium atom when it is shot towards a nucleus of the liquid drop type. When the helium atom is approaching the nucleus the electrons are decelerated by the electrical pressure from the outer electron shells around the target atom. At the same time the positive protons are accelerated. When the helium atom splashes into the target atom, its shape has been altered to look something like diagram (d) of Figure p-21.

A DESIGN FOR THE NUCLEUS

The line-of-sight bonding of the electron-proton sets makes the whole structure of the nucleus quite simple in its basic design. Every proton must have a line-of-sight corridor with one electron to allow the interconnecting photon to exist. One possible model is hollow, like a hollow spherical ball. The Helium atoms would remain on the surface of the ball with all of the protons exposed. The other model, first suggested by the need to explain only two electrons in the first quantum level, is a layered barrel shape or wheel shape. The number of Helium atoms in each layer coincides with the numbers of electrons.

The most compact shape that the four nucleons in the Helium atom would take would be a tetrahedron. However, the diamond shape shown in Figure p-22 permits the protons to take positions away from each other. The most likely configuration is one which permits the greatest exposure for the protons. The diamond shape allows each individual proton to establish a photon link with a separate electron.

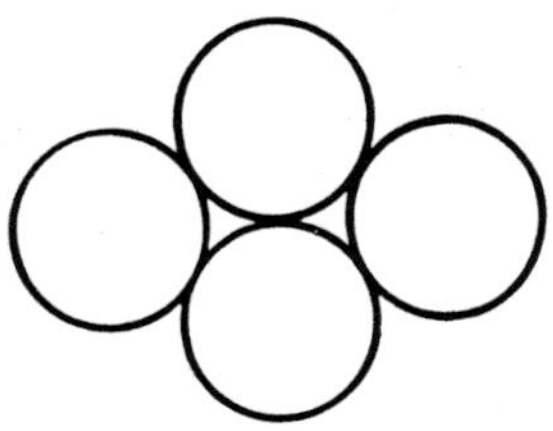

Figure p-22: The four nucleons in the Helium nucleus may take a diamond shape as a result of the positive charges mutual repulsion.

When the diamond shape is turned sideways to give the protons the maxium exposure, the side view of the Helium nucleus can be represented by a

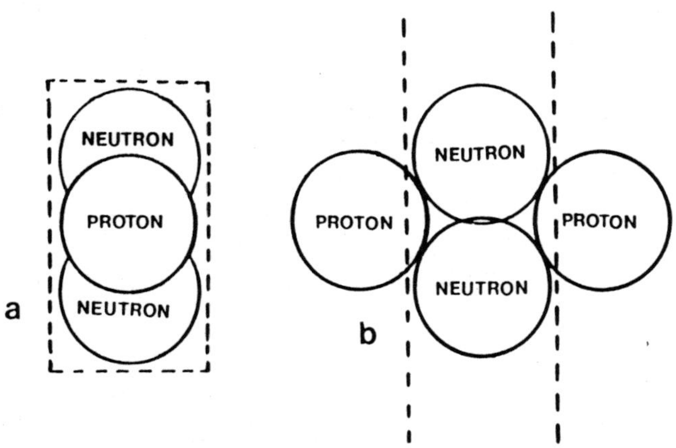

Figure p-23: Diagram (a) shows the position the protons in the Helium nucleus must take to permit the maximum exposure for photon connections. The side view at (b) occupies a rectangular space, approximately 1 to 2 in proportion, shown by the dotted outline.

rectangle, shown by the dotted line in Figure p-23 (a). In the diagram of Figure p-24, each rectangle represents a helium nucleus. The nuclear structure would be a wheel shape or disc shape with the protons sticking out of the sides like studs, in the manner shown in Figure p-23 (b). Such a nucleus provides the maximum space necessary for the electron/proton exchange particles to be established.

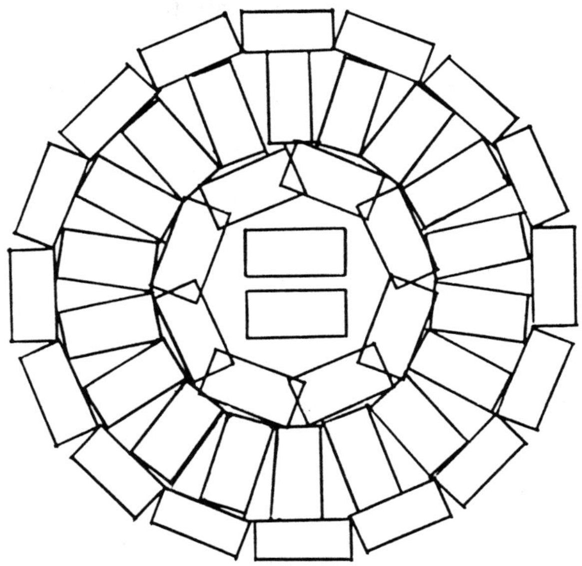

Figure p-24: In this diagram, each rectangle represents a helium nucleus. The number of rectangles in each level is the same as the number of electrons in the first four quantum levels, that is 2, 8, 18 and 16. Since each Helium nucleus carries two electrons, all of the electrons in the atom are accounted for. This is the nucleus of Radium, with 88 electrons.

Each of the electrons has a line-of-sight corridor to the positive charge in the proton. The electrons, gripped on the ends of their exchange particles, stand away from each other and also stand away from the nucleus, like pins in a pin cushion.

At the center two helium nuclei occupy the core. The two electrons on the first quantum level as well as the two outer electrons weakened by "energy shadows" are attributed to the core pair of helium atoms. The beat frequency that is established between an electron and its parent proton sets the frequency for the second electron. This concept is shown in the diagram of Figure p-25.

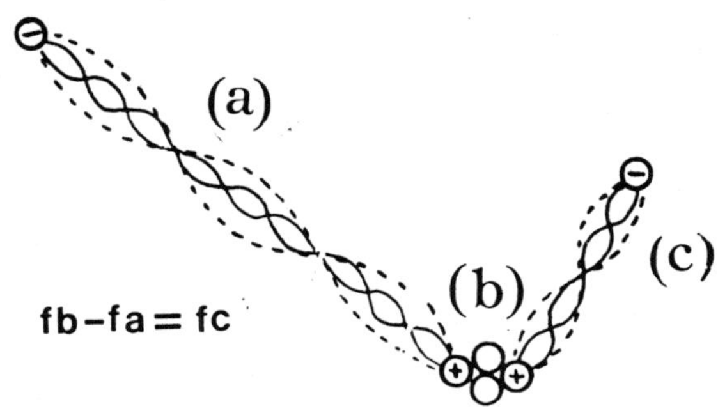

Figure p-25: The energy sharing attribute of the nucleons is the basis for the disparity in the two photons associated with one Helium nucleus. When one of the photons has a high frequency, the other photon connected to the same nucleus has a low frequency. The constant frequency of the whole nucleus is the sum of the frequencies of the two photons.

There are 8 Helium nuclei on the second level. These are responsible for the 8 electrons on the second level and the 8 outer electrons. Again, the difference in the quantum levels for the two electrons of one nucleus is due to the shared energy. The photon frequency blends with the natural frequency of the Helium nucleus, generating a third frequency which controls the more distant electron of the pair. This energy sharing relationship is related to the Pauli exclusion principle, in which no electron can occupy a quantum state that is occupied by another electron.

There are 18 Helium nuclei on the third level. These control both sets of 18 electrons, in the same manner as described above. In the top level there are 16 Helium nuclei, and these are responsible for the 32 electrons in the middle quantum level.

The rows of Helium nuclei alternate between standing on end and laying on one side. The positive charges on the protons may be the reason for this effect. Each Helium nucleus may take up a tilted position within the alotted rectangle, in such a way that the protons are exposed to the greatest degree on both sides of the disc. By coincidence, the positions taken by the positive protons are those most likely to be effective for establishing the line-of-sight photon bonds.

INTERSTICIAL BONDING

An important concept that may explain the strong forces within the nucleus is intersticial bonding. This happens when groups of basic charges within one nucleon establish one-on-one photon-like attachments to the same basic charges in the proximate nucleon. Matter is mainly space. The spaces between particles is immense when compared to the sizes of the particles themselves. It is quite possible for one nucleon to enter into another nucleon to some degree and overlap their charge structures. When this happens the photon-like attachments between the basic charges inside the nucleons become extremely powerful. The diagram in Figure p-26 shows the partial overlap of two nucleons with the basic charges interconnected by photon-like attachments.

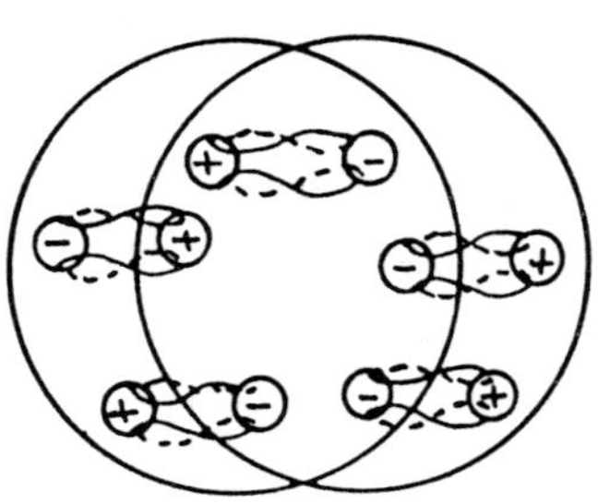

Figure p-26: Matter is mainly empty space. The strong force of the nucleons in the atom may be the result of the intrusion of one nucleon into another, as diagrammed above. The photon-like bonds would then be extremely short and powerful.

NEWMAN'S GRAVITY

Joseph Newman developed an idea to explain gravity. His idea is simple to understand and the way it works is demonstrated in the diagram of Figure p-27. Newman's reasoning is that within any group of electric charges (which includes all matter) the positive charges attract the negative charges. These charges will then move as closely packed as physical space allows. On the other hand, negative charges will repel negative charges, and positive charges will repel positive charges. So these charges will take up positions within their sphere of influence as far apart as possible. The overall result is that there will be a slightly greater force of attraction than there will be of repulsion. The charges that attract are slightly closer to each other and so exert slightly more attractive force.

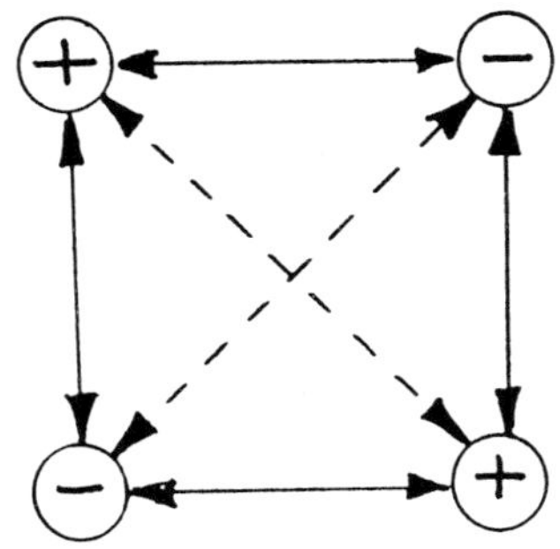

Figure p-27: In any group of equal numbers of charges, the charges that attract each other will move closer together and the ones that repel each other will move further apart. The result is that the force of attraction will be slightly greater than the force of repulsion.

ELECTRON PATTERNS

In the atom, electrons are strongly attracted towards the nucleus because of the positive charge associated with each proton. The electron is prevented from approaching too close, however. It is held in an energy node generated by the standing waves of energy in the exchange particle. In a complex atom, one with several electrons around the

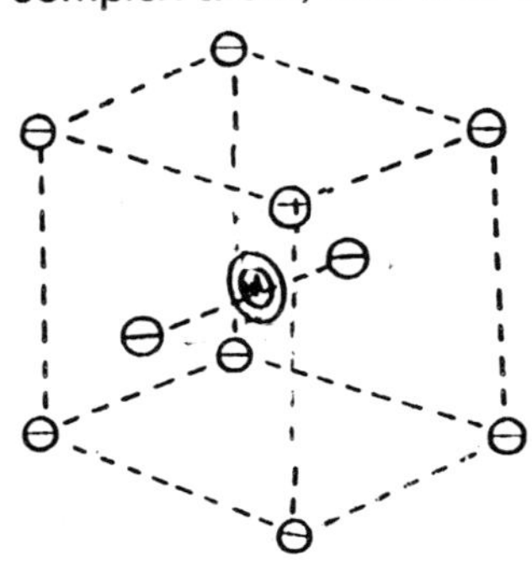

Figure p-28: The space occupied by the electrons in the atom is immense compared to the space occupied by the nucleus. The locations of the electrons in the neon atom are show in the diagram. The center "target" is the location of the nucleus.

nucleus, the electrons are organized in accordance with the amount of energy in the attendent exchange particle.

Electrons repel each other with forces of similar magnitude to that of the attractive forces. The combination of attractive force and electron repulsion then dictate where the electron can be located within the region around the nucleus. The electrons remain as far apart from each other as the available space permits. Another factor is the line-of-sight imperative. Each electron must have a clear channel to one positive charge.

Figure p-28 shows the pattern of electrons in the neon atom. The two inner electrons are opposite

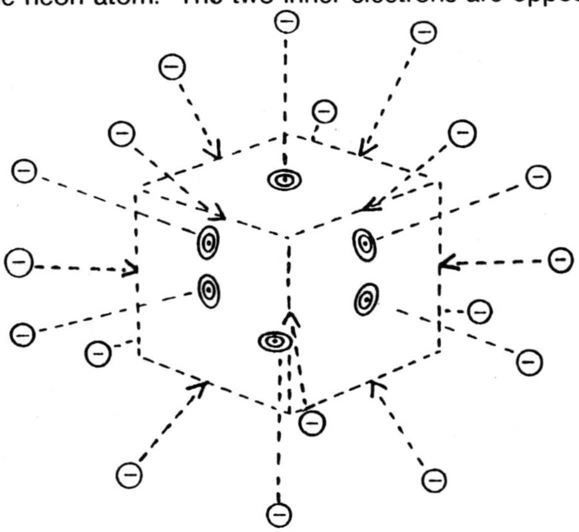

Figure p-29: A cube has 6 faces, 8 corners and 12 edges. The structure patterns adopted by electrons in the atom are related to these numbers. An electron "shell" with 18 electrons is derived from the centers of the faces of a cube and from the centers of the edges of a cube. As shown in the diagram, these 18 positions permit the electrons to be the maximum distance apart while each electron retains a line-of-sight with the home nucleus. The targets in the diagrams are the nuclei of component helium atoms.

the central nucleus. The other 8 electrons are spaced approximately at the corners of a cube. The cube is not perfectly cubical, but is narrow in one direction. This shape is the result of the electrons all moving as

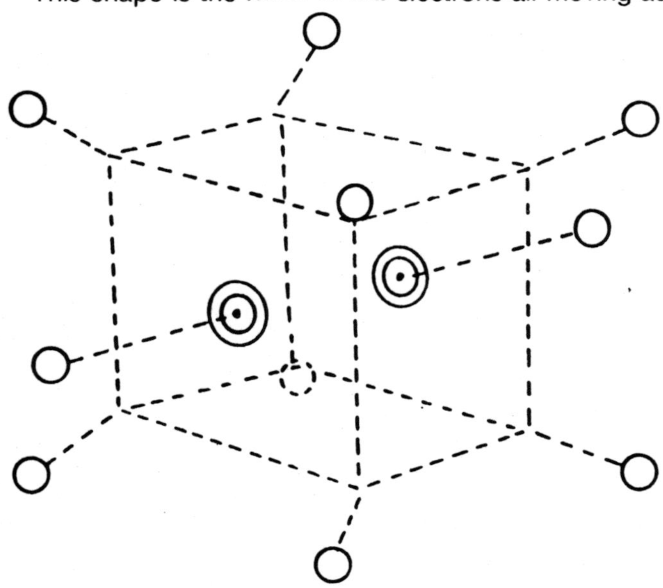

Figure p-30: The chemical bonds between atoms and the periods of the periodic table are revealed by the positions adopted by the electron shells of the atom.

far apart as possible, while remaining gripped by their respective exchange particles.

A cube has 6 faces, 8 corners and 12 edges. When the next outer layer or "shell" of electrons begins to form, the new electrons take up positions where they are far from the existing electrons, yet they can "see" the nucleus, in accord with the line-of-sight imperative. As shown in Figure p-29, the new electrons avoid the corners already occupied by the neon atom. They take up positions opposite the 6 faces and the 12 edges, as shown. This totals the 18 electrons in that layer.

The periodic table is founded on groups of 8 elements and groups of 10 elements. These groups relate to the chemical composition of each element, dependent on the number of electrons in their outer shells. Figure p-30 is a diagram showing the probable positions of the outer electrons. The cube shape is narrow in one direction, permitting the 9th and 10th electrons to take up the positions shown.

EVENTS FROM THE PAST

This scene of "people power" is in the city of Boston, Ma., in the year of 1872. An epidemic of grippe had attacked horses in United States and only a few survived. Tramways and streetcars depended on horses for motive power. New York and Philadelphia switched to teams of oxen, but in Boston, conductors and drivers and volunteers were paid one dollar a day to pull the tram cars through the streets.

This artist's view of the event was made over a hundred years ago for the popular press of the day.

WHAT IS THE UNIVERSE LIKE

DID THE BIG BANG EVER HAPPEN

The universe is made up of matter and space. The universe is infinite in size. It goes on forever in every direction. It always existed. There was no original big bang and there will never be an end to the universe. These statements are nearly self evident.

The universe is comprised of either matter or empty space. If empty space is nothing, then there is no conceivable barrier to the extent of the universe. Whatever end point you can imagine, beyond that point there is either something or nothing. In either case, the universe goes on. If space is curved, the dimensions of the universe remain unchanged. Whatever direction you go, whatever distant point you can imagine, beyond that point the universe goes on and on.

If the universe has no end in terms of distance or dimension, that means that there was never any beginning to time and there will never be an end. When you look into distant space, you are looking back in time. If there is no end to space, then, by the same reasoning, there is no end to time. There was never a beginning and there will never be an end.

THE BIG BANG THEORY

The big bang theory was originated in order to explain why the whole universe appears to be expanding in every direction. The theory predicts that if you roll time backwards, all of the matter in the universe would travel back to a single point. Therefore, the evidence appears to support the idea of a tremendous explosion that started it all. Presumably there was no universe as we know it before the big bang happened. But what if the evidence is simply misinterpreted?

The reason scientists believe that the universe is expanding is because there is a red shift associated with spectroscopic graphs taken of light from distant matter. The further away any particular star is from the earth, the greater is the red shift. The red shift is usually interpreted to mean that the object is moving away from the earth. Is it possible that some other phenomenon causes the red shift?

One theory that has been advanced is that old light gradually loses energy to space. When light loses energy, it shows up as a lowering of the frequency. The farther light travels to reach the earth, the more energy it has lost, and hence the red shift is greater for more distant objects.

Another theory proposes that the red shift is caused by the motion of earth through space. Since we, the observers, are travelling with the earth, the red shift is caused by our observations.

When you look up to observe the stars in the night sky, your line of vision must correct the image to accomodate the velocity at which the earth is moving. The orbital speed of the earth through local space is 19000 miles per second. The effect is similar as when the star quarterback throws a pass downfield to a receiver who is running across the field. The quarterback throws the ball to the place where he believes the receiver will be when the ball arrives. The angle at which the earthly observer must look to correct star images is 21 seconds of arc.

Relatively, if the earth was not rotating, most of the stars would be whirling around the earth at speeds greater than the velocity of light. The more distance there is between the earthbound observer and a star, the greater is the turning effect in order to maintain the angle of observation. The effect of this is shown in the diagrams of Figure U-1. In diagram (a), because of the earth's motion, the star apparently travels across the sky at a high velocity. The individual waves of radiant energy leaving the region of the star are shown as circular wave fronts, each one originating at a point where the star actually was when the wave was released.

If the part of the wavefront intercepted by the observer's eye is the portion of the wavefront closest to the observer, the wavelength of the arriving radiation would be like the intervals shown on the left side of diagram (a). However, what actually happens is shown by diagram (b). The part of the wavefront that reaches the eye of the observer comes in at an angle, as shown. The result is that the train of waves arrives at the observer's eye with a definite doppler effect. The further away the star is that is producing the radiation, the more pronounced the effect is.

An experiment could be performed in which the space shuttle, travelling in orbit at speeds in excess of 23000 miles per hour, would be used as an observation platform. During a small portion of the orbital path, the shuttle will be moving against the direction of the motion of the earth through local space, thereby reducing the cause of the red shift by a small amount. Observations would be timed to coincide with the retrograde motion. If the red shift is caused by the motion of the earth through space, then the red shift from distant sources would be reduced by 0.0336 percent. This amount of change could be detected by todays sophisticated spectrographic interferometer.

HOW IS AN ELECTRIC CHARGE MADE

The universe appears to be mainly empty space with globs of matter speckled here and there with immense distances in between. The matter portion of the universe is made up from fantastically small bits and pieces. Solid matter is comprised of molecules which are comprised of atoms. The atoms are made up out of electric charges. The electric charge is therefore the basic building block of the universe. If the way in which an electric charge is put together could be understood, that would be a step toward understanding the whole universe.

An electric charge really is just a point where a bundle of lines of force converge. It is nothing more than that. The characteristics of the charge are all caused by the lines of force which converge at the point. So in fact the real basic unit of the universe is the electric line of force. Without the lines of force and the way in which they perform, there would be no electric charges.

An electric line of force is made from moving magnetic fields. In Figure U-2 the magnetic fields are shown as circles which collapse toward the line, but pass right through the line and expand again with opposite polarity. The polarity has not changed, of course. Only the direction of motion has changed. There is no reason why the moving magnetic fields have to be circles. Any moving magnetic fields will contribute to the line of force. The criteria is that in order for the magnetic fields to disappear, the sum of the vectors of such moving fields must be zero. When that condition exists, the line of force remains in evidence, but the magnetic fields which produce it are not in evidence.

If a short length of the line of force is examined, one end of it will be exhibiting negative electricity and the other will show positive electricity. The magnetic fields that cause the line of force cannot be seen or measured in any way, but they still exist. The line

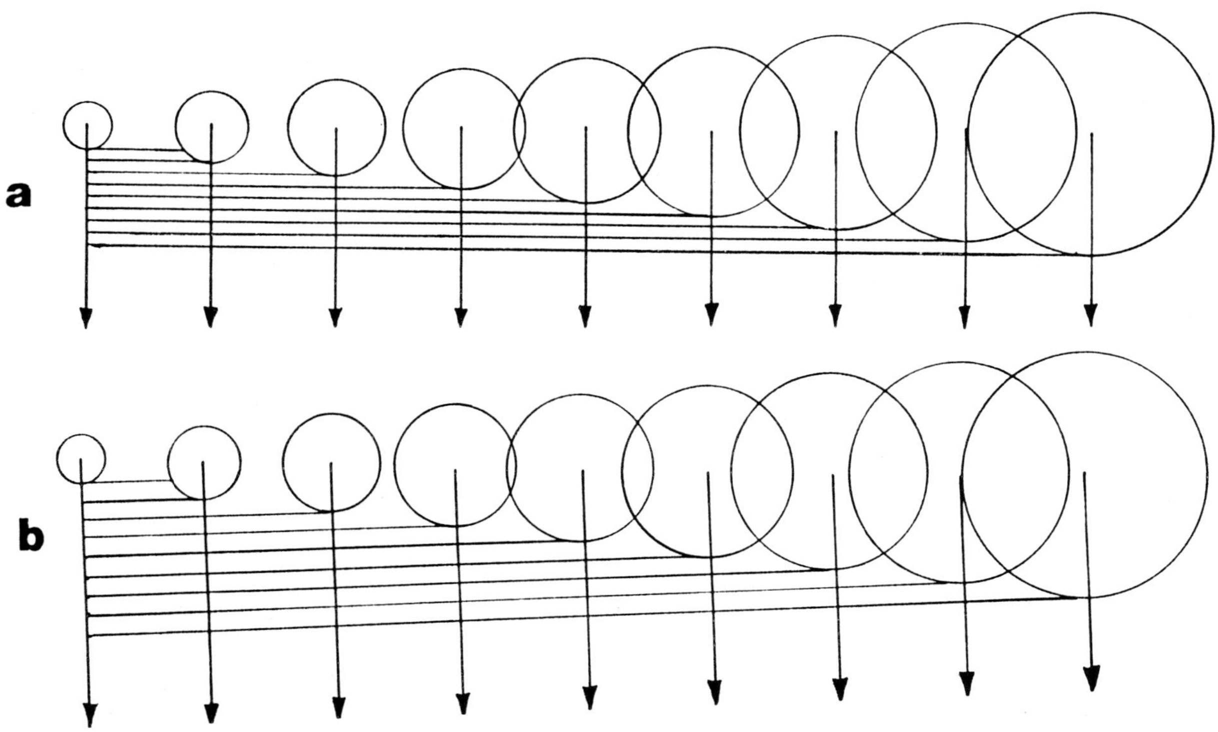

Figure U-1: In the diagram, each circle is a wavefront that was released by a radiating star which is moving rapidly across the field of view from left to right. In diagram (a), if the near side of the waves would reach the eye of the observer in succession, the wavelength of the arriving radiant energy would be like the intervals shown at the left of diagram (a). However, because the earth is moving through space and rotating, the wavefronts that reach the eye of the observer come in at an angle as shown in diagram (b). A doppler effect is generated. The effect is greater for stars at greater distances.

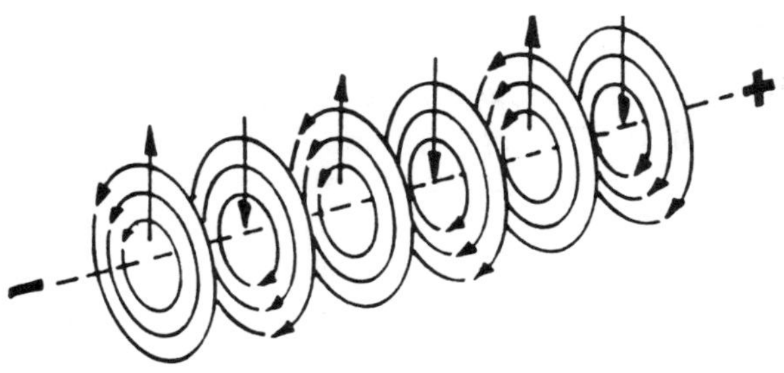

Figure U-2: Expanding and collapsing magnetic fields result in lines of electric force. A magnetic field that collapses to the center line passes through the center and then must expand out again instantly to comply with the conservation of energy. The vector sum of these magnetic fields is zero.

of force is the proof that the moving magnetic fields are still there.

THE STRAIGHT LINE IMPERATIVE

The most important characteristic of a line of force is that it can only exist as a straight line. Another quality that it possesses is to shorten itself to the shortest possible length, which makes it become the shortest distance between two points (which is the definition of a straight line). When permitted to shorten itself entirely, it disappears into a puff of radiant energy.

In all of the manifestations of electric fields, there are usually many curved lines. As Maxwell pointed out, the curved lines of force are merely vector additions of underlying straight lines.

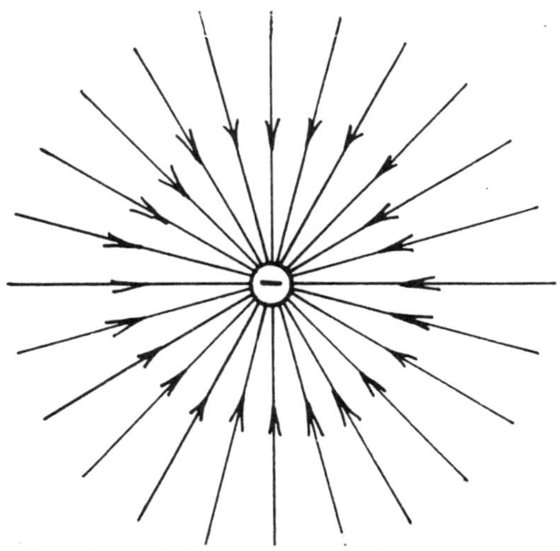

Figure U-3: The radial pattern of force lines around a lone charge. The charge is three-dimensionally radial rather than the two flat dimensions shown here.

THE POINT OF CONVERGENCE

Lines of electric force of the same polarity if placed parallel will repel each other. If two lines of force are placed side by side with opposite polarity, they will form into a continuous line shaped like a sausage and collapse toward the center where they disappear and release a photon of radiant energy. The main difference between a line of force and a photon is that the photon undulates with a certain frequency and moves off at the velocity of light.

Lines of force are extremely volatile. What keeps them all from converting to radiant energy is the fact that they are held captive by millions of tiny points of convergence that we call charges. All of the positive ends converge to form negative charges and all of the negative ends converge to form positive charges. What is it about a point of convergence that holds the lines of force captive?

When lines of force have the same polarity, they repel each other. That explains why the lines radiate away in every direction from the point of convergence, because they try to separate as far as possible away from each other. But at the very ends of the lines where they converge, they do not fly apart. There is some kind of glue holding them there. The nature of that glue is an intense concentration of the moving magnetic fields which make the lines of force exist.

ELECTRIC GLUE

One of the amazing things about electric charges is that when two electrons are accelerated side by side, they don't repel each other as strongly, and at a certain velocity, a pair of electrons do not repel each other at all. The reason is that the circular magnetic fields associated with acceleration tend to blend the charges together and make the pair of electrons act as one.

The circular magnetic fields that rise and fall around a line of force are identical to the acceleration fields of moving charges. Therefore, near the points of convergence of the lines of force in an electric charge, the magnetic fields blend together. The closer the converging lines of force are to the center of convergence, the stronger is the binding force of the surrounding magnetic fields. The diagram in Figure U-4 shows the combining of the magnetic fields around two lines of force. This effect is multiplied a million times in an actual charge structure.

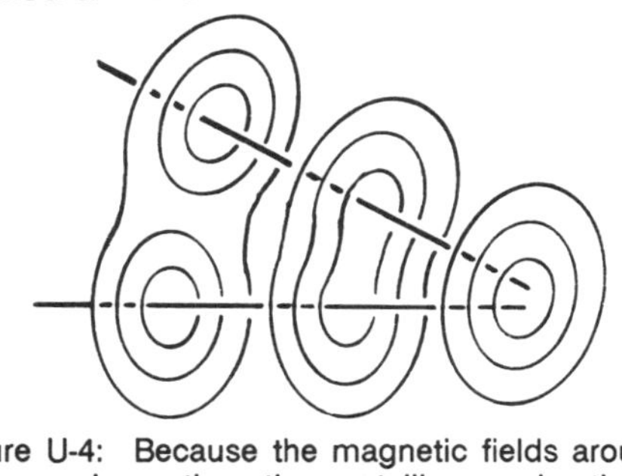

Figure U-4: Because the magnetic fields around a line of force are in motion, they act like acceleration fields and combine together when in close association. Two lines that converge become bound together near the convergence. The closer the converging lines approach each other, the stronger the binding force.

Take a closer look at the illustration of a line of force shown in Figure U-2. Then try to imagine this incredible fact. The whole universe is made out of nothing except these lines. All of the electric charges are made from lines of force. All of the atoms are made up from electric charges. Molecules are made from atoms. Matter in the large, like you and me and the earth and the stars are made up from molecules and atoms. Now, if somebody happens to ask you if you know anything about "string theory", you can say "yes, if the string is a line of electric force".

WHAT IS A MAGNETIC FIELD

The great puzzle that remains is the magnetic field itself. Why does it act the way it does? How does it make electricity flow? James Clerk Maxwell tried to find an answer to these question nearly a century ago. He made many drawings of spinning vortices, wheels and rollers, and elastic tubes full of liquid and whatever else he could think of. Eventually he discarded all of the models and wrote down the formulas that are still in use today. Maxwell's formulas show the relationship between an electric field and a magnetic field and how they interact. What causes the magnetic field to exist is still a mystery.

JOSEPH NEWMAN'S MAGNETIC FIELD

Joseph Newman (of mass-equivalence electric car fame) came the closest to visualizing an actual magnetic field. His theory arose out of the fact that when you accelerate a charge, the magnetic field forms in a circle around the accelerated charge. In the same way, when you accelerate a gyroscope, the gyroscope turns and aligns itself around the axis of acceleration. The gyroscope then has the same polarity as a circular acceleration field.

This relationship convinced Newman that an electric charge is surrounded by millions of tiny objects that are spinning in randon directions. When you accelerate the charge, the spinning points turn like gyroscopes to align their spin with the axis of acceleration. The result of the many gyroscopic points being aligned in the same direction is that they show up as a circular magnetic field.

MAGNETIC SPACE

It can be said without any doubt that a magnetic field is made up from whatever kind of particles occupy space. A magnetic field will arise anywhere, whenever an electric charge is accelerated. That means that the tiny particles are evenly distributed throughout space. Their main occupation is to form into rising and collapsing circles of magnetism around electric lines of force.

What physical attributes would be necessary to make millions of tiny particles organize themselves into circles that expand and collapse? If a circle of tiny magnets are placed on floating chips of wood, they will first turn and align themselves with unlike poles facing, as shown by the arrows in Figure U-5 (a). Then they will converge to the center exactly like a collapsing circle of magnetism, as shown in Figure U-5 (b).

It seems very likely that Newman's tiny gyroscopic particles do behave like magnets. This does not mean that they are magnets. What it does mean is that the spin factors inherent in each particle, and their inherent spinning energy, causes them to align themselves in certain preferred directions.

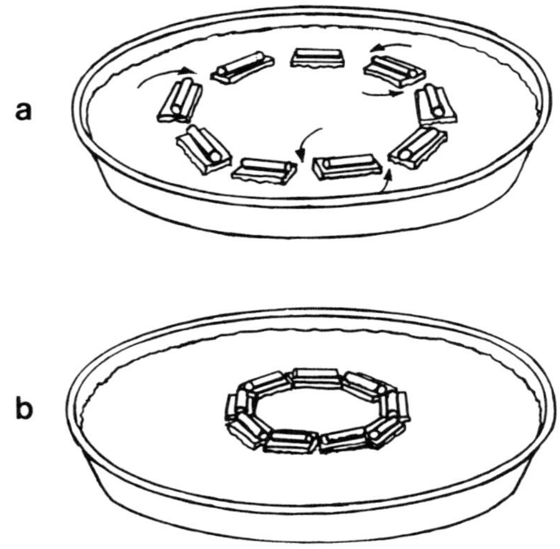

Figure U-5: Small magnets floated on a pan of water first swivel until the unlike poles are adjacent, as shown at (a), and immediately converge on the center of the pan as at (b). This indicates that whatever fine particles make up a magnetic field, they probably exist as magnets.

Groups of magnetic lines that are placed in an array as shown in Figure U-6 (a) do not automatically combine to make one large magnet. The vector additions of the overlapping circles produce the pattern shown at (b), which is actually of opposite polarity to (a). In order for a large number of tiny magnets to form into one large magnet, there must be a magnetic path present, like an iron core.

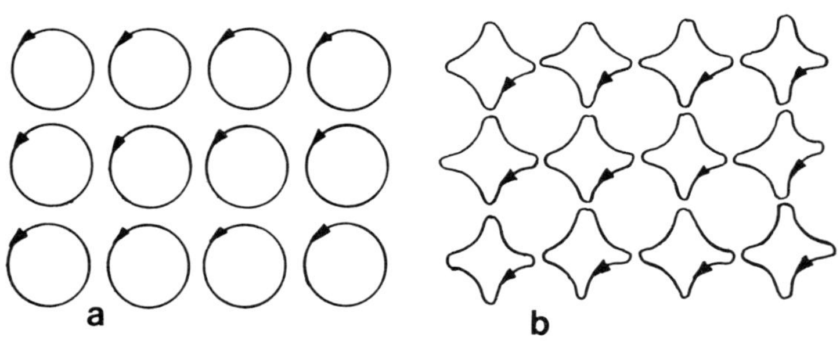

Figure U-6: Groups of small circular fields as shown in diagram (a) do not automatically unite to form a large field. Vector additions of a group of fields yields the pattern in diagram (b), which has a reversed polarity. However, groups of small circular field lines form one large field when a magnetic path is introduced, such as an iron core.

Perhaps what we call magnetism is really made from millions of miniature vortices or whirls of force in space. The vortices or whatever they are do act like miniature gyroscopes and they do combine to form circles of magnetism, and that is evident in the acceleration field of a moving electric charge.

HOW DENSE IS SPACE

Perhaps twenty years ago there was an article in the Scientific American in which the authors described a series of experiments they were making. The authors were experts on magnetism, and their main experimental thrust was to discover if there was a limit to the strength of a magnetic field.

This is an interesting idea because magnetism is not something you can compress into a tank like compressed air. If there is a limit to the strength of a magnetic field, does that mean that there is a limit to the density of the tiny gyroscopic particles? Are there X number of particles per cubic centimeter?

Many years have passed since these experiments were performed so the details are difficult to remember. However, the result was that there is no limit to the strength of a magnetic field. What this means is that if there is a limit to the density of space, it was not reached, and there was no indication that there would ever be a saturation point where a limit would be reached.

THE VELOCITY OF LIGHT DEFINED

A photon is very much like an electric line of force. The main difference is that a photon is an organization of such lines in which many lines of electric force are held parallel. Captive undulations travel within the photon at an incredible velocity, the velocity of light. The captive bonding consists of toroidal (doughnut) shaped magnetic fields with opposite polarity at intervals equal to the frequency of the radiation. The photon has the durability of a particle and only releases its energy when a sympathetic charge in its path matches its vibrations exactly. At that time the energy is transferred to the charge.

The electric lines of force in the photon and the associated magnetic fields are comprised of fine particles. As previously proposed, such particles exist in immense numbers and are either magnetic or else spin in such a way that the preferred directions they take have the same characteristics as magnetism.

The fine particles of space, considered separately from the intense structured organizations they form, are like a special kind of gas. In fact because they are magnetic and attract each other in a particular way, they are very much like a plasma. (A plasma is a gas made from extremely hot atoms in which the electrons are disassociated from the nuclei by the intense concentration of heat energy. The atmosphere of the sun is an example of a plasma). Waves travel through the magnetic plasma at the velocity of light.

In all forms of energy propagation the upper limit of velocity is controlled by the density of the wave conductor and the velocity of motion (or of vibration) of the contributing atoms and molecules. The least complex forms of sound propagation are waves moving through liquids and gases. Sound waves in air have been thoroughly studied.

The velocity of a series of waves through a medium is related to the velocity of the particles in that medium. From every point in the medium any energy disturbance travels outward in a circle from the point of origin. The velocity of the wave front is then the average velocity of the individual particles in that direction. The average velocity of the particles is their root mean square velocity. The result is the formula

$$\text{Velocity of wave} = \frac{(\text{rms})\,V}{\sqrt{2}}$$

The basic idea from this relationship is that the particles of the medium travel faster than the wave front by a factor of more than a third. For example, when the velocity of light is given as 3×10^{10} centimeters per second, the velocity of the particles in the magnetic field is therefore 4.2426×10^{10} centimeters per second. This does not mean that the particles are flying around randomly in every direction. But it does mean that within whatever organization they form, the average speed of the particles in their vibrations match that velocity.

THE CURVATURE OF SPACE

Einstein expected that a ray of light would be bent by gravity because of the mass-equivalence relationship. It was proposed to examine the light rays which pass close to the sun. The angle of the bent rays was measured during an eclipse because stars near the dazzling sun could not be seen when the sun was visible The angle was measured by groups of scientists observing from different locations. A predetermined angle was deducted from the measured angle in order to account for the effects of the sun's atmosphere. The residue angle was determined to have been caused by the gravity of the sun. The amount of the angle was very nearly what Einstein had predicted.

DENSITY GRADIENT IN MAGNETISM

When light rays pass through layers of different density, refraction occurs, which causes the light rays to bend. When the moon or the sun is rising or setting, atmospheric refraction is visible and can be measured. When light rays are curved and there is no atmosphere present, there is the possibility that there may be a density gradient in space itself near massive bodies. The density gradient would apply to the magnetic particles of space. If such a density gradient exists, it would effect the intensity of a magnetic field for a given coil and current. The same coil with the same amount of current should exhibit a slightly less intense magnetic field when measured out in space as compared to when it is measured at the surface of the earth.

THE NATURE OF TIME

One of the great riddles of the universe and of

our existence in the universe is the nature of time. Time appears to flow by in an orderly sequence of intervals of duration. Our lives are ruled by time; our very lifeforce is called longevity to remind us that we are captives of time. We are born into the current of time and although time is like a river in many ways, we cannot swim upstream or downstream. We remain fixed in the current until the river releases us into limbo in our old age. Meanwhile we are free to exercise our will and do with our time whatever we wish. We can squander it away in day dreams or we can apply it in many ways to good or evil. It has been pointed out that it is the minutes that we waste in idle or useless pursuits to which most of our life is lost. Some people with organized habits have the ability to go through college with the same amount of time allotted as the time that most of us fritter away in indecision.

One of the things that mankind does very well is to measure time. When we measure time we discover that there is only one way to measure it, and that is by comparing it to something else. The rotation of the earth gives us a basic unit of a day. A vibrating atom gives us a very small unit. But there are no units for time itself. Time can only be observed by keeping track of the motions of objects, either celestial or manmade.

Why does time exist? The answer to this question is based in the conservation of energy. Time exists only because energy cannot be created or destroyed. The result is duration. Conservation is duration. Time is duration.

TIME AND ACCELERATION

The rate of the flow of time changes when a mass is accelerated or decelerated, in accordance with Einstein's predictions. When an object is accelerated, the atomic oscillators within the mass are affected. The range of the vibration rates of the internal resonators is altered. Since the rate of the flow of time is related to the speed at which the object is moving through Galactic space, the Galaxy can be used as a frame of reference against which to measure the flow of time.

A change in the rate of time flow is called time dilation. It happens to people as well as to inanimate objects. When a spaceman goes on a trip in a spaceship, the flow of time changes as the ship is accelerated. Time slows down as the velocity of the star-ship increases. A few years later, when the spaceman arrives at some distant planetary system, he has remained youthful because time passed so slowly during the trip. Meanwhile, his earthbound friends aged at the normal rate and are now perhaps twenty years older than he is.

Since the flow of time is related to speed through space, on the return trip the spaceman will again experience a slowing of the flow of time. Again he may lose twenty years as compared to his earthbound friends.

When the spaceman returns to earth some years later, the sweetheart that he left behind is now an old wizened grandmother, while he remains vibrant with youthful vigor.

The evidence that time dilation actually happens was demonstrated by Rudolph Mossbauer in 1958, when he devised a nuclear clock for studying such phenomenon. A quartz crystal with a known frequency was clamped to the edge of a rapidly rotating disc. The rate of time flow in the quartz crystal slowed down the exact amount prediced by Einstein's theory. The change in the rate of vibration of the crystal is easily detected with Mossbauer's method (So, if you want to stay young, spend your life on a merry-go-round)!.

LIVE MATTER AND DEAD MATTER

Theories about the origin and eventual destination of the universe are dependent on the most accurate knowledge that is available about the distant stars. It is a natural result of distance that objects far away are much more difficult to define correctly than objects that are near at hand.

One of the tidbits of information needed by cosmologists is the density of matter in the universe. In the early days star counting was a relevant procedure. The amount of matter accumulated in this way was far short of the amount of matter needed to support most theories about the universe.

An early investigator who took note of all of the stars closest to the earth and their density reached the conclusion that if the stars continued in every direction with the same density as local stars, then the whole sky would be ablaze with light night and day. Somebody else came up with idea of "dead matter", that is matter that does not have its own source of illumination, like the earth. In this case the "blazing sky" would be blocked out and shaded by all of the dead matter in the universe. An argument ensued in which the blazing sky advocate assured his detractor that there was no dead matter in the universe, because it would gradually absorb radiation from all of the hot sources until everything was the same temperature. While making this argument, the individual was standing on a cold, non-illuminated earth. When this fact was noted, it turned out he was talking about dust particles.

Right here in our own solar system we have an illuminated part (the sun) and a few non-illuminated parts (the planets). In between the planets there are many smaller cold bodies including asteroids and meteor showers. All together, the amount of illuminated matter in the solar system is much greater than the amount of dead matter. This proportion is probably normal for systems like the solar system in the outer wings of the Galaxy.

Closer to the center of the Galaxy, it should be natural that more hot matter should exist. However, whatever direction is considered, whereever we aim our telescopes, great masses of dead obscuring matters are visible throughout the Galaxy. Considering that most dead matter cannot show itself except by the fact that it obscures illumination, there is likely to be much more dead matter in the universe than what is visible. Certainly there is sufficient matter unidiscovered and unknowable to pad out the requirements of any cosmological theory based on matter density.

THE STEADY STATE UNIVERSE

In 1948 Fred Hoyle and a couple of his friends (Bondi and Gold) originated the idea of a steady state universe. It is quite different than the Big Bang theory, advocated by George Gamow, which claims that the universe began as an enormous explosion of a primeval glob of matter. The expanding universe that we live in today is the result of the explosion, Gamow claims. All of the stars and galaxies throughout the universe are still flying apart from the original explosion.

In his book about the steady state universe, Hoyle accepts the expanding universe. But he believes that the universe was always expanding. It always was and always will be. The basic idea of the steady state universe is that the universe never changes. The thinning out of the matter in the universe by expansion is replaced by the creation of one hydrogen atom per year in every ten million cubic feet of space. According to Hoyle, this minor creative activity would hardly be noticed, but would be sufficient to counterbalance the matter lost to deep space by the constant expansion of the universe.

It is interesting to note that both of these theories, the Big Bang and the Steady State, are based on the apparent expansion of the universe occasioned by the red shift (doppler effect) observed in distant matter. If it happens that the red shift is caused by some effect other than expansion, then both of these theories would be of little consequence. Meanwhile, either one could be true. One day the truth will be known.

THE VELOCITY OF LIGHT

The fact that the velocity of light is the same for any observer regardless of the motion of the source or the motion of the observer is the evidence of a mechanical effect. What this evidence confirms is that the constancy of the velocity of light is not related to any particular observer, but to the matter out of which the observer's light measuring instrument is constructed.

The apparent contradictions (Einstein's assertion that the velocity of light is a constant for everyone) are removed by expressing the velocity of light in a mechanical way. An experiment to illuminate this characteristic of the interaction between a ray of light and matter begins by procuring a long thin elastic thread (long thin strands of elastic can be obtained by removing the cover from an old golf ball). The elastic strand is stretched from point A to point B in a room, fastened in place, and the velocity of sound through the elastic thread is measured and recorded. The elastic is then detached as it is no longer needed.

The next step is to make five or six small blocks from soft wood and place the blocks at different positions in a room. Ask some friends to cooperate and hold each block in position while long elastic threads are connected from block to block, using thumb tacks to secure the elastic strands to the wood blocks. The wood blocks represent observer stations. Each wood block is connected to all of the other observer stations by elastic threads.

The co-operating observer's are now asked to move the blocks of wood about. While the blocks are in motion, the speed of sound in the elastic attached to each block is measured. What is noted is that when a block of wood is in motion, the velocity of sound in the elastics attached to that block is a constant relative to the block, measured near the block. Each observer notes that the velocity of sound in the elastic strands is a constant relative to any one observer station regardless of the motion of any or all of the observer stations.

This is exactly the same as the Einsteinian postulate for light rays. The mechanical analogy to the velocity of light is completed by substituting the velocity of light for the velocity of sound. The elastic strands then represent the elastic electric lines that connect between the charges in centers of mass.

The result from these experiments confirms that the speed of sound in an elastic thread is a constant. The conclusion from the above demonstration for light rays can be stated with one mechanical cause-and-effect explanation: "Radiant energy travels along electric field lines at the velocity of light".

One has to wonder if Einstein was aware of this mechanical explanation. Why did not Einstein say that the paths along which light rays travel are elastic, and that they might be stretching or relaxing in length during the time that the radiant energy is apon them? Perhaps the reason it was never mentioned is that the mathematics were already of considerable complexity. To introduce a flexible elastic path for the velocity of light would have made the mathematics of relativity virtually impossible. Better to merely state the truth, that the velocity of light is a constant for every observer.

INVENTIONS FROM THE PAST

This cartoon of a family outing on a tricycle appeared in a popular English news magazine many years ago when cycling became popular. The cartoon is an antique.

BASIC ELECTRICITY

STATIC ELECTRICITY

(1-5) Ancient Greek philosophers observed and recorded the effects of static electricity. When a piece of amber was rubbed, they saw that the amber was imbued with a special attractive power. The amber could attract and pick up small bits of leaves and seeds. What happened was that the amber was given a static charge by rubbing contact with another material. Soft materials like cloth or fur are the most effective. From these early experiments with amber, our word for *electricity* was originated. The Greek word for amber is *elektra.*

Glass, plastic, compressed rubber and other materials will accept a static charge when they are rubbed with fur. A piece of glass rod or plastic rod are useful for demonstrating the resulting forces. Make two small balls of kleenex or tissue paper and suspend them on threads. They can be charged by touching them with the end of the glass rod (or plastic) on which a charge has previously been induced.

It will soon be noted that different materials hold different kinds of charges. The paper balls will fly apart when both have been touched with the glass rod. However, they will pull together if one of them is now touched with the charge on a piece of bakelite or compressed rubber. The reason is that glass carries a negative charge whereas compressed rubber holds a positive charge.

An interesting observation is that similar static charges repel each other and different static charges attract each other. This is the basic fact about electricity. *Like charges repel; unlike charges attract.*

ELECTROSCOPE

A simple electroscope for comparing the charges on different substances can be made by suspending two small balls of tissue paper on narrow strips of thin foil inside a glass jar. The tissue paper balls are suspended from a piece of metal or wire. The wire is pushed through a seal of insulating material like cork, wax or a rubber eraser. This insulates the electroscope from the metal lid of the jar. See the illustration in Figure S-1,(a).

A more sensitive electroscope can be made with two pieces of gold leaf suspended close together inside a jar, drawing (b). The charge in the electroscope is registered by the increased distance between the two paper balls or the two gold leaves.

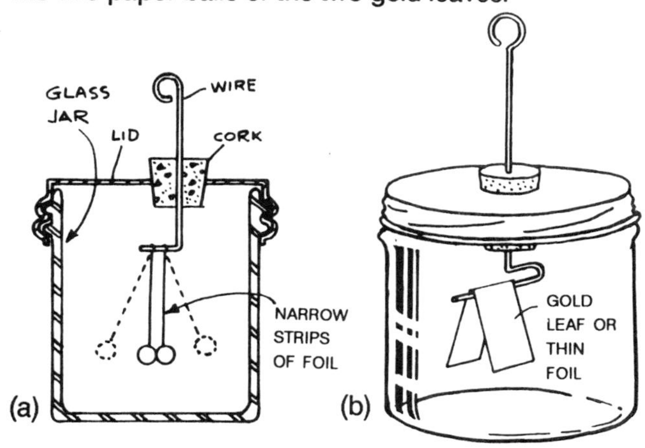

Figure S-1: Making an electroscope out of a small glass jar.

ELECTRIC ATOMS

(2--5) Where does static electricity come from? Everything is made out of atoms and molecules. Static electricity must be attributed to the structure of atoms and molecules. The first design for an atom was conceived by Ernest Rutherford and developed by Niels Bohr. It was decided that each atom must be like a miniature solar system, with planetary electrons circling around a central core. What the core could possibly be made of was unknown at the time, but then Rutherford's associate, James Chadwick, discovered the existence of the neutron. With the neutron as the core, the planetary atom was complete. Although the planetary atom is not a plausible concept today, it was a useful idea that contributed greatly to the advancement of atomic knowledge.

The simplest atom known is the hydrogen atom. For a nucleus it has one proton with a positive charge and one planetary electron with an equal negative charge. The attraction between the two unlike charges holds the atom together. A diagram of the hydrogen atom is shown in Figure S-2, (a).

Helium is next after hydrogen. It has two

electrons and two protons, but it also has two neutrons in the core, which gives the Helium atom four nucleons, as shown in the diagram, Figure S-2, (b). Next is Lithium. It has three electrons, three protons and three neutrons.

Each atom has a series of levels like the floors in a department store. Each level can accomodate only an exact number of electrons. The first floor, or inner level, can hold only two electrons. Since the Lithium atom has three electrons it follows that it must place one of its electrons on the second floor. However, the second floor will hold up to eight electrons. The first two levels together will hold ten electrons.

The ten atoms whose electrons occupy the first two levels are listed as follows:

Hydrogen (H)	1 electron	1 proton	o
Helium (He)	2 electrons	2 protons	2 neutrons
Lithium (Li)	3 electrons	3 protons	3 neutrons
Beryllium (Be)	4 electrons	4 protons	4 neutrons
Boron (B)	5 electrons	5 protons	5 neutrons
Carbon (C)	6 electrons	6 protons	6 neutrons
Nitrogen (N)	7 electrons	7 protons	7 neutrons
Oxygen (O)	8 electrons	8 protons	8 neutrons
Fluorine (F)	9 electrons	9 protons	9 neutrons
Neon (Ne)	10 electrons	10 protons	10 neutrons

Figure S-2.

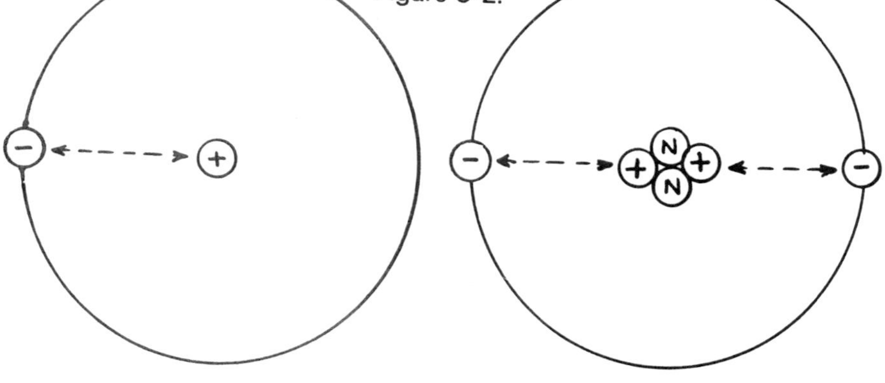

(a) Diagram of the Hydrogen atom. (b) Diagram of the Helium atom.

Some atoms are *inert*. That means they hold onto their electrons securely. Some atoms actively exchange electrons with other elements. These atoms are said to be *chemically active.*

For example, Sodium (Na) has eleven electrons. It has two electrons on the first level and eight electrons on the second level. There is one electron left over which must go alone onto the third level. With all that space for the lone electron to occupy, it easily gets lost, and it often does get lost.

On the other hand, both Fluorine and Chlorine each have one less electron than needed to fill up a level. They each have a vacancy. Atoms like Sodium which have a lone electron are very likely to combine with atoms which have an electron vacancy. Sodium and Chlorine often combine to produce common salt (NaCl).

The word used to describe atoms with extra electrons or extra electron vacancies is *valence.* An atom with a giveaway electron has a valence of +1 after it has lost its electron. It has a positive charge. An atom that has gained an electron has a valence of –1, and it has a negative charge.

Atoms are extremely small. A Hydrogen atom (one electron and one proton) has a mass of only 1.674×10^{-24} grams. The dimensions of the Hydrogen atom are tiny beyond any real comprehension of its size, about 0.5×10^{-8} centimeters. The mass of one electron is 9.11×10^{-28} grams.

GENERATING STATIC ELECTRICITY

(3-5) Electrons are easily rubbed off of certain materials which causes those materials to lose their electrically neutral state. When electrons are shifted from one substance to another, the object that loses electrons becomes positively charged. The object that gains electrons becomes negatively charged.

Atoms or molecules which have an electric charge are called *ions.* When a charge leaks away from the charged substance into the surrounding air, the air is said to have become *ionized.* Sometimes a charge difference between substances will become so strong that the air between them will become ionized and form a conductive path. The electrons will jump across the intervening space and neutralize the charge. When this happens, a display of sparks is produced.

When your pet cat is softly stroked in a darkened room, a glitter of small sparks will be seen. The electric charges on the cat's fur can be felt by your hand. Also, in the winter time, when you walk across a room with a broadloom rug on the floor, and then reach for a light switch, you will see a lively spark and feel it as well.

Summer lightning is a similar phenomenon on a larger scale. An enormous static charge is built up between cloud layers. When the charge leaps from cloud to cloud, or from the clouds to the earth, a display of lightning is produced. The resulting clap of thunder is a measure of the energy involved in a lightning display.

INSULATORS AND CONDUCTORS

(4-5) Charges of static electricity gradually leak away and all of the affected substances become electrically neutral again. This shows that air is not a perfect insulator, although it is a good one. There are some materials that can hold a charge for a long time. They do not permit a charge to pass through them and they inhibit charges from moving on their surfaces. Examples are glass, bakelite and porcelain. These substances are called *dielectrics* or *insulators.*

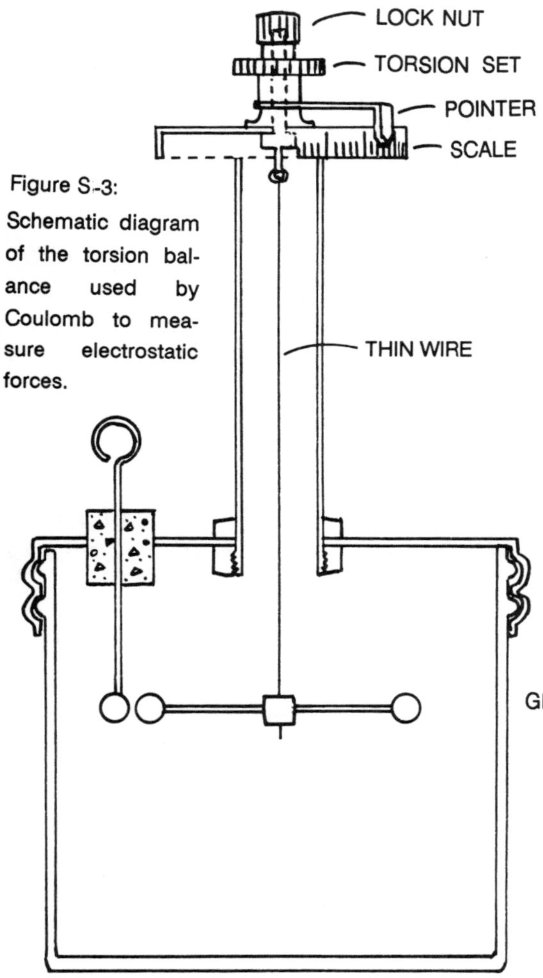

Figure S-3: Schematic diagram of the torsion balance used by Coulomb to measure electrostatic forces.

Other materials encourage a charge to run through them fast and easy. Silver, copper and other metals allow charges to slip easily along their surfaces. These materials are called *conductors.* Every substance known has a certain characteristic ability to conduct electricity. The dielectrics are very poor conductors and the metals are excellent conductors. The earth, being damp and full of minerals, makes quite a good conductor. When a charge is *grounded,* it quickly runs into the earth.

THE FORCES EXERTED BY STATIC CHARGES

(5-5) For the early experimenters it was easy to see that there were two kinds of charges. It was also apparent that like charges repelled each other and unlike charges attracted each other. The really difficult thing was to try and quantify the forces involved.

The measurement of these forces was first recorded by Charles Coulomb (1736--1806). He used a torsion balance placed within an insulated container, something like the illustration, Figure S-3. The results of his investigations showed that the forces exerted between charges, whether attraction or repulsion, are inversely proportional to the square of the distance between the charges. The strength of the force is proportional to the strength of one charge multiplied by the strength of the other charge.

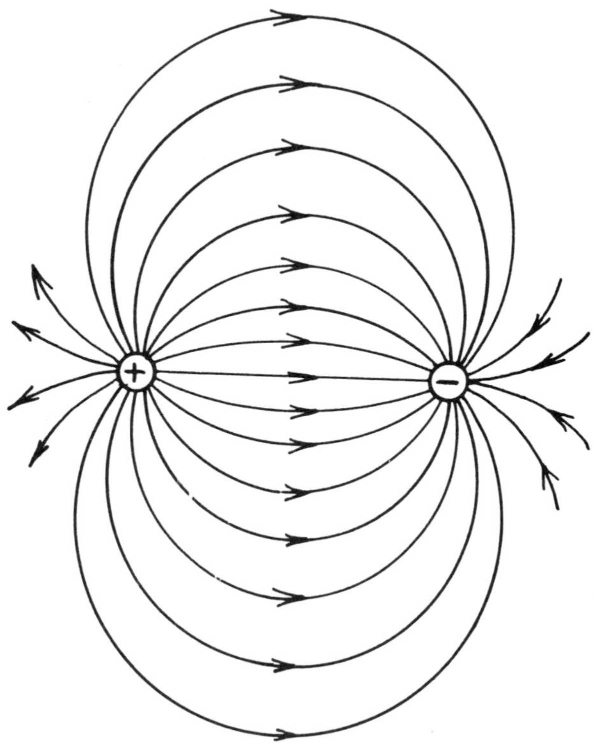

Figure S-4: The shape of an electric field around two unlike charges.

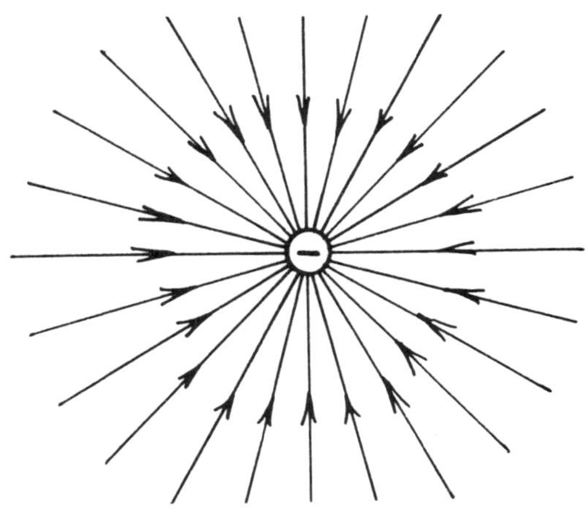

The shape of the field around one lone charge.

When two charges Q_1 and Q_2 are separated by a distance r, the force F between them is:

$$F = \frac{Q_1 Q_2}{r^2} \qquad (1)$$

This result is slightly different in a vacuum as compared to air. A factor ϵ, called the permitivity of the surrounding medium, is usually included in the formula

$$F = \frac{Q_1 Q_2}{\epsilon r^2} \qquad (2)$$

The factor ϵ is equal to one (1) in a vacuum, and in air it is slightly larger, approximately 1.0006.

THE ELECTRIC FIELD

(6-5) The space around a charged object exerts forces on other charges that are nearby. Such a space is said to be occupied by an electric field. The electric field cannot be seen nor felt. It can only be sensed by other charges.

The strength of an electric field can be measured at various points and mapped out on a chart. The resulting chart has a shape like the example shown in Figure S-4. The field intensity at any point is measured to be

$$\mathcal{E} = \frac{Q}{\epsilon r^2} \qquad (3)$$

Field intensity is measured in DYNES. One dyne of force results when two equal electrostatic units of charge (esu) are placed one centimeter apart. The charge on one electron is 4.8×10^{-10} esu. The number of electrons in one esu of charge is therefore:

$$ESU = \frac{1}{4.8 \times 10^{-10}} = 2.083 \times 10^9 \text{ electrons.}$$

Electrostatic lines of force have the folowing characteristics. In the direction along the electrostatic lines of force, the field tries to shorten itself to the shortest possible length. But in the direction across the field lines, the lines try to get as far apart as possible. The two opposing forces create the shape of the overall electrostatic field, Figure S-4.

ELECTROSTATIC POTENTIAL

(7-5) Since an electric field exerts a force, it is capable of doing work. The work that can be performed by an electric field is that of moving a charge or a charged object from one place to another within the field. The work done when a charge is moved in an electric field is expressed in ergs. An erg is a unit of energy or work.

Potential is actually the amount of work that would be necessary to bring a charge from an infinite distance away right up to its present location within the field of another charge. Suppose two charges Q_1

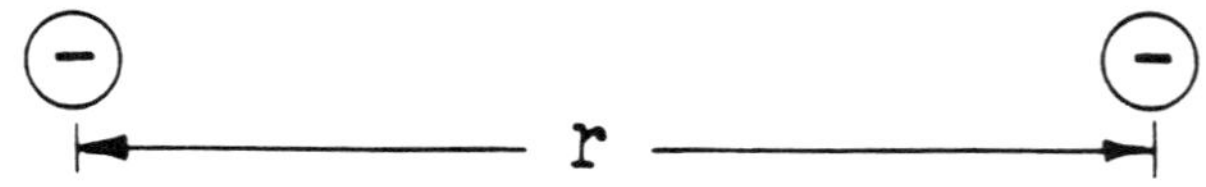

Figure S-5.

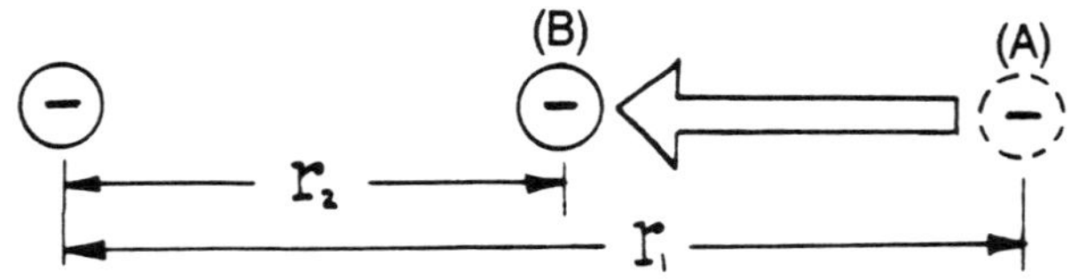

Figure S-6.

and Q_2 are located r centimeters apart as in the drawing, Figure S-5. The potential of Q_2 (because it is in the field of Q_1) is:

$$P = \frac{Q_1}{r} \qquad (4)$$

If the charge Q_2 is now moved from point A to point B in Figure S-6, the potential of Q_2 has been changed by the amount:

$$P_2 - P_1 = \frac{Q_2}{r_2} - \frac{Q_1}{r_1} \qquad (5)$$

If Q_1 and Q_2 each had a charge of 500 esu the distance r_1 is 10 cms. and r_2 is 5 cms, the amount of work done when Q_2 is moved from point A to the point B is:

$$\text{Work} = Q_2 (P_2 - P_1) = 500 \left(\frac{500}{5} - \frac{500}{10}\right) =$$
25,000 ergs.

If the event described above took place within a copper wire, the work done would be the work necessary to establish a current in the wire.

CONDUCTION OF ELECTRONS THROUGH GASES

(8-5) Figure S-7 shows a length of glass tubing with an electrode sealed into each end and a means for pumping out the air within the tube. The first step is to induce a positive charge on the electrode A. Since it is positive, it is now called the annode. A negative charge is then induced on electrode B. It is now the cathode. Because the charges on A and B are unlike, they try to find a path from the cathode to the annode.

When there is ordinary air in the tube at normal pressure, nothing happens because the air is an insulator and forms a barrier. However, as the air is pumped out of the tube, a fine glow of light begins to shine. A thread of light begins to run through the tube from the cathode to the annode. Sometimes the arc is a bluish-purple line, and sometimes the whole tube glows faintly as if it was hot. What is happening in the tube is that a fine stream of high velocity electrons from the cathode is knocking electrons off of the remaining air molecules, and ionizing the air that is left

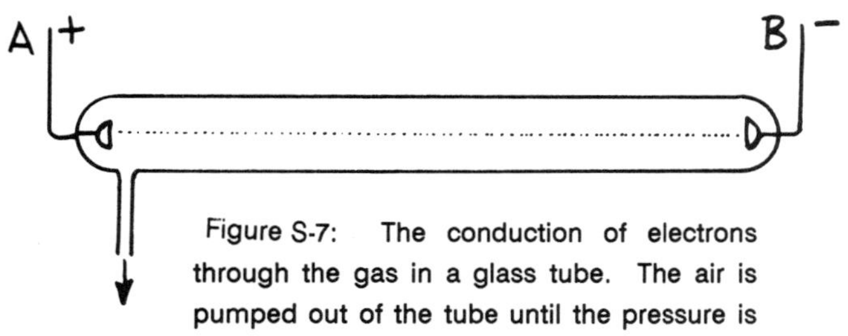

Figure S-7: The conduction of electrons through the gas in a glass tube. The air is pumped out of the tube until the pressure is very low. The remaining air molecules in the tube become ionized and act as a conductor.

in the tube.

In the ionized state, air is a good conductor of electricity. When the pressure in the tube is low enough the electrons flying off the cathode are able to penetrate far enough through the tube to establish ionization. The electrons from the cathode are called cathode rays.

THE STATIC ELECTRICITY GENERATOR

(9-5) With the early textile mills came running machinery, all driven by running belts. It soon became necessary to protect against the fire hazard and high voltage injury caused by the build-up of static electricity. Metal grounding became an important safety precaution.

Robert Van de Graaff observed the static effect of running belts and decided to build a high voltage generator based on the running belt principle. He soon demonstrated a charge build-up of over 400,000 volts using a belt machine running at 4000 feet per minute.

A diagram of a running belt voltage generator is shown in Figure S-8. A large hollow metal sphere is mounted on the top of an insulated supporting column. A pulley at the top and the bottom guides a belt so that the charge on the belt is carried upwards into the metal sphere. At the bottom pulley, the belt passes close to a row of pointed metal prongs on which a continuous negative charge is fed.

Static charges can exist only on the outer surfaces of objects because of the self-repulsion. For this reason, the negative charge is not resisted as it is fed to the inner conductive surface of the metal sphere, where it immediately goes to the outer surface build-up on the sphere. Having lost its charge to the sphere, the belt continues downward to pick up another charge at the bottom.

MEASURING THE CHARGE ON THE ELECTRON

(10-5) In 1913 Robert A. Millikan decided to build an apparatus that would measure the charge on the electron. He tried several ideas and finally succeeded by spraying a fine mist of charged oil droplets into the space between two horizontal metal plates. Millikan arranged to carefully control the electric potential between the two horizontal plates. He could prevent the oil droplets from falling due to gravity. Millikan knew the strength of the electric field on the suspended oil drop. With that he could compute the electric charge on the oil drop.

By repeating the same experiment many times he found that the charge on the oil drops was always a multiple of 4.8×10^{-10} esu. Apparently the smallest possible charge had to be equal to the exact multiple that he observed. He deduced that it must be the charge on the electron that he was measuring. Since that time, no smaller unit of electric charge has ever been discovered.

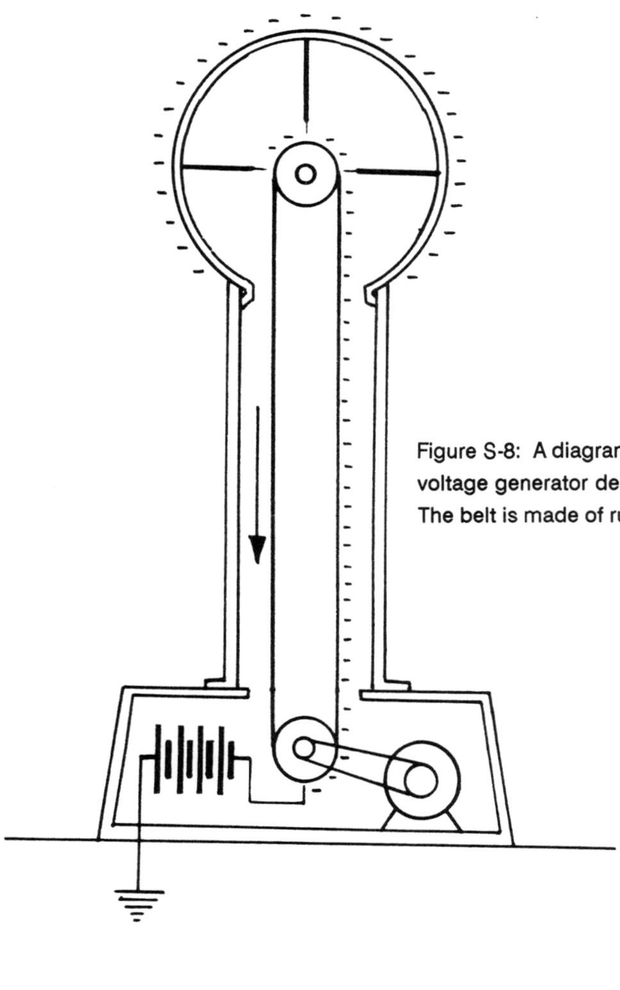

Figure S-8: A diagram of the running belt high voltage generator designed by Van de Graaff. The belt is made of rubberized fabric.

EVENTS FROM THE PAST

The date is 1837. Thomas Rogers has completed the manufacture of his first locomotive. Locomotives had names in those early times. Rogers decided to name his first locomotive the Sandusky.

The first locomotive and train of coaches to operate on a length of railroad track in New York State. Note the similarity between the railroad coaches and the horse-drawn stage coaches of the day.

MAGNETISM

THE MAGNETIC POLES

(1-6) Some natural minerals have an attribute which causes them to attract objects made of iron or steel. A blackish mineral which has this quality was called lodestone (leading stone) by the ancient ones. A piece of lodestone could be suspended on a thread and it would always turn to show the direction towards a sacred place. Such minerals, like lodestone, are called magnetic substances.

A bar of iron or steel can be made into a magnet by rubbing it with a piece of lodestone. A new magnet can be made by placing a piece of iron against a strong magnet and striking it sharply. Hard steel retains its magnetism for many years and is often called a permanent magnet.

Magnets have two attractive poles of equal strength. When a magnet is suspended on a thread or floated on a dish of water, one of the poles gradually turns to point north. The other pole then points to the south. Every magnet has a north pole (N) and a south pole (S). A bar magnet and a horseshoe (or "U") magnet are shown in Figure M-1.

When two magnets are brought near each other, the (N) poles repel each other and the (S) poles repel each other. However, the (N) poles are strongly attracted to the (S) poles. The important fact to remember about magnets is that similar poles repel and unlike poles attract each other.

When Charles Coulomb was measuring the strength of electrostatic charges, it occurred to him to use the same torsion balance to measure the forces between magnetic poles. He discovered that when two magnetic poles, N_1 and N_2 are placed r centimeters apart, the force acting between them is

$$F = \frac{N_1 N_2}{r^2} \qquad (7)$$

The force is slightly different in a vacuum, and the difference is usually shown by a factor μ. The force is then expressed by the equation

$$F = \frac{N_1 N_2}{\mu r^2} \qquad (8)$$

The difference is so slight that it is usually ignored in all practical applications, but it is of significance to scientists.

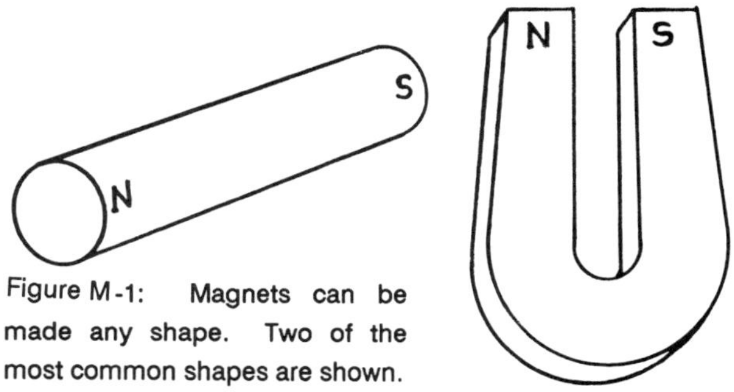

Figure M-1: Magnets can be made any shape. Two of the most common shapes are shown.

THE UNIT MAGNETIC POLE

(2-6) To evaluate the strength of a magnetic pole, the following criterion is applied: A unit magnetic pole is one which exerts a force of one dyne on another magnetic pole of equal strength when the two magnetic poles are one centimeter apart.

THE MAGNETIC FIELD

(3-6) A magnetic pole has a magnetic field surrounding it. A magnetic field cannot be seen or felt, but it can be sensed by a small magnet or by a moving electron. The intensity of a magnetic field is measured in *oersteds.* The oersted is a magnetic field intensity which exerts a force of one dyne on a unit magnetic pole. The oersted is named for Hans Christian Oersted (1777--1851), who discovered electromagnetism.

MAGNETIC LINES OF FORCE

(4-6) When a small magnet is placed within the field of a larger magnet, the small magnet will turn to align its poles with the larger magnetic field surrounding it. The directions that the small magnet chooses can be plotted on a chart and mapped out to form the magnetic lines of force around a magnet. The chart is shown in Figure M-2.

Magnetic lines of force have the following characteristics. In the direction along the lines of force, that is from north to south, the lines try to make themselves as short as possible, like elastic under tension. However, in the directions across the field lines, the lines of force stay as far apart as possible.

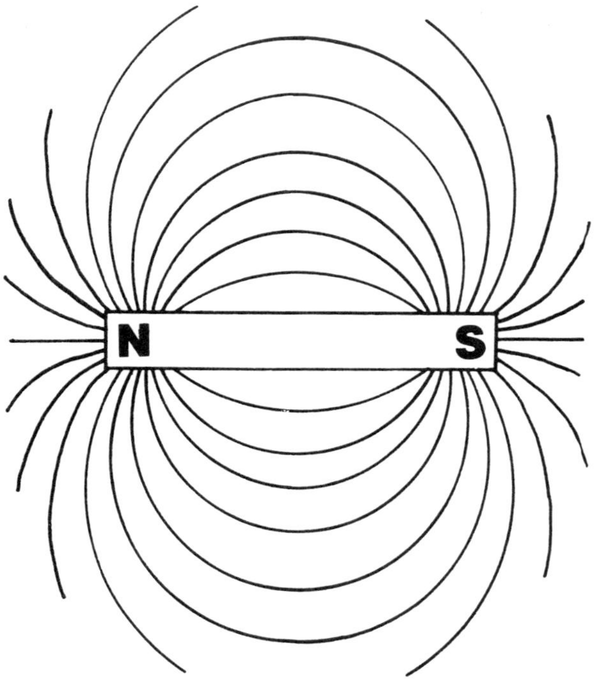

Figure M-2: The shape of the field around a bar magnet is developed as the magnetic lines of force act to put the greatest possible distance between the lines, and at the same time act to make the lines as short as possible.

The combination of push and pull in the field causes it to be distributed evenly around the magnet.

Magnetic lines become very visible if some iron filings are sprinkled onto a piece of cardboard under which a magnet has been placed. The iron filings quickly align themselves with the magnetic field lines. Tapping the cardboard edge with the finger tip causes more iron filings to scramble into alignment. The shape of the magnetic field around a bar magnet will be like the diagram in Figure M-3.

ATOMIC MAGNETS

(5-6) If a magnet is cut in half across its length, two magnets are formed. Each one of the new magnets is just as strong as the original magnet. If the two magnets are now cut in half, four magnets are produced. If the magnets are cut again and again into smaller and smaller pieces, eventually only individual atoms will remain. But each tiny element will still have the characteristics of a magnet.

From this experiment, which was actually carried out by the early experimenters, it is evident that a magnet is made out of atoms or molecules which are themselves magnets. When these tiny basic magnetic elements are lined up with each other, they form a large magnet. But when they are not aligned, and scrambled, they do not form a magnet. The tiny magnets cancel each other out because they are facing in random directions.

Magnetism can be removed from a magnet by heating it red hot. When it cools, the magnetism

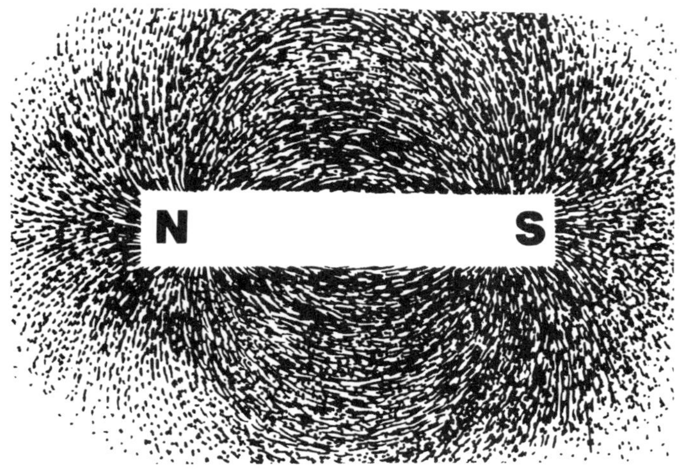

Figure M-3: The characteristic magnetic field around a bar magnet is formed when iron filings are sprinkled onto a piece of cardboard with a bar magnet below it. To make this pattern a piece of masking tape was placed over the position of the magnet and later removed.

will be gone. It can be quickly restored by stroking it with another magnet, or it can be placed in a strong magnetic field and then struck sharply with a hammer. The sudden jolt acts to realign the magnetic particles in the iron.

TERRESTRIAL MAGNETISM

(6-6) When a lodestone or a magnet is suspended on a thread, it will turn and come to rest with a certain direction or orientation. This action provides the evidence for the existence of a background magnetic field. The earth is completely surrounded by a magnetic field, which is termed terrestrial magnetism.

The configuration of the earth's magnetic field is the same as if the earth had a huge bar magnet imbedded in its core. The long direction of the bar would point north and south. This is where the terminology for North magnetic pole and South magnetic pole originated.

USING A MAGNETIC COMPASS

When the magnetic field of the earth reaches latitudes near the north pole it dips steeply and goes down into the core of the earth. It emerges near the south pole. This means that if there was an immense bar magnet buried in the core of the earth, the north pole of the magnet would be at the south end of the world, and the south pole would be located near the north pole of the earth. This fact is apparent if you look at the drawing of the bar magnet on page 96.

The earth's magnetic field is rather weak. The horizontal component of the field ranges from 0.10 to 0.25 oersteds. The strength of the field depends on the location where it is measured. A magnetic compass is used for finding directions, but because the earth's magnetic field points in many random directions, the method is not very reliable. To make it more reliable a magnetic line chart can be acquired for each township or locality, which shows the deviation from true north in the local magnetic field. In Canada these charts are available from the Department of Lands and Mines. In United States, they are available from the Department of Commerce.

The magnetic map shown in Figure M-4 was adapted from a map supplied by the United States Department of Commerce. The irregularities in the earth's magnetic field are very obvious, and also, there is a small change in the direction of the magnetic field lines of the earth from time to time.

For accurate compass use, an up-to-date map is necessary. If a magnetic compass is used in any one of the cities marked on the map, the deviation for that city can be measured approximately from this map. The deviation is measured in degrees, which can be added or subtracted from your compass reading to give a close approximation to true north.

The earth's magnetic field is level only at the equator. Everywhere else the field dips into the earth. The further north or south, the greater is the dip. The amount of the dip in degrees can be measured using a dip needle, which is like a compass turned sideways with a special bearing for the needle.

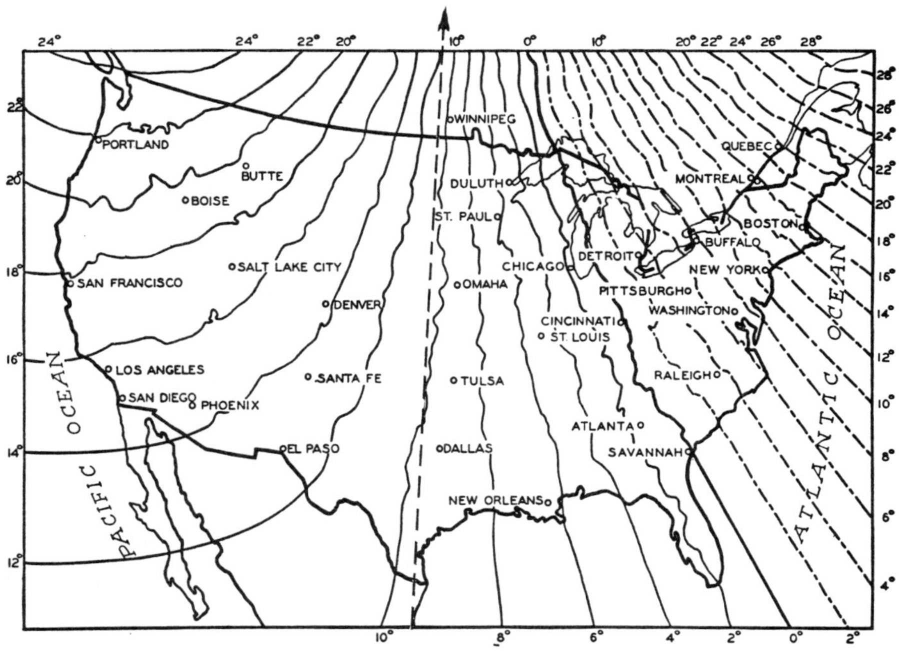

Figure M-4: The wide variations in the direction of the earth's magnetic field are shown on this map of United States. Evidently, the directions that the lines of magnetism follow are influenced by the location of the continents and their great mineral deposits. The principal cities are indicated. True north is shown by the dotted line with an arrow head at the top of the map.

EVENTS FROM THE PAST

The evolution of the railroad is evident in these three antique illustrations from the old times. Can you spot an interesting innovation in the middle example? Is this the first piggy-back car? Like, take your own coach on a rail excursion?

ELECTRIC CURRENTS

ELECTRONS IN MOTION

(1-7) Electrons will flow along a conductor if the potential at one end of the conductor is different than the potential at the other end. The flow of electrons will continue until the difference in potential has been equalized. Such a flow of electrons is called a current of electricity.

A wire conductor can be bent around into a loop and the electrons made to flow around it. Such a loop is called a circuit. A continuous difference in potential can be produced in a circuit by placing a battery or an electric generator in the circuit. The electrons flow from the negative terminal of the battery to the positive terminal.

MEASURING ELECTRIC CURRENT

(2-7) When electricity became a commercial commodity, it was necessary to measure it and meter it. Because the electron is so small, a larger unit of electricity was chosen. It was called a coulomb, after Charles Coulomb. A coulomb of electricity contains 3 x 10^9 esu. Therefore, a coulomb contains 6.25 x 10^{18} electrons.

The energy of electricity travels through the conductor as waves of potential. The current of electricity is measured as the rate of electron flow past a given point in a conductor. A flow of one coulomb per second is called an ampere. The ampere was named for Andre M. Ampere (1775-1836). He became famous as one of the original experimenters in electricity.

When Q coulombs of electricity flow past a given point in a conductor in t seconds, the current I in amperes at that point is

$$I = \frac{Q}{t} \qquad (9)$$

ELECTROMOTIVE POTENTIAL

(3-7) In chapter 5 it showed that work has to be performed on a charge to move it from a region of low potential into a region of higher potential. In a conductor the work of moving a charge through a circuit is performed by electromotive force (emf). The *emf* is provided by a battery or by an electric generator.

When a charge Q has an amount of work W performed on it, the strength of the applied emf is

$$E = \frac{W}{Q} \qquad (10)$$

Potential difference in a circuit is measured in volts. The volt is named after Volta, the man who made the first battery. A volt is defined as the potential difference between two points in a conductor that will cause one coulomb of electricity to expend one joule of energy. The joule is a unit of work. It is important and will be the subject of a later paragraph.

In equation (10) E is in volts when W is in joules and Q is in coulombs.

ELECTRIC POWER

(4-7) The work done by an electric circuit is usually expressed in terms of the current I. An ammeter is used for measuring current. The amount of current is shown by a needle which is arranged to swing over a curved segmented bar. The needle is attached to a small magnet. A fine coil of wire passes around the magnet without touching it. When current flows through the coil, the magnet moves in proportion to the amount of current. See the diagram in Figure C-1.

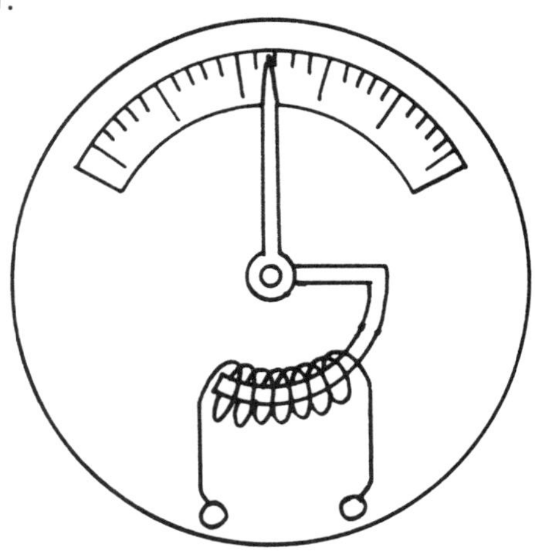

Figure C-1: Operating principle of an ammeter. The position of the needle indicator is controlled by the amount of current passing through the coil.

The amount of work done by an electric circuit is expressed by this relationship:

$$W = E I t \qquad (11)$$

W is in joules, E is the potential drop in volts, I is the current in amperes and t is the time duration in seconds.

The term power is used when the rate at which energy is used up is measured. Power is defined as joules per second. The joule per second is also called a watt, named for James Watt, the inventor of the steam engine. When the power P is expressed in watts, a factor for time is already included, and therefore

$$P \text{ (power in watts)} = V I \qquad (12)$$

In other words, one watt is equal to one volt multiplied by one ampere.

RESISTANCE AND OHM'S LAW

(5-7) The rate of current flow in a circuit is always exactly related to the potential difference, or emf, of the circuit. George S.Ohm (1787-1854), noted that the current was slightly different in some metal conductors as compared to copper. It was greater in silver and less in iron. It became evident to Ohm that an inherent characteristic of the conductor was acting to hold back the current flow. After many experiments Ohm determined that the following relationship was true in every case:

$$I = \frac{E}{R} \qquad (13)$$

Ohm's formula confirms that the current is equal to the voltage divided by the resistance R. The R stands for all of the resistances in the circuit.

The unit of resistance was named the *ohm*, for George Ohm. The ohm is defined as the amount of resistance that will cause a voltage drop of one volt when the current is one ampere.

JOULE'S LAW

(6-7) James P. Joule (1818-1889) was a thorough investigator into the nature of energy. His many experiments provided the evidence for quantifying many energy states which had been a mystery. His experiments were painstaking and accurate.

One of the phenomenon that Joule worked on was the apparent generation of heat in a conductor when a current is passing through it. He carefully measured the heat produced by electric currents in different conditions. Joule expected that when he doubled the rate of current flow, the rate of heat production would also double. But the heat generated was much greater than he had expected. After many extremely delicate experiments he was able to determine that the rate of heat production in a conductor is proportional to the square of the current flow. So twice as much current produced four times as much heat energy, in the same period of time.

Since different conductors gave different results for the same amount of current, Joule realized that the resistance of the conductor to the passage of current must be an important factor in the production of heat energy.

Joule found that the quantity of heat energy W produced by a current I in a time t, within a conductor with a resistance R, would be equal to

$$W = R I^2 t \qquad (14)$$

When a current of one ampere is sustained for one second in a conductor with a resistance of one ohm, the amount of heat produced is one joule. One joule equals 0.239 calories.

The power that is lost as the heat energy escapes into the air is

$$P = R I^2 \text{ watts.} \qquad (15)$$

RESISTANCE OF STANDARD CONDUCTORS

(7-7) To standardize the electrical industry and to assure safe practices, a system for calculating the resistance in wire conductors was established. The method for determining the resistance of a conductor is to measure the length in feet and give the cross section in circular mils. Circular mils is abbreviated CM. One CM has a cross section area the same as a circle 0.001 inches in diameter. One mil is 0.001 inches.

The area of a circle is proportional to the square of its radius, and the diameter is twice the radius. Therefore it is only necessary to square the diameter of a wire to arrive at the number of circular mills in its cross section. A wire with a diameter of 4 mils has a cross section of 16 mils (4 squared).

The only other factor that has to be known is the resistance of the material the wire is made out of. Since almost all conductors are copper, that matter is also simplified. The chart in Figure 7-3 lists the resistance (resistivity) of some common materials.

The CM system permits the resistance of a wire conductor to be expressed in ohms and CM's, by the foot of wire. A length of copper wire 10 feet long and one mil diameter will have a resistance of

$$10 \text{ feet} \times 1 \text{ CM} \times 10.4 \text{ ohms} = 104 \text{ ohms.}$$

Wires of low resistance are used when the objective is the efficient transmission of electric power. However, wires of high resistance are used when heat energy is needed for electric ovens and heaters.

METAL OR MATERIAL	RESISTIVITY IN OHMS PER C.M. FOOT
ALUMINUM	17
BRASS	45
CADMIUM	46
CHROMIUM	16
COPPER	10.4
IRON	59
SILVER	9.8
ZINC	36
NICHROME	650
CONSTANTAN	295
MANGANIN	290
MONEL	255

Figure C-3.

CALCULATION OF WIRE SIZE

(8-7) The non-hazardous use of conductors makes it necessary to specify minimum wire sizes for carrying electric currents. Preferably a wire size is used that controls the potential drop along its length. Lighting circuits should not have a potential drop greater than 3%, or 3 volts per one hundred volts. Power circuits are allowed 5 or 6 volts potential drop per hundred volts. These guide lines should be adhered to in order to prevent excessive heating of the circuit and to avoid power losses along the lines. It is an advantage to always use a slightly larger wire for a particular application rather than a smaller one.

SERIES AND PARALLEL CIRCUITS

(9-7) In the diagrams of circuits shown below, (A) illustrates a *series circuit*, which means that the electric components occur one after the other. Diagram (B) shows a *parallel circuit* in which case the lamps or whatever are wired in side by side.

The important difference between series and parallel circuits is that in a series circuit all of the current passes through every electrical component; in the parallel circuit, the current is divided up between the electrical elements that are wired in parallel. Each lamp or whatever only gets a share of the currrent. A circuit can also be partly in series and partly parallel, like the example shown at (C) in Figure C-4.

The symbol for resistance in a circuit is a short sawtooth line. Circuits (A), (B) and (C) are shown in symbols in Figure C-5.

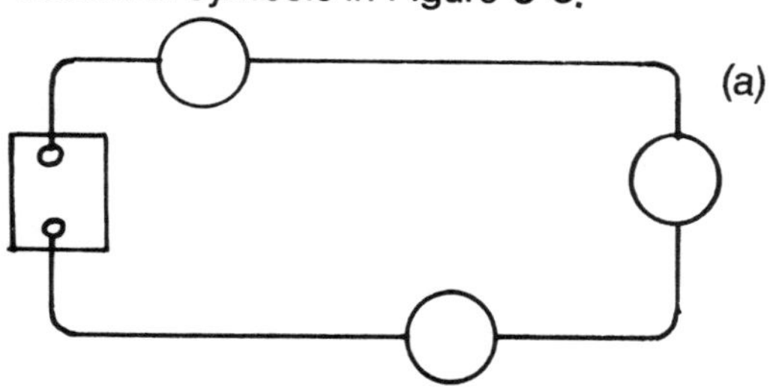

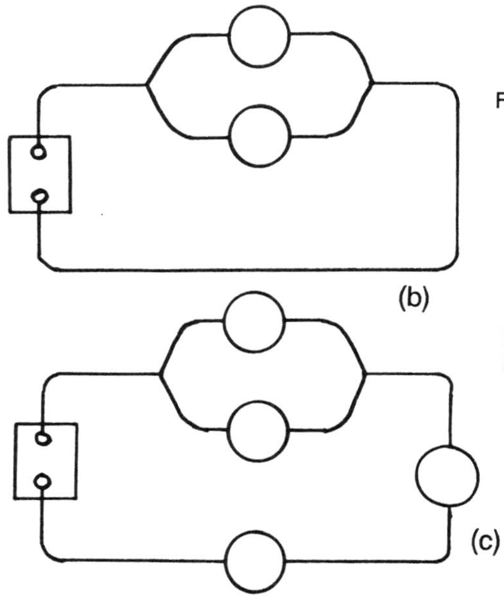

Figure C-4: Circuit (a) shows three lamps wired in series. Circuit (b) shows two components wired in parallel. Circuit (c) is an example of both series and parallel circuits combined into one circuit.

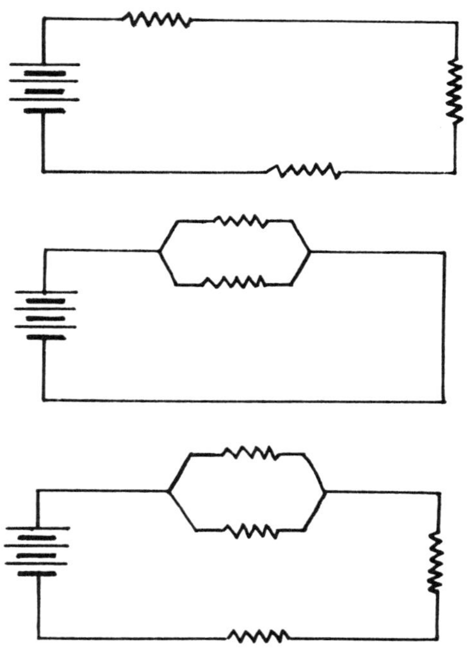

Figure C-5.

Figure C-5: Lamps in an electric circuit are usually shown as a resistance. The symbol for resistance is a short sawtooth line. The three circuits (a), (b) and (c) are shown as they would look in symbols.

Each circuit has a source of emf, such as a battery or an alternator or generator. The source of the emf has its own internal resistance. The result is that the emf across the output terminals is slightly reduced.

When R is the actual resistance inside the source of emf, I is the current and E is the emf delivered to the circuit, the voltage drop within the source due to internal reistance is

$$V \text{ (voltage drop)} = E - IR \qquad (17)$$

RESISTANCE CALCULATIONS

(10-7) When trouble shooting circuits (trouble shooting means to locate a problem and solve it) it is often necessary to isolate all or part of a circuit to determine the actual resistance of suspected components, or to pinpoint a defective element.

To simplify a circuit, any group of resistors can be replaced with a single substitute resistor in the following manner: For resistances in series, measure the resistance of each component and then add the resistances together. The total resistance will indicate the size of the substitute resistance to use. Resistances that cannot be measured can be copied from the circuit diagram. The calculation is

$$R \text{ (substitute)} = R_1 + R_2 + R_3 + R_4 + R_5$$

When adding resistances in parallel, a different method has to be used. First convert the resistance values to their reciprocals (to find the reciprocal of a number just divide it into the number 1). When the reciprocals are added, the total is the reciprocal of the total resistance. It must be converted back to a normal number, and to accomplish that, divide it into the number 1 again.

For resistances in parallel, each resistance acts to reduce the total resistance. The more conductors there are in parallel, the more paths there are for the current to take. Each new path reduces the resistance in all of the paths that are in parallel. The formula for adding resistances in parallel is

$$\frac{1}{R}\text{(substitute)} = \frac{1}{R_1} + \frac{1}{R_2} + \frac{1}{R_3} + \frac{1}{R_4} + \frac{1}{R_5} \qquad (19)$$

There is another way of adding resistances in parallel, and that is to take them two at a time and apply the following formula

$$R = \frac{R_1 \times R_2}{R_1 + R_2} \qquad (20)$$

When you use this method, the operation converts two resistances into one. Just multiply the values of any two resistances together and then divide by their sum. Keep in mind that this operation only applies to resistances in parallel. Use a calculator to speed up the process. Repeat as many times as necessary to reduce all of the resistances in parallel to one resistance.

To simplify the job of troubleshooting circuits, an inexpensive meter can be purchased which will measure currents, voltages and resistances.

RESISTANCE IS INFLUENCED BY TEMPERATURE

(11-7) The resistance of a conductor is higher when there is an increase in the temperature. Resistance is less when it is cold. For most practical applications, the ambient temperature considerations can be ignored.

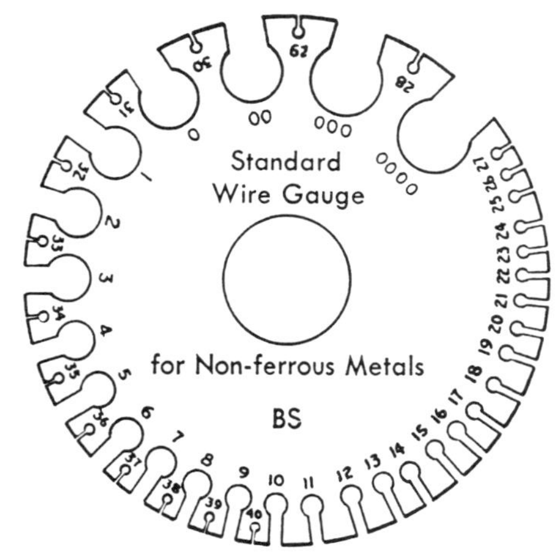

Figure C-6: The wire gauge illustrated is made out of tough metal so that it will not change size with use. To check a wire for size, the bare wire is fitted into one of the slots. The proper wire will just slide into the slot without forcing. The number at the top of the slot gives the number of the wire.

ELECTRIC NETWORKS

(12-7) Electric circuits can be quite complicated. When there are several components in parallel and in series, the analysis of the circuit can present a difficult problem. The problems associated with solving networks were greatly simplified by Gustav Kirchhoff (1824-1887), who formulated the laws which affect electric networks. The Kirchhoff rules are:

(1) At any junction of two or more wires in a network, the sum (addition) of all of the currents arriving at that junction is equal to the sum of all the currents leaving that junction. See Figure C-7. In the form of an equation:

$$I_1 + I_2 = I_3 + I_4 + I_5 + I_6 \qquad (21)$$

(2) Follow any wire through a network and return to the starting point by a different route; the algebraic sum of the voltages add up to zero (the algebraic sum means that some of the quantities you are adding may have minus values, and you subtract instead of adding). The formula is

$$V_1 + V_2 + V_3 + V_4 + V_5 + V_6 = 0 \qquad (22)$$

The voltages are measured from connection to connection. An example loop is shown by the heavy line in diagram C-8.

The two Kirchhoff rules can be applied, along with other methods of troubleshooting a circuit, to find the value of the current, the voltage and the resistance for any component or element in an electric circuit.

After you have applied all of the strategems, including Kirchhoffs laws, you may still be in doubt about a component. The final strategem is to isolate a component by disconnecting it from the circuit. Then measure the resistance with a meter. This will give a known value to insert into an equation or into an algebraic addition. The component can then be reconnected into the circuit (if it proves to be satisfactory and not defective).

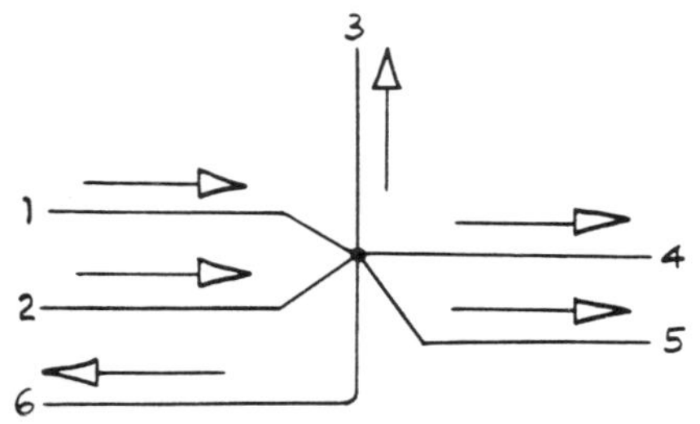

Figure C-7: Kirchhoff's first rule is that all of the currents arriving at a junction must equal all of the currents leaving the same junction.

COPPER WIRE CHART

WIRE NO.	DIAM. IN MILS	CIRCULAR MIL AREA	FEET PER POUND		OHMS PER 1000 FEET	CURRENT CAPACITY 1500 C.M. PER AMP	DIAM. IN M.M.
			BARE	COVERED			
1	289.3	82690	3.947	----	0.1264	55.7	7.348
2	257.6	66370	4.977	----	0.1593	44.1	6.544
3	229.4	52640	6.276	----	0.2009	35.0	5.827
4	204.3	41740	7.914	----	0.2533	27.7	5.189
5	181.9	33100	9.980	----	0.3195	22.0	4.621
6	162.0	26250	12.58	----	0.4028	17.5	4.115
7	144.3	20820	15.87	----	0.5080	13.8	3.665
8	128.5	16510	20.01	19.6	0.6405	11.0	3.264
9	114.4	13090	25.23	24.6	0.8077	8.7	2.906
10	101.9	10380	31.82	30.9	1.018	6.9	2.588
11	90.74	8234	40.12	38.8	1.284	5.5	2.305
12	80.81	6530	50.59	48.9	1.619	4.4	2.053
13	71.96	5178	63.80	61.5	2.042	3.5	1.828
14	64.08	4107	80.44	77.3	2.575	2.7	1.628
15	57.07	3257	101.4	97.3	3.247	2.2	1.450
16	50.82	2583	127.9	119	4.094	1.7	1.291
17	45.26	2048	161.3	150	5.163	1.3	1.150
18	40.30	1624	203.4	188	6.510	1.1	1.024
19	35.89	1288	256.5	237	8.210	0.86	0.9116
20	31.96	1022	323.4	298	10.35	0.68	0.8118
21	28.46	810.1	407.8	370	13.05	0.54	0.7230
22	25.35	642.4	514.2	461	16.46	0.43	0.6438
23	22.57	509.5	648.4	584	20.76	0.34	0.5733
24	20.10	404.0	817.7	745	26.17	0.27	0.5106
25	17.90	320.4	1031	903	33.00	0.21	0.4547
26	15.94	254.1	1300	1118	41.62	0.17	0.4049
27	14.20	201.5	1639	1422	52.48	0.13	0.3606
28	12.64	159.8	2067	1759	66.17	0.11	0.3211
29	11.26	126.7	2607	2207	83.44	0.084	0.2859
30	10.03	100.5	3287	2534	105.2	0.067	0.2546
31	8.928	79.70	4145	2768	132.7	0.053	0.2268
32	7.950	63.21	5227	3137	167.3	0.042	0.2019
33	7.080	50.13	6591	4697	211.0	0.033	0.1798
34	6.305	39.75	8310	6168	266.0	0.026	0.1601
35	5.615	31.52	10480	6737	335.0	0.021	0.1426
36	5.000	25.00	13210	7877	423.0	0.017	0.127
37	4.453	19.83	16660	9309	533.	0.013	0.1131
38	3.965	15.72	21010	10666	672.6	0.010	0.1007
39	3.531	12.47	26500	11907	848.1	0.008	0.0897
40	3.145	9.88	33410	14222	1069	0.006	0.0799

Figure C-9: Chart for copper wires from numbers 1 to 40 provides basic information about standard conductors.

Figure C-8: Kirchhoff's second rule is that the sum of the voltages around any loop in a network add up to zero. The voltages are measured from connection to connection. It is important to give each voltage a plus or a minus value depending on the local voltage gradient. The values are then added algebraically.

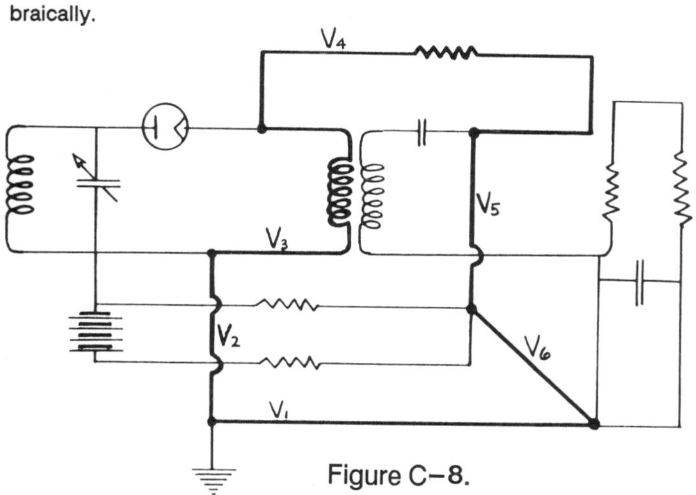

Figure C–8.

EVENTS FROM THE PAST

The evolution of automobile design is apparent in these three antique illustrations. The top model is an example of the "horseless carriage". Buggy wheels and spring-suspension were used. Drawings in the middle and at the bottom show the complete change from open air driving to closed-in comfort.

BATTERIES AND ELECTROLYTES

ELECTROLYTIC CELLS

An electrolytic cell has three parts. These are, (a) an insulated container (like a glass jar), (b) a pair of conductive electrodes, and (c) an ionic solution in the jar into which the electrodes are immersed or suspended. The solution is called an electrolyte and the container is called an electrolytic cell.

Electrolytic cells are usually separated into three categories, which are:

(1) When the cell is used for generating a voltage it is called a voltaic cell, or a battery.

(2) When the cell is used for producing one or more gases the cell is called a gaseous electrolytic cell and the process is called electrolysis.

(3) When the cell is used to produce or to plate metals it is called an electro-plating cell.

The electrolyte can be any solution, gel or paste in which there is a tendency for ions to be created and sustained. The ions can be positive or negative. All acids, bases and salts are effective electrolytes when in solution. An electrolyte will conduct a current of electricity.

DECOMPOSITION OF WATER

(2-8) One way to decompose water into its component gases is to use an electrolyte made of a weak solution of sulphuric acid and water (H_2SO_4). The electrodes should be chemically inert, such as platinum. An electric current passed through the cell will break-down the solution into component gases. See Figure B-1.

The negative electrode attracts positive hydrogen ions. When the ions reach the electrode, they combine with electrons on the electrode and are converted to hydrogen gas. The gas can be seen rising as a series of bubbles. They can be collected at the surface by an inverted funnel with a plastic tube to guide the gas to a container.

Meanwhile, the sulphur dioxide ions (which are negative) drift towards the positive electrode, the annode. When the sulphur dioxide ions touch the annode, electrons are released onto the annode. The action causes the sulphur and the oxygen to disassociate. The oxygen rises up as a series of bubbles and can be collected by a suitable collector and container. During this process, more sulphuric acid is formed, so the process becomes automatic.

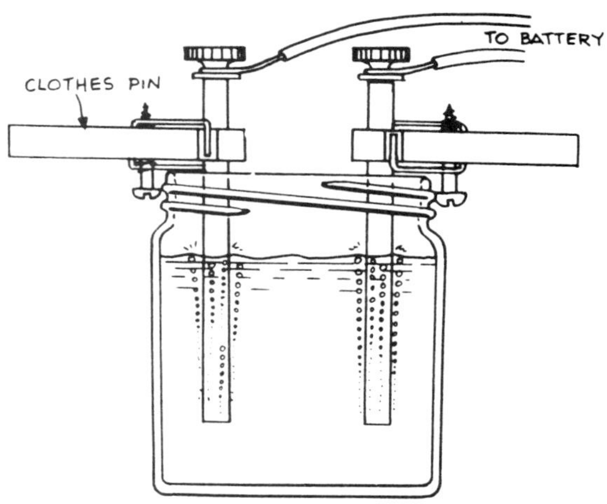

Figure B-1: When a current is passed through a weak solution of sulphuric acid and water, the water is broken down into its component gases, oxygen and hydrogen. The gases bubble up from the electrodes. The electrodes should be made from an inert metal such as platinum.

AVOGADRO'S CONSTANT

(3-8) Amadeo Avagadro (1776-1856) was a diligent investigator. His great interest was the nature of gases and their properties. One of his most fruitful intuitions about gases was that if you take any kind of gas and enclose it in a container, you could determine the number of molecules or atoms in the container. If you arranged for the gas to be a certain volume, pressure and temperature, then the number of molecules or atoms would be the same for any other gas at the same volume, pressure and temperature. He surmised that this fact would be true for any gas, regardless of the size of the individual molecules or atoms. Many experiments proved Avogadro to be right.

The proof of that idea gave him a deeper insight about matter. He then stated that if you take the atomic weight of any substance, gas, solid or liquid, and carefully measure out the same number of grams of the substance as its atomic weight, the resulting mass would always contain exactly the same number of atoms. That also was proven to be true. This quantity of any substance is called a gram-atom of the substance.

The number of atoms in a gram-atom is always the same number. It is called Avogadro's constant. The number is 6.02×10^{23} atoms. The gram-atom (atomic weight in grams) for some common elements is shown in Figure B-2.

FARADAY'S LAWS OF ELECTROLYSIS

(4-8) When electricity is passed through an electrtolytic cell, a process occurs in which individual electrons attached to atoms or molecules will drift from the cathode to the annode. This fact makes it possible to measure exact amounts of electricity. For example, when a current of one ampere is passed through an electrolytic cell for one second, the number of electrons transported throught the cell is 6.25×10^{18}. This quantity of electricity is called one coulomb.

Consequently, in any electrolytic process, the mass of the substance transferred through the cell in one second at one ampere of current will always be 6.25×10^{18} atoms or molecules (if the valence of the substance is one). If the valence of the substance is two, like copper or oxygen, then only half the amount will be transferred, 3.125×10^{18} atoms or molecules.

Based on these facts, Faraday's first rule is: The same quantity of electricity will liberate an amount of a substance in proportion to the number of electrons passing through an electrolytic cell.

Faraday's second rule is: The same quantity of electricity when passed through an electrolytic cell will liberate an amount of substance in proportion to its atomic weight.

To liberate a gram-atom of a substance the current required is

ELEMENT	GRAM-ATOM	VALENCE
HYDROGEN	1.008 gm	+1
OXYGEN	16.000 gm	−2
COPPER	63.5400 gm	+2
SILVER	107.8800 gm	+1
CHLORINE	35.4600 gm	−1

Figure B-2: The weight in grams of some substances that will be produced by an electrolytic cell when 96,500 coulombs of electricity is passed through the cell. (One faraday of electricity is 96,500 coulombs).

$$\frac{6.02 \times 10^{23}}{6.25 \times 10^{18}} = 96{,}500 \text{ coulombs of electricity.}$$

This quantity of current, 96,500 coulombs, is called one *faraday*.

Example: How many grams of copper can be liberated by an electrolytic cell when the current is ten amps maintained for 24 hours? The mass of copper is

$$\frac{10 \times 24 \times 60 \times 60}{96{,}500} \times \frac{63.54}{2} = 284.44 \text{ grams}$$

VOLTAIC CELLS

(5-8) When an electrolytic cell is used for producing a voltage, it is called a voltaic cell, or in our language, a battery. A simple kind of battery can be made with a glass jar containing salt water or dilute sulphuric acid. When a copper and a zinc electrode are suspended in the solution, a voltage can be measured across the terminals of the electrodes. If the solution is sulphuric, the copper plate will acquire a positive charge and the zinc electrode will become negative. The copper is the annode and the zinc is the cathode.

What happens is that positive hydrogen ions in the solution drift towards the copper electrode partly due to the low solution pressure of copper. At the zinc electrode some of the zinc atoms dissolve into the solution but leave their electrons sticking to the zinc. The negative charge around the zinc tends to encourage positive hydrogen ions to move away towards the copper electrode. The combination of events causes a voltage to appear across the exterior terminals of the electrodes. When the connection is made, a current can be measured going from the cathode to the annode. But as hydrogen collects on the copper it forms a barrier which blocks further battery action. The annode is said to have become polarized.

To make a battery function continuously, a *depolarizer* is used. In ordinary dry cells, manganese dioxide is used to prevent polarization. A typical dry cell is shown in Figure B-3. A carbon electrode is used instead of copper. The zinc electrode forms the container. The interior of the zinc electrode is coated with a fibrous filter (blotting paper is often used).

A dry cell is not really dry. What they mean by dry is that it is not runny like liquid. The solution used is a paste of mixed carbon particles and manganese dioxide. The liquid used to make the paste is ammonium chloride and a drop of zinc chloride. The cell will operate for several hours, and then will recharge itself during periods of rest.

It is dangerous to heat dry cell batteries with the intention of restoring their energy. When a battery is heated it may explode. The possibility of eye injury or other injury is always to be uppermost in mind.

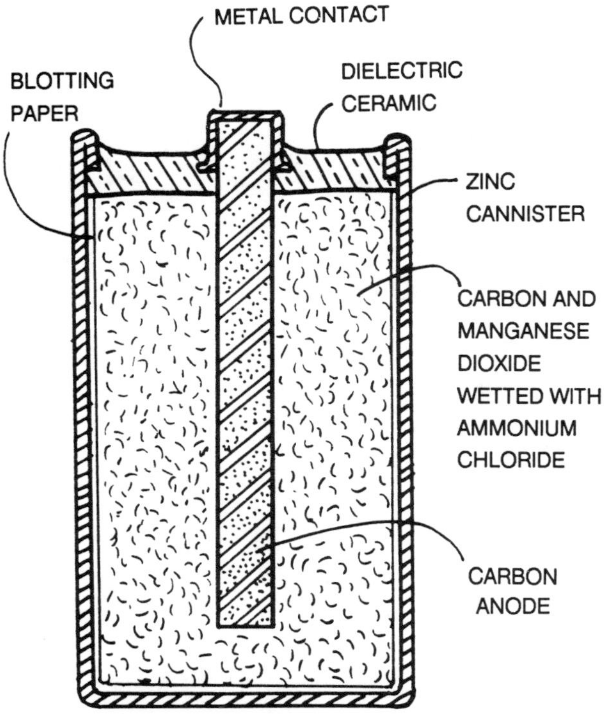

Figure B-3: Diagram in cross section of a dry cell battery.

VOLTAIC CELLS IN PARALLEL AND IN SERIES

(6-8) To get a higher voltage out of batteries, they can be connected in series. To get a higher current at the output, the batteries are connected in parallel.

When batteries are connected in series, the total emf is equal to the addition or sum of the voltages of each cell. When connected in parallel, the voltage is equal to the voltage of the cell with the greatest voltage. The current produced in parallel connections is the sum or addition of the currents produced by each battery.

Batteries can also be connected in series and in parallel at the same time, as in Figure B-4, to generate whatever voltage and current is wanted for a particular application.

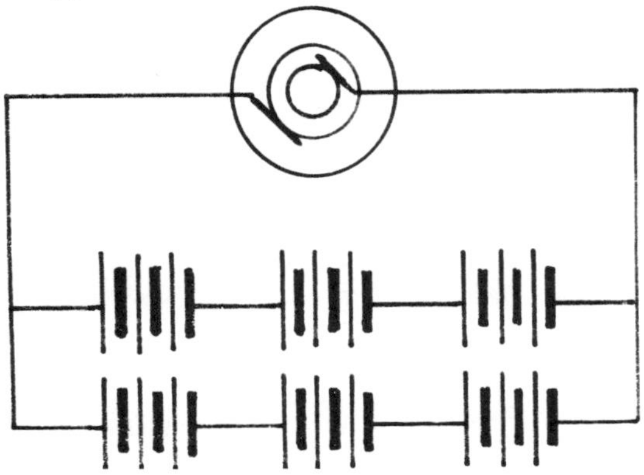

Figure B-4: Six 4 volt batteries connected in series and in parallel to operate an experimental 12 volt electric motor.

ELECTROPLATING

There is no advantage to plating a low value metal like copper onto a costly metal like platinum. Usually, a decorative metal like brass or chromium are plated onto less attractive metals like cast iron or copper. However, the process for plating copper onto platinum is a good example to describe.

Start with an electrolyte of dilute copper sulphate. For the electrodes use one of copper and one of platinum. The copper atoms dissolve into the electrolyte and drift to the platinum where they become plated on the surface of the platinum. In the circuit use enough voltage to initiate the process.

Other metals can be plated onto metal parts by making the part that is to be plated an electrode immersed in the solution of the electrolyte. The anode metal is the metal that is to be plated on the part. The electrolyte is a solution containing at least one of the salts of the metal. Zinc and copper can be plated at the same time, using a cyanide solution. The plating deposited on the part is a combination of copper and zinc, which is brass.

IMPORTANT

Electricity has been a great blessing to mankind. In the myriad ways electricity is applied to advantage, it is like the magic Genie who has been released from his bottle to grant us all of our wishes. Without electricity, mankind would still be in the dark ages. There would be no burgeoning electronic technology, no television, no street-lighting.....our lives would be drab and dull, devoid of flavor. The magic would be gone from our modern world.

From another point of view, electricity is so powerful that it is incredibly dangerous. Hundreds of lives have been lost to its whimsical fury. The only protection there is against the lightning bolt sudden death of electricity is to be ever alert. Those who choose to work with electricity must be forever watchful. Never be contemptful of its power.

Inventors who choose to experiment with electricity must be responsible for their own safety and the safety of anyone near....friends, companions, family members, all are at risk. Safety procedures should be set-up and then carefully adhered to. The magic and power of electricity makes it very easy to work with and fun to experiment with. But don't ever take chances. You only live once.

INVENTIONS FROM THE PAST

Inventors discovered many ways to redesign and improve the machinery for fighting fires. These antique illustrations of fire engines show a variety of ideas, including a bugle or trumpet for clearing the road of traffic.

ELECTROMAGNETISM

CONCENTRIC MAGNETIC FIELDS

(1-9) In 1820 Oersted discovered that there was a circular magnetic field around a conductor in which a current was flowing. Experiments revealed to Oersted that the lines of magnetism around a live wire were rings, like the diagram is Figure E-1. Oersted noted that the strength of the magnetic field diminished with distance from the live conductor.

Although the circular magnetic lines have no visible magnetic pole, the polarity of the rings can be seen by placing a small magnetic compass needle near a live conductor. The needle will swing to show a north and a south pole. A strange result is that the polarity of the circular magnetic field is always the same direction. If you point with the thumb of your right hand, as in Figure E-2 , your fingers will curl around in the direction of the circular magnetic field. Your thumb will be pointing in the direction of the current.

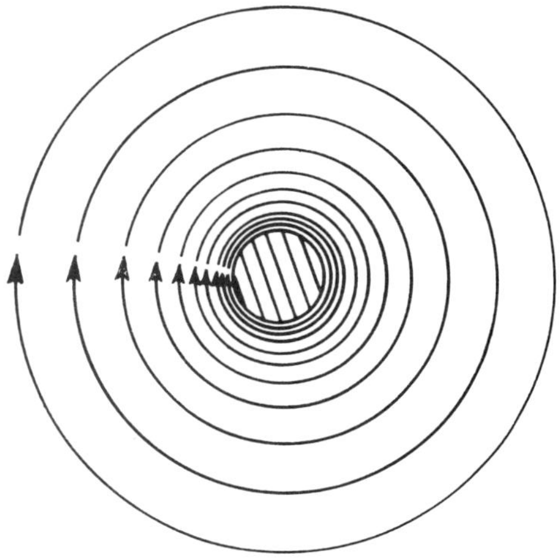

Figure E-1: Diagram of the concentric rings of magnetism around a live conductor.

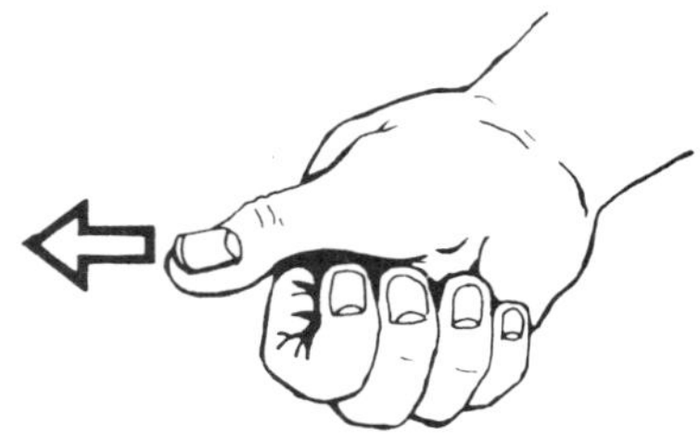

Figure E-2: The thumb and fingers of the right hand are used to show the direction of the current and the curl of the magnetic field around a conductor. The north pole is in the direction indicated by the fingernails.

GYROSCOPIC PARTICLES

(2-9) It has always been a real puzzle for scientists since the days of Oersted why the concentric magnetic field around a live wire always turns in the same direction. What made the circular magnetic field choose one direction instead of the other. It is predictable, yet strange.

Joseph W. Newman was studying electricity about the year 1965 when his mind came to the same problem. It occurred to Newman that the field formed exactly when the current started, and he recognized a similar phenomenon in the toy gyroscope he had been playing with. When the gyroscope is given a small acceleration or push in any direction, the spinning flywheel immediately aligns itself with the direction of motion. The spin in the flywheel is always in the same direction as the polarity of the magnetic field around a live conductor. Newman imagined that the space around the conductor is occupied by millions of tiny spinning particles, like miniature gyroscopes. When the electrons in the conductor are accelerated into motion, the tiny gryroscopes turn in accordance with the gyroscopic principle. When they are all lined up in the same direction, the circular magnetic field is created. Of course this is just a theory, but it is the cause-and-effect action it describes that gives it the ring of truth.

MAGNETISM IN COILS

(3-9) The concentric rings of magnetism around a live conductor can be greatly intensified by winding the wire into a coil. If the coil has ten turns, the magnetism is ten times stronger. If the coil has 100 turns of wire, the magnetism is 100 times stronger. When the wire is wound uniformly around a cylindrical form, it is called a solenoid. The magnetic effect is greatly increased by placing an iron or steel core inside the coil of wire. Solenoids have a wide application in technology and industry. Solenoids are the muscles in machines and appliances. They open valves or close switches, or perform whatever is necessary, automatically.

The force exerted by a solenoid can be determined by placing one pole of a long permanent magnet into the center of a coil of wire. The number of turns of wire are counted and a measured amount of current is sent through the coil. The force exerted by the solenoid can be expressed mathematically if everything is carefully measured.

If the coil has a radius r, the current is I, the number of turns of wire is indicated N, and the unit magnetic pole is P. The force F is:

$$F = P\frac{2NI}{10r} \qquad (24)$$

The magnetic field around a coil of wire with a current passing through it is exactly like the magnetic field around a bar magnet. A diagram of the magnetic field around a live coil is shown in Figure E-3. The field around a bar magnet is illustrated in Figure E-4.

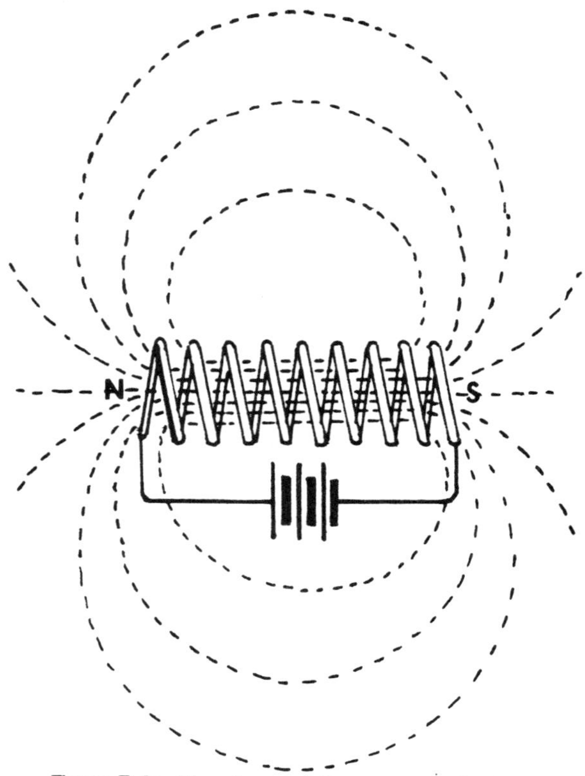

Figure E-3: The shape of the magnetic field around a live coil of wire is the same shape as the magnetic field around a bar magnet.

RESULTANT MAGNETIC FIELD

(4-9) When two magnetic fields with different strengths and directions are merged together, they blend and form one field. The new field strength and direction is found by the vector addition of the two merged fields. In the example, Figure E-5. two fields (A) and (B) have the directions and strengths indicated by the vectors (a) and (b). When fields (A) and (B) are merged, field (C) is created. Its strength and direction are shown by the vector (c). The fields (A) and (B) have totally disappeared, but they still remain intact and invisible. The important point is that magnetic fields are often composed of several fields blended together.

MOVING MAGNETIC FIELDS

(5-9) When a magnetic field is in motion through a region, it has the attributes of an electric field in that region. If two or more magnetic fields are moving through a region with such strengths and directions that their vector addition adds up to zero, the magnetic fields vanish entirely. However, the electric field caused by their motion through the region remains unaffected.

It is important to note that the electric field produced by two or more moving magnetic fields is not different in any way from the electric field around an electron. Experiments do not reveal any difference whatever. A natural question that arises is this: Why does nature have two ways of making an electric field? The answer is unknown. But perhaps the real answer is that all electric fields, including the electric field around an electron, are really magnetic fields in motion.

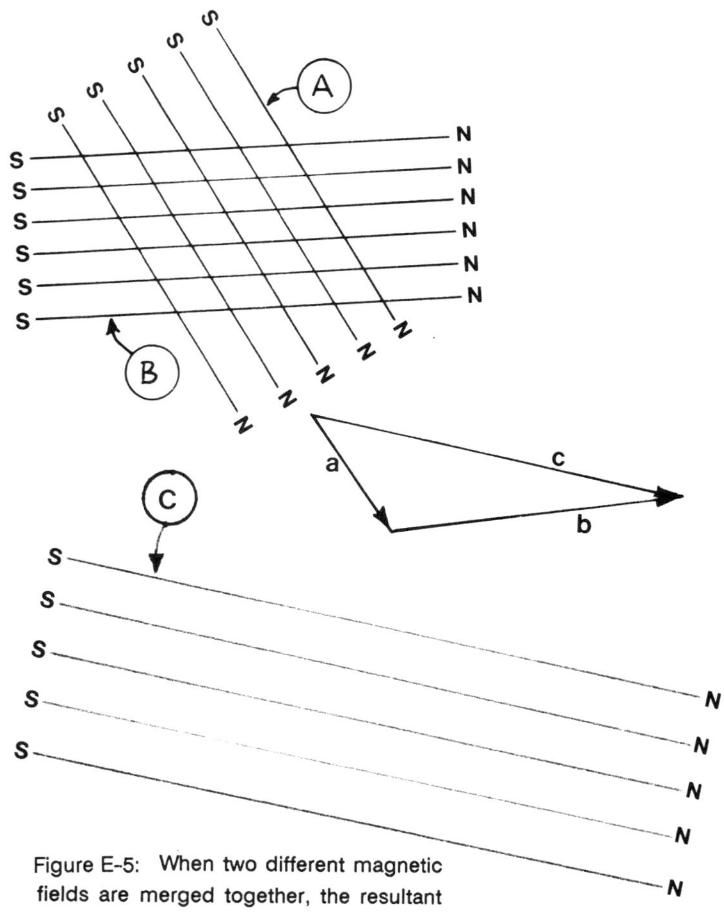

Figure E-5: When two different magnetic fields are merged together, the resultant field is the vector addition of the two fields

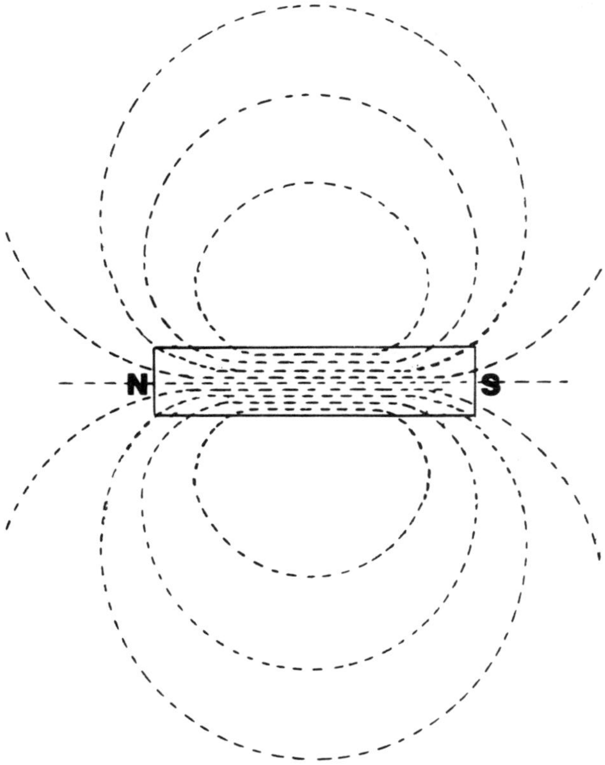

Figure E-4: The magnetic field around a bar magnet.

FORCE BETWEEN CONDUCTORS

(6-9) When two live conductors are parallel, the magnetic field of each conductor interacts with the magnetic field of the other. Magnetic field lines are like stretched elastic. They try to make themselves shorter, so the force between the conductors is attractive.

When the distance between the two conductors is r, the currents flowing in the two conductors are I_1 and I_2 flowing in the same direction, and l is the length of the wires, the attractive force F is:

$$F = \frac{\mu I_1 I_2 l}{50r} \qquad (25)$$

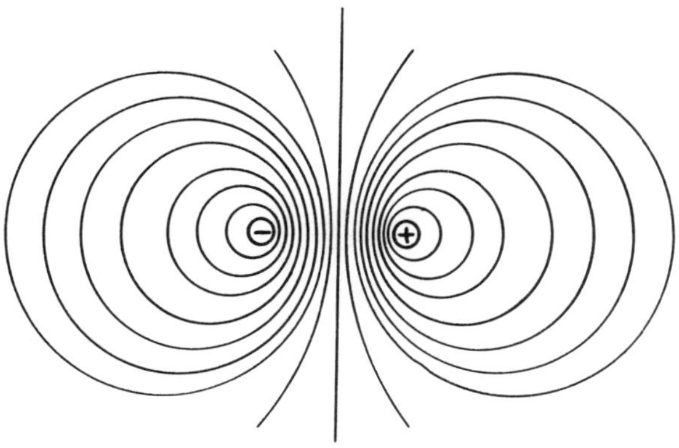
Figure E-6: The magnetic flux around two parallel conductors with the current going in opposite directions. There is a force that tends to push the conductors apart.

The opposite effect occurs when the current is going in opposite directions in two parallel conductors. The force acts to push the conductors away from each other. The magnetic field around two parallel conductors with currents going in opposite directions is shown in Figure E-6.

FORCE BETWEEN ELECTRONS

(7-9) When two electrons are travelling side by side in parallel paths (often happens inside a cathode ray tube or oscilloscope) the circular magnetic field attending each electron has an influence on the other. The combined fields of two moving electrons is shown in the diagram, Figure E-7. There is an attractive force which tends to reduce the repulsion between the negative charges. The strength of the attractive force alone is:

$$F = \frac{e^2 V^2}{r^2 C^2} \qquad (26)$$

In formula (26), e is the charge on the electron, V is the velocity of the pair of electrons, r is their distance apart in centimeters, and C is the velocity of light.

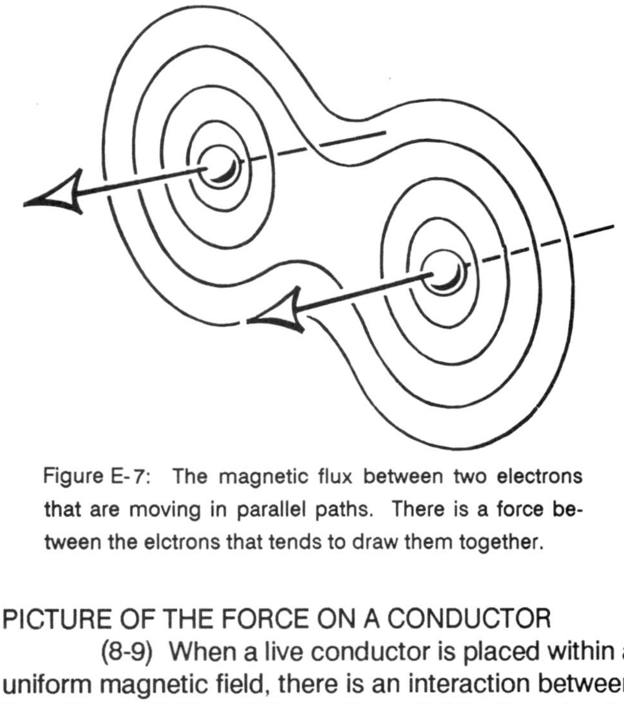
Figure E-7: The magnetic flux between two electrons that are moving in parallel paths. There is a force between the elctrons that tends to draw them together.

PICTURE OF THE FORCE ON A CONDUCTOR

(8-9) When a live conductor is placed within a uniform magnetic field, there is an interaction between the circular field and the uniform field. The circular field around the conductor is often called the magnetic flux. The magnetic flux of the conductor absorbs the uniform field on one side and increases it on the other. The result is that the conductor is given a strong push in one direction. It is this interaction that represents the magnetic muscle of all electric machinery.

Figure E-8 illustrates the shape of the two fields before they are merged. Figure E-9 shows what happens to the two fields when they are merged together. The force on the conductor, and the direction of the force, is shown by the vector.

Figure E-8: A circular magnetic field and a uniform magnetic field illustrate the optimum conditions for the efficient operation of electrical machinery.

INDUCED ELECTROMOTIVE FORCE

(9-9) In the year of 1831 Michael Faraday discovered that when a magnet is in motion near the wires of a circuit, a current and an emf are generated in the circuit. The induced emf continues as long as the magnet is in motion. It later became clear to Faraday that to produce a voltage in a circuit using a magnet, the field lines of the magnet must cut across the circuit. Today all electric generators and alternators operate on this principle.

Faraday built many experimental generators and learned something from each one. He used a coil of wire to produce a magnetic field and he made several experiments with a rotating copper disc. Typical examples of Faraday's experimental apparatus are shown in Figure E-10) and Figure E-11.

THE UNIVERSAL MAGNETIC FIELD

(10-9) The early astronomical investigators like Kepler and Galilleo made their observations of the planets against a motionless backdrop called the fixed stars. The stars seem to turn and travel across the night sky, but actually it is the earth that is turning. The star field remains motionless. As it turned out, many of the fixed stars that we can see belong in the *milky way*, which is our local *galaxy*. It moves so slowly it seems to be motionless.

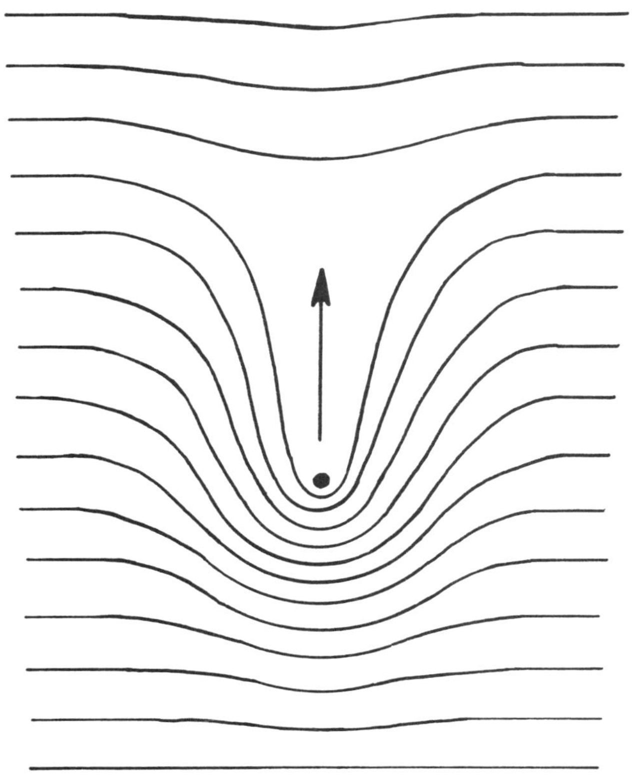

Figure E-9: When the circular field and the uniform field are merged, the uniform field exerts a strong force on the circular field in the direction shown by the vector. The circular field is the magnetic flux of a live conductor.

Scientists of 100 years ago were fascinated by the fact that the inertial system within which the earth and the whole solar system operates is controlled by the stars.

In 1851, J. B. L. Foucault built a celestial pendulum, designed with the purpose of demonstrating the powerful influence of the stars. He made a heavy iron ball and attached it to a strong wire 200 feet long. He suspended the ball from the top of the center dome of the Pantheon. Fine sand was smoothed over a concave floor. A pointer attached to the bottom of the ball was adjusted until it traced a line in the sand. To start the pendulum in motion, it was pulled to one side and tied with a thread. Then the thread was burned off to release the ball, which eliminated any unwanted motions in the pendulum. (A duplicate of this pendulum can be seen in some museums. There is one in the museum in the park at San Francisco).

Foucault was aiming to prove that all inertial forces on earth are dependent on the great star fields of the heavens. His pendulum continues to swing back and forth along a straight line dictated by the stars, while the earth turns. The Foucault pendulum

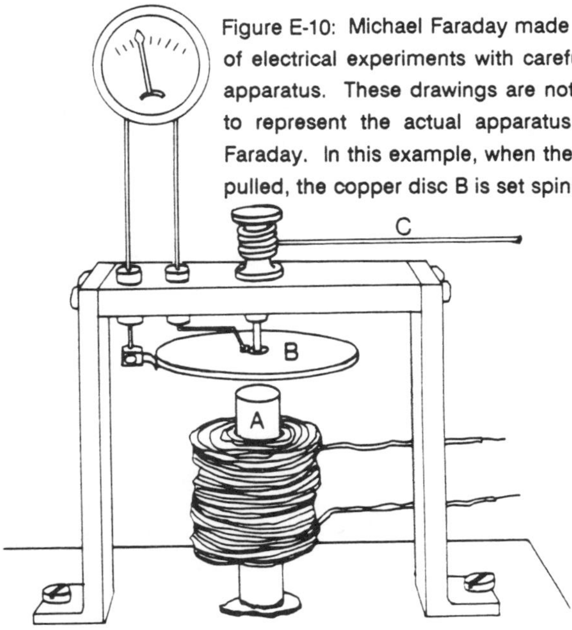

Figure E-10: Michael Faraday made hundreds of electrical experiments with carefully made apparatus. These drawings are not intended to represent the actual apparatus used by Faraday. In this example, when the cord C is pulled, the copper disc B is set spinning.

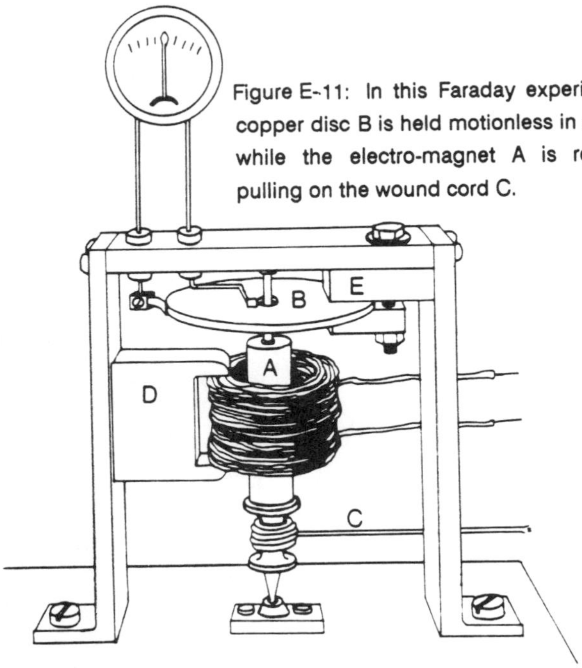

Figure E-11: In this Faraday experiment, the copper disc B is held motionless in a clamp E while the electro-magnet A is rotated by pulling on the wound cord C.

appears to make one revolution each 24 hour day. But it is really the earth that is turning once every 24 hours.

In 1966, experimental evidence convinced Peter Horsley that there is a universal magnetic field constant with the stars in the same way that the inertial field is fixed to the stars of the galaxy. Some of the experiments that confirm the existence of the universal magnetic field are described in the following.

The first experiment considers the mutual attraction that occurs between electrons that are moving in parallel paths within a cathode ray tube. Each electron in motion is affected by the magnetic field around the other. See the diagram of two electrons, Figure E-7, page 58. For an electron to be influenced, it must pass through the lines of a magnetic field. The two electrons that attract each other are both passing

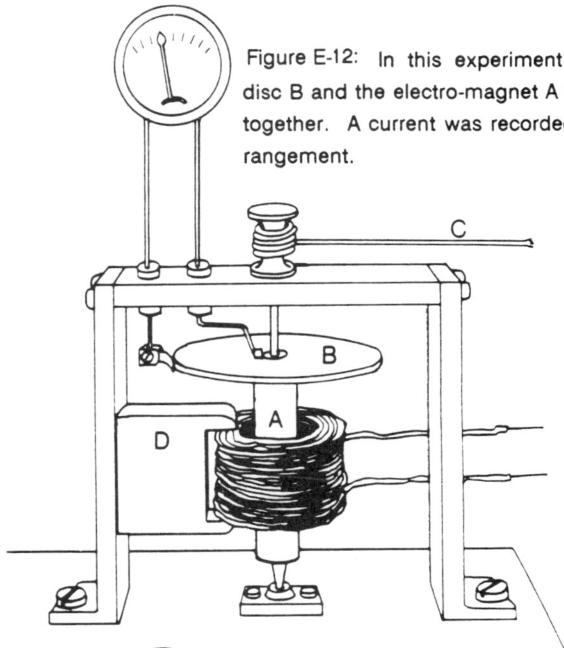

Figure E-12: In this experiment the copper disc B and the electro-magnet A were rotated together. A current was recorded for this arrangement.

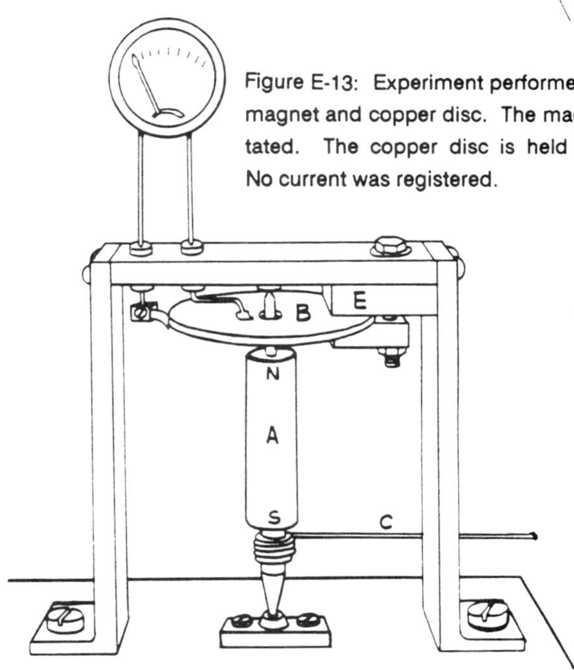

Figure E-13: Experiment performed with a bar magnet and copper disc. The magnet A is rotated. The copper disc is held motionless. No current was registered.

through the magnetic field of the other. That means that the magnetic fields are not moving with the electrons. If they were, there would be no interaction and no attraction. This indicates that magnetic flux around a moving charge rises and falls like a water wave, but the actual magnetic field remains fixed in space.

The second experiment is shown in Figure E-14. In this diagram a battery (A) is connected in the circuit (B). Also connected in the circuit is a long electrolytic cell. The electrolyte in the cell is chosen to deliver only positive ions as the conductive element in the cell. When the circuit is connected, the battery sends a current through the circuit and a circular magnetic flux rises.

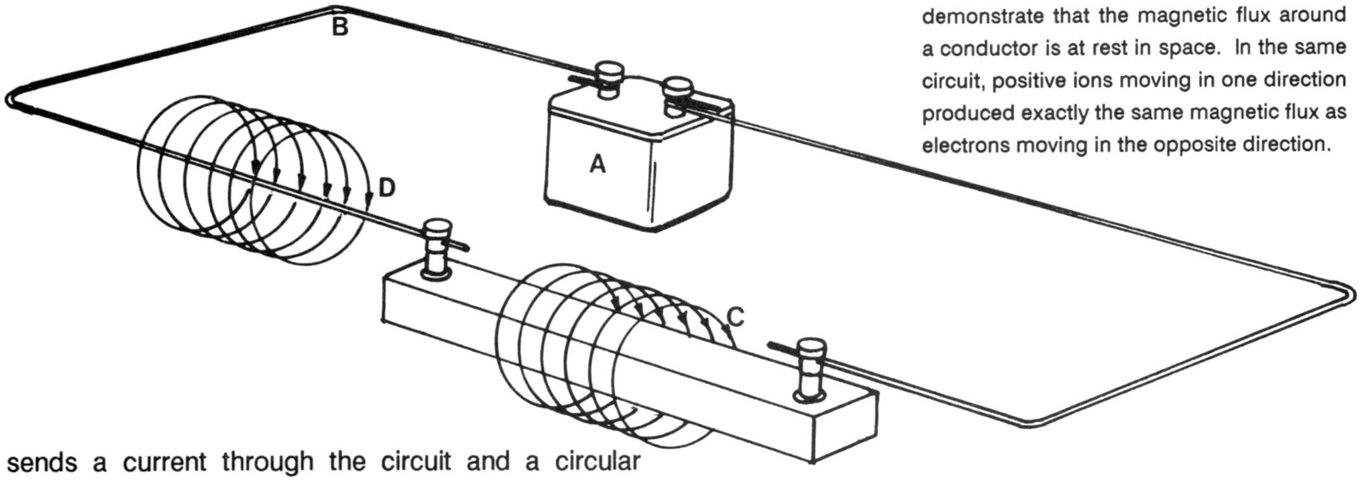

Figure E-14: Experimental circuit to demonstrate that the magnetic flux around a conductor is at rest in space. In the same circuit, positive ions moving in one direction produced exactly the same magnetic flux as electrons moving in the opposite direction.

The magnetic flux around the wire conductor is now compared to the flux around the electrolytic cell. They are found to be exactly the same. Although the current in the cell is in the reverse direction with positive ions, the magnetic flux is identical and motionless. Again the magnetic flux is shown to be fixed in space.

The third experiment was originally performed by Michael Fraday. He made no conclusions, but wrote it in his notes.

Faraday found that by spinning a copper disc over the end of a magnet, a current would be generated. The current appeared between the hub of the disc and the outer edge, as shown in Figure E-12 on page 60. He tried holding the copper disc motionless and spinning the magnet instead. In this case no current was evident.

When the same experiment is repeated using a permanent magnet, as in Figure E-13, again no current is generated. The magnetic field does not rotate with the magnet, but remains motionless in space.

The results of these and other experiments show that there is a universal link between the inertial field of the fixed stars and a universal magnetic field. They appear to be co-extensive and function in unison. When an electron is accelerated, for example, the inertial resistance of the acceleration and the rise and fall of the magnetic flux around the electron occur together. The two fields are interdependent. Or perhaps there is only one field......the universal inertial-magnetic field.

MAGNETIC FLUX LINES

(11-9) The lines of magnetic flux around a live conductor are called maxwells, named after the famous physicist who was the first to define electricity and magnetism in mathematical terms. One imaginary line of flux is called one maxwell. To induce one volt of EMF in a circuit, a conductor must pass through or be cut-across by 10^8 maxwells per second. The number 10^8 is equal to 100 million.

When a circuit conductor is wound into a coil, the magnetic flux passes through all of the windings, so that the same conductor is cut-across many times. Each time the magnetic flux passes through one conductor, the emf in the circuit is increased. Therefore, the more windings there are in a coil, the more voltage can be generated when a magnet is moved across it.

A substantial voltage can be produced by inserting a bar magnet into a coil as shown in Figure E-15. The faster the magnet is inserted, the greater is the voltage produced. Evidently the speed of the magnetic lines as they cut across the conductor is a factor.

The emf generated in a circuit by a moving magnetic field is

$$E = \frac{B l V}{10^8} \qquad (27)$$

In this equation, E is in volts, B is in maxwells per square centimeter, l is the length of the conductor acted on by the magnetic field, and V is the velocity of the flux lines as they cut across the conductor. The velocity of the magnetic field controls the number of maxwells per second, and the voltage is then proportional to that velocity.

The terminology used regarding the magnetic flux cutting across the conductor is the usual way of expressing electric generator action. In actual fact, any change in the strength of a magnetic field will induce a current in a nearby circuit.

If a magnet moves quickly near a circuit or if a coil is moved near a magnet, a sensitive galvanometer will respond to show that there is a momentary current in the circuit.

THE DIRECTION OF THE INDUCED EMF

(12-9) Emil Lenz (1804-1865), made many experiments in electricity. He discovered and recorded an important relationship now called Lenz's Law. Lenz determined that whenever a current is induced in a circuit by a moving magnetic field, the current will always be in the direction so that the magnetic flux of the current will oppose the action that caused it.

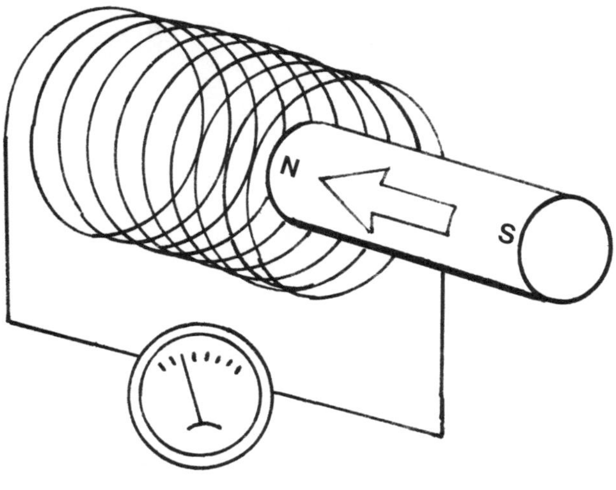

Figure E-15: When a magnet is moved near a closed circuit, a voltage appears in the circuit. If there is a coil near the magnet, the effect is multiplied. The maximum voltage is produced when a bar magnet is rapidly inserted into a coil, as in the diagram.

Lenz's law is demonstrated by the diagram in Figure E-16. When the bar magnet (A) is inserted into the coil (B), the current generated in the coil sets up an opposing magnetic field. This is shown by the dotted outline of a bar magnet inside the coil. Because similar poles repel each other, the imaginary magnet within the coil resists the insertion of magnet (A).

ELECTRON VOLTS

(13-9) When an electron is acted upon by a voltage of one volt, the amount of energy received by the electron is called one electron-volt. One erg of work per esu is equal to 300 electron-volts. The charge on the electron is 4.8×10^{-10} esu. Apply a bit of algebra to show that one electron-volt is equal to 1.6×10^{-12} ergs. This statement also defines the value of the erg.

MAGNETIC FLUX DENSITY

(14-9) One imaginary line of magnetic flux is called a maxwell. The number of maxwells passing through one square centimeter is called the flux density of magnetic induction. The unit of flux density is called the gauss, named after Karl F. Gauss (1777-1855). A gauss is defined as a flux density of one maxwell per square centimeter.

Ampere's law states that a live conductor in a magnetic field will experience a force that is proportional to the flux density, proportional to the current and proportional to the length of wire that is immersed in the magnetic field.

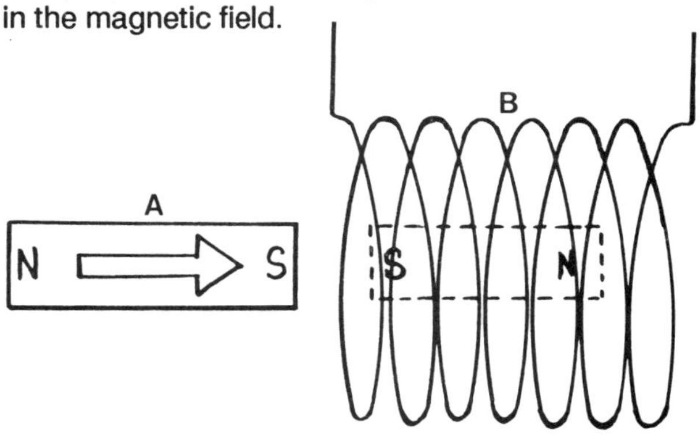

Figure E-16: Inserting a bar magnet into a coil generates a current in the coil. The direction the current takes in the coil is always such that the magnetic field produced by the current opposes the magnetic field that caused the current. This fact was discovered by Lenz about the year 1835.

THE NATIONAL ELECTRICAL CODE

The National Electrical Code is sponsored by the National Fire Protection Association under the auspices of the United States of America Standards Institute.

The National Fire Protection Association was organized in 1896 to promote the science and improve the methods of fire protection. Memberships in the Association are open to anyone who may be interested. For complete information, and to order the latest edition of the Code, the address is:

National Fire Protection Association,
60 Batterymarch Street,
BOSTON, MA., 02110.

EVENTS FROM THE PAST

When the railroad was just beginning, the real lifeblood of the nation was the steamboats that plied the big rivers and lakes of North America. Loaded with cargo and passengers, paddle-wheelers carried on the trade and commerce of the New World.

INDUCTANCE AND CAPACITANCE

MUTUAL INDUCTION

(1-10) When two circuits are placed near each other, a change in the flow of current in one circuit will cause a current to be induced in the other circuit. This effect is called *mutual induction*. Mutual induction occurs when the magnetic flux lines of a live circuit cut across the conductors in a nearby circuit. Two coils that are close to each other will show mutual induction when the current flow is varied in either coil. Mutual induction occurs when the current is increased, decreased, stopped, started, or varied in any way. There is no induction effect when the current flow has a steady value.

Mutual induction is measured in *henrys*, named after Joseph Henry (1797-1878). When a change of one ampere per second in one circuit causes a change of one volt in the other circuit, the mutual induction is said to be one henry.

The degree of closeness between circuits is called *coupling*. The closer the coupling, the greater the mutual induction. Mutual induction is at a maximum when two coils are wound together on the same iron core.

SELF INDUCTION

(2-10) When a live conductor is wound into the shape of a coil, the magnetic flux from the current in the coil cuts across many of its own conductors. Because of Lenz's law, there is a reverse voltage or *counter emf* generated in a coil because of its own magnetic flux. This effect is called *self induction*. A coil has one henry of self induction when a current change of one ampere per second causes a counter emf of one volt. The more windings a coil has, and the closer the windings are together, the greater will be its self induction.

An iron core inside a coil has the effect of increasing the self induction of the coil. What self induction does in a circuit is to resist the rise of current in a coil when a voltage is applied. It also acts to resist the fall of current in a coil when the voltage supply has been removed. When the voltage is removed, the magnetic flux in the coil begins to collapse. The collapsing field cuts through the conductors of the coil and generates a voltage in the circuit which tries to sustain the current flow. The self inductance of a coil is

$$L = \frac{N\Phi}{I}10^{-8} \qquad (28)$$

The symbol Φ is the Greek letter Phi. It is used to indicate the density of the magnetic flux that is cutting across the conductors of a coil.

ENERGY OF AN INDUCTIVE CIRCUIT

(3-10) When a voltage is applied to an inductive circuit the self inductance produces a reverse emf. The applied voltage has to overcome the resisting emf of induction in the circuit. That means that it takes some extra energy to establish a current in an inductive circuit. However, when an inductive circuit is shorted, the extra energy in the circuit is returned in the form of increased current output. The amount of energy that can be stored in an inductive circuit is

$$W = \frac{L I^2}{2} \text{ joules.} \qquad (29)$$

The current I is in amperes and the inductance L is in henrys.

THE INDUCTION SPARK COIL AND VIBRATOR

(4-10) An induction coil comprises two coils of wire (A) and (B) wound on a soft iron core (C). One of the coils is energized by a battery (D). See the diagram in Figure D-1.

A flexible metal leaf (or spring loaded leaf) is placed near one end of the iron core. The leaf is made from a magnetic material so that the magnetized iron core will draw the leaf towards it and thus open the contact (F). When the contact is broken, the induction flux collapses, generating a counter emf in the live coil and producing a surge of current in the second coil (B). The voltage in coil (B) is sufficient to cause a spark to jump the spark gap (G). The condenser (H) acts to prevent an unwanted spark from jumping the open contact (F).

The spring-loaded leaf now returns to its former position, again closing the contact (F). This

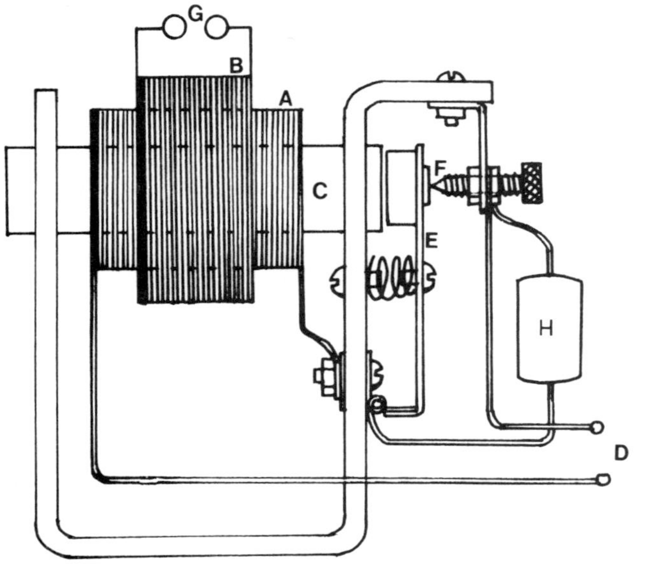

Figure D-1: Induction coil and vibrator are operated by a battery (not shown). When the coil is magnetized, the contact is opened, breaking the circuit. The energy in the coil is released through the spark gap in a second coil.

causes the action to repeat. The cycle repeats itself over and over in a rapid vibrating mode. If the vibrating action is all that is needed, the second coil with the spark gap is not included. Vibrators are used for sanding machines and for massaging machines. A vibrator also is used to activate a transformer in a d-c circuit.

THE TRANSFORMER

(5-10) The most widely used electrical device that operates on mutual induction is the transformer. Like the induction coil, two coils are wound on the same iron core. The coil to which the voltage is applied is called the primary coil. The output coil is called the secondary coil.

The output voltage of a transformer is proportional to the number of windings in the secondary as compared to the primary. If there are twice as many windings in the secondary, the output voltage will be twice the input voltage. The operation of a transformer is described in the section on alternating currents.

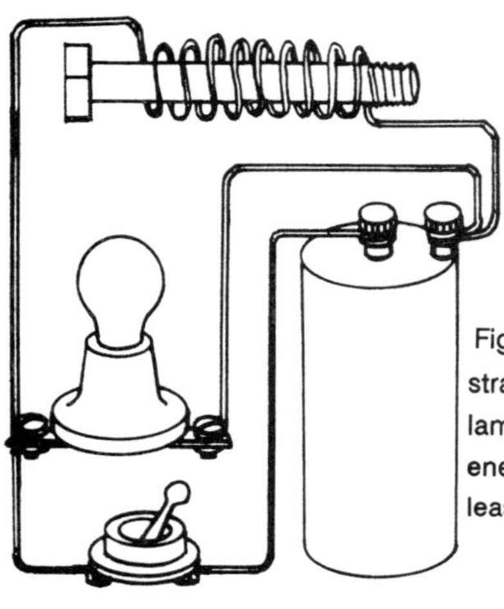

Figure D-2: Circuit for demonstrating self induction. The lamp burns brightly when the energy of inductance is released.

DEMONSTRATING SELF INDUCTANCE

(6-10) The circuit shown in Figure D-2 is an easy method for demonstrating the effects of self induction. A battery, a switch, a lamp and socket, a piece of iron and some insulated wire are connected together as shown. Wind the wire around the iron bolt at least fifty times.

When the switch is closed, the lamp burns brightly at first because the self inductance and counter emf of the coil is forcing the current to go through the lamp. As the current is established through the coil, the lamp grows dimmer. When the switch is turned off, the lamp emits a bright flash as the magnetic flux in the coil collapses. The open switch prevents the inductive current from escaping that way so it has to go through the lamp. The bright flash of the lamp represents the energy that was stored by the inductance of the coil.

CAPACITANCE

(7-10) When an object is charged with electricity, the charge represents energy. The amount of energy is related to the actual work that would be necessary to move an electron from its "home" atom to a resting position on a different substance. In an earlier chapter, a display of lightning was used as an example of tremendous energy stored in electrically charged objects such as clouds.

An object charged with negative electricity can be brought near to an object that is charged with positive electricity. Unlike charges are attracted to each other strongly, and the attractive force is stored energy. Evidently, electricity could be stored by bringing unlike charges close together, but preventing them from actually connecting. Such a plan would give charged objects a "capacity" to store electricity.

THE CAPACITOR

(8-10) The capacitor is a device for storing electricity. The basic idea is to place two flat plates together with a thin layer of dielectric material between them to prevent them from ever "leaking". When the capacitor is put into an electric circuit and a voltage applied to the circuit, the capacitor accepts and stores a charge equal to the voltage applied. One wire of the circuit goes to one of the plates of the capacitor, and the other wire goes to the other plate, as shown in Figure D-3.

The dielectric material that is chosen for the barrier in the capacitor is very important. As shown by the diagrams in Figure D-4 (a) shows a dielectric material when the capacitor is not charged and (b) shows the dielectric when the capacitor is charged. The electrical structure of the dielectric atom or molecule is stretched out of shape like a spring. The deformed atom represents stored energy. The atom will bounce back to its original shape when the stored energy is released.

Some dielectrics are better than others for

storing electric energy, and Figure D-5 is a chart showing the characteristics of some common dielectrics.

DIELECTRIC MATERIAL	$\frac{\epsilon}{\epsilon_v}$
WINDOW GLASS	6 TO 8
PYREX	4.5
MICA	5.4
PORCELAIN	7.0
QUARTZ	4.2
HARD RUBBER	2.8
TEFLON	2.1
SULPHUR	3.8
BLACK BAKELITE	5.5
TRANSFORMER OIL	2.2
LUCITE	2.6
POLYSTYRENE	2.55
ACETATE	3.7
PURE WATER	81

Figure D-5: The capacitance of a test capacitor with a calibrated voltage is measured with the test dielectric in place. The same capacitor is then measured with a vacuum as the dielectric. The proportion of the two results is the dielectric constant.

APPLICATIONS OF CAPACITORS

(9-10) Capacitors are used in many ways, but perhaps the most important application is in eliminating or controlling undesirable inductances. Capacitors are used for tuning radio circuits, and for controlling radio "noise" in filter circuits. Another application is for eliminating arcing in switches and contacts.

To make a large capacitor available in a small space, capacitors are often made with two sheets of thin foil separated by a thin sheet of oiled or waxed

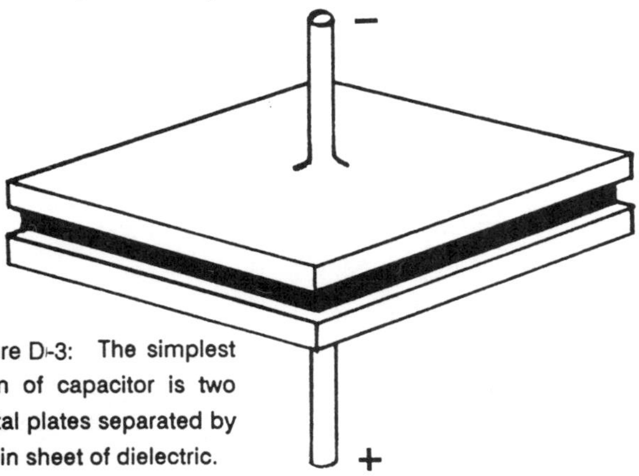

Figure D-3: The simplest form of capacitor is two metal plates separated by a thin sheet of dielectric.

paper. This "sandwich" can be tightly rolled and sheathed with insulation such as bakelite. A capacitor in the process of being made is shown in Figure D-6. Capacitors are also made to be adjustable and variable. A variable capacitor used in tuning a communications circuit is comprised of a set of movable plates with a set of fixed plates in the spaces between the plates. Air is used as the dielectric.

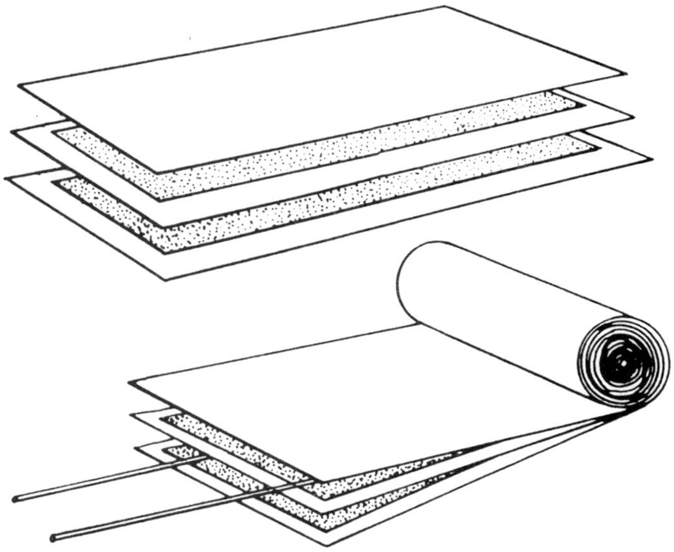

Figure D-6: Layers of foil and oiled paper are rolled up together into a tube to make a compact capacitor.

CAPACITANCE DEFINED

(10-10) The capacitance of a capacitor is a measure of the energy it will store when a voltage is applied. The amount of the voltage is not critical in determining capacitance. What is critical is the ratio of the amount of stored energy compared to the voltage applied. This relationship is a constant regardless of the voltage.

The stored energy is the electrostatic charge,

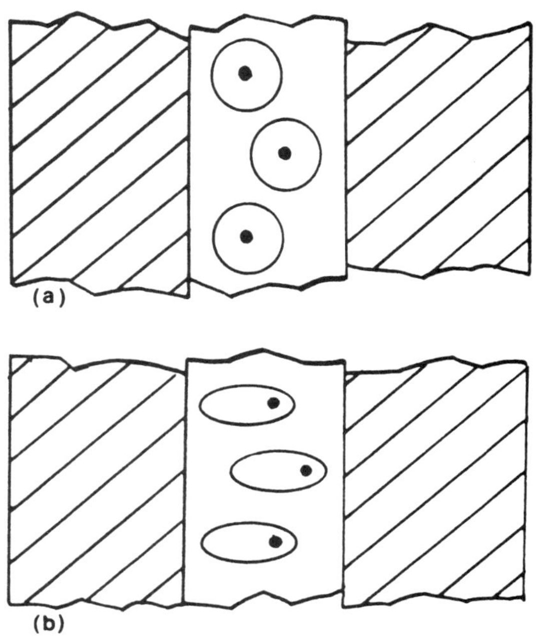

Figure D-4: Enlarged section through a capacitor showing the atoms or molecules in the dielectric. When the capacitor is not charged, (a), the atoms are normal. When the capacitor is charged, (b), the atoms are deformed.

Q. The applied voltage is V and the capacitance is C. The relationship that defines capacitiance is

$$C = \frac{Q}{V} \qquad (30)$$

The unit of *capacitance* was named after Faraday and so it is called a farad. When C is in farads, V is in volts and the charge Q is in coulombs, in the foregoing equation.

The original name for a capacitor is a "condenser". The name condenser is still used in particular capacitor applications. A condenser or a capacitor has a capacitance of 1 farad when 1 volt of emf produces a charge on the capacitor of 1 coulomb of electricity. The farad as a working unit of capacitance is too large for general use, and therefore the *microfarad* is more commonly used. One farad is the same as 10^6 microfarads. 10^6 is one million.

Electricity travels at the velocity of light, 186,000 miles per second. Therefore, when a voltage is applied to the terminals of a capacitor, the capacitor becomes fully charged instantaneously.

The amount of energy stored in a capacitor when fully charged is

$$W = \frac{C\,V^2}{2} \qquad (31)$$

In this equation, the capacitance C is in farads and V is the voltage. W stands for work or energy.

CAPACITORS CONNECTED IN PARALLEL AND SERIES

(11-10) Groups of capacitors act like one large capacitor when they are all conected in parallel. This is shown in the diagram of Figure D-7. The dotted line connecting the capacitors together illustrates that the effect would be the same if they were all one capacitor. Therefore the total capacitance of a group of capacitors in parallel is the total or sum of all the capacitances. The equation is

$$C = C_1 + C_2 + C_3 + C_4 \qquad (32)$$

Capacitors connected in series act entirely different. A diagram of capacitors connected in series is shown in Figure D-8. The voltage measured at a capacitor is actually a voltage drop in the circuit, like a waterfall in a river. The highest waterfall the circuit is capable of delivering is limited by the voltage of the battery. Now if you add another capacitor downstream, the voltage has to be shared between the two capacitors in the circuit. It is the same as adding another waterfall. Some of the height has to be borrowed from the first waterfall.

The total capacitance of a group of capacitances in series is found by taking the sum of the reciprocals of the individual capacitances. That is

$$\frac{1}{C} = \frac{1}{C_1} + \frac{1}{C_2} + \frac{1}{C_3} \qquad (33)$$

The reciprocal of any number is that number divided into the number one.

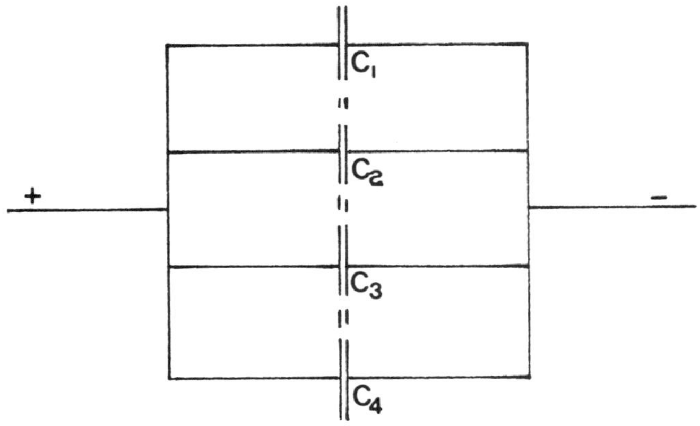

Figure D-7: Capacitances connected in parallel in a circuit have the same effect as if they were one large capacitor. The four capacitors are shown with a dotted line connecting them to indicate that if they were touching each other, the capacitance would be the same as when they are not touching.

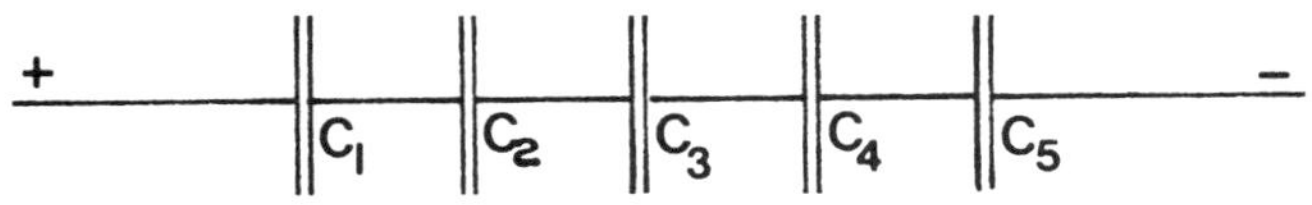

Figure D-8: Capacitances connected in series have to share the available voltage supplied by the battery. The capacitance on any one capacitor cannot be greater than the voltage drop at that capacitor. When the total voltage is shared, the capacitance is reduced every time another capacitor is added in series.

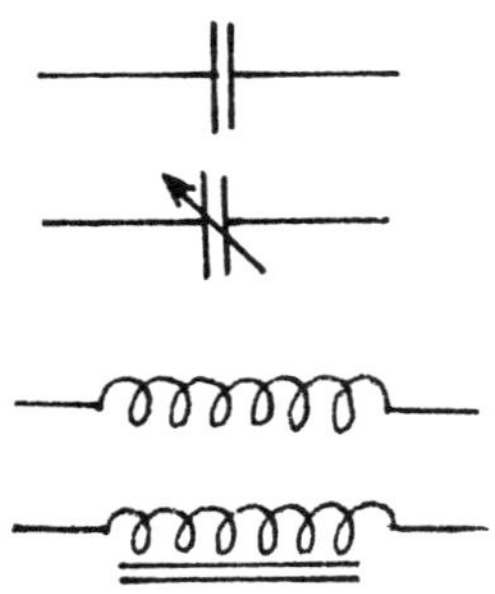

SYMBOL FOR CAPACITOR

VARIABLE CAPACITOR

SYMBOL FOR INDUCTANCE

IRON CORE INDUCTANCE

TUNING CONDENSER

The variable air capacitor has a number of movable plates positioned between an equal number of stationary plates. The movable plates slide between the still plates without touching them. The plates are usually stamped out of aluminum or alloy. In the example shown, Figure D-9, the dotted line outlines the shape of the hidden part of the plates. A small amount of turning of the control knob makes a large difference in the amount of area between the plates. The dielectric is air at normal pressure.

The disadvantage of the variable air capacitor is that the design has inherent inductances which counteract the desired effect of the capacitor. The disadvantages can be partly avoided by making cross pieces, which are at right angles to the plates, non-metalic. The turning axle and through-rods can be of nylon or other plastic. The plates have to be connected together somewhere in order for them to function as a capacitor. The two connections across the two sets of plates can be positioned far apart so that there will be a minimum of direct inductive linkage between them.

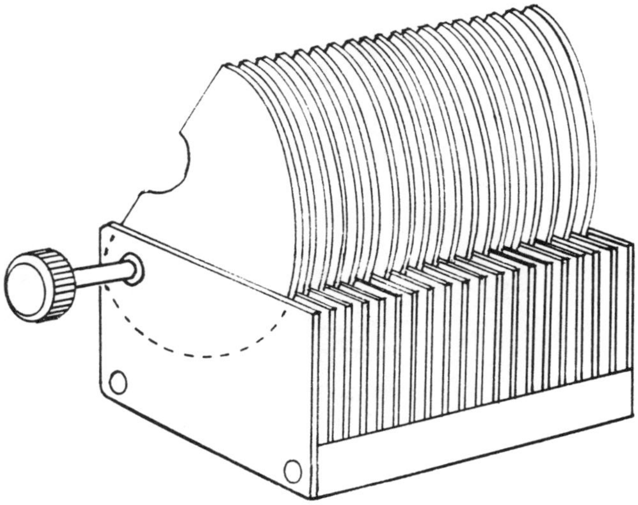

Figure D-9: The variable air capacitor has a set of movable plates and a set of stationary plates. The plates do not touch each other as there are air spaces in between. This type of capacitor is used for fine-tuning radio reception.

EVENTS FROM THE PAST

Horse-drawn wheeled vehicles came in all sizes and types. Inventors dreamed up new uses for wagons with automatic mechanical loaders and off-loaders. Designers produced beautiful carriages for the rich and powerful. It was into this clean scene that the automobile intruded, roaring thunderously and belching smoke. Most people thought that the automobile was just a fad that would pass away eventually.

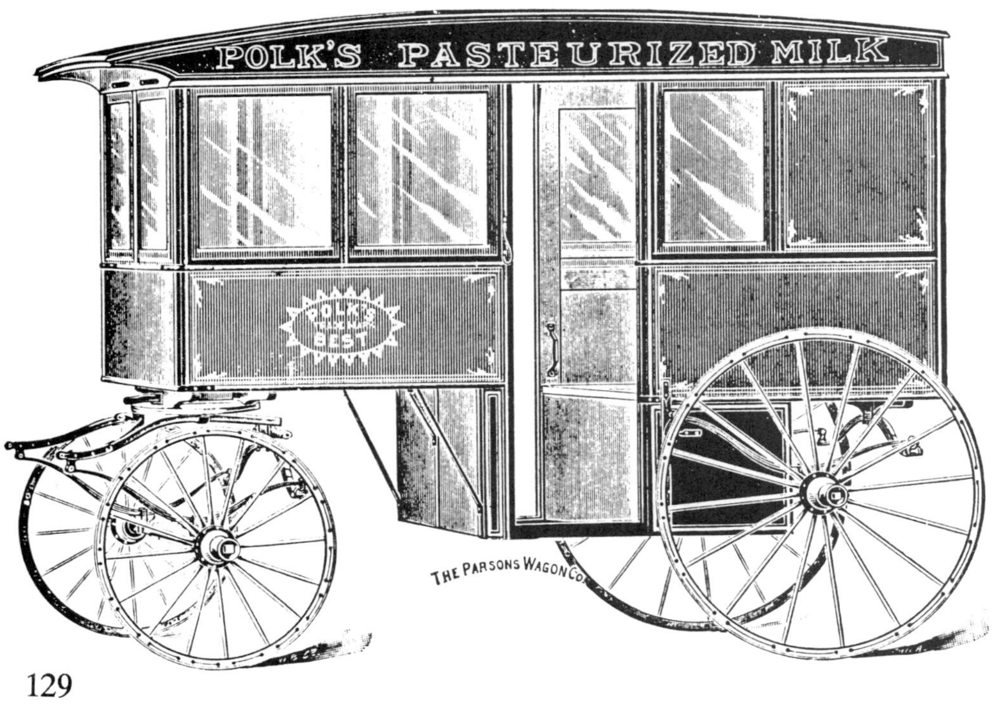

ALTERNATING CURRENTS

THE TECHNICAL WORLD OF ELECTRICITY

Alternating current, usually abbreviated a-c, is used in our homes and factories and in the electrical systems of industry across the nation. Almost all of the electrical circuits and equipment applied in the technical world are operated by alternating current.

This section on alternating currents is the most important part of basic electrical knowledge. It should be studied carefully and thoughtfully. The practical application of electricity depends on an understanding of a-c circuits and devices.

FREQUENCY OF THE ALTERNATIONS

(1-11) In the development of the first moving pictures it was apparent that the pictures would flicker unless the film was speeded up to a minimum of sixteen frames per second. About that time the very first electric generators were being built. They were faced with the very same problem. They wanted to use alternating current for lighting circuits, but the flicker of the alternations gave people a headache and bothered their eyes. On the other hand they wanted the alternations as slow as possible because it was a lot more expensive to deliver higher alternations. They finally settled on 25 cycles per second. That was somewhat higher than the frequency that the movie industry had decided on.

As it turned out, at 25 cycles per second there were lots of people who couldn't stand the flicker. Also, certain disadvantages to the number 25 became apparent in some technical applications of alternating current. Eventually, a new standard of 60 cycles per second was chosen. Alternations at 60 cycles has a relationship to clock time which became very important as the new communications industry burgeoned into existence.

HERTZ

To simplify electrical language, and to honor Hertz, the genius who first demonstrated the existence of radio waves, the three words "cyles-per-second" have been changed to "Hertz". Instead of saying 60 cycles-per-second, the correct technical usage is now 60 hertz.

THE VELOCITY OF ELECTRICITY

(2-11) To produce alternating current, a voltage is applied first in one direction and then in the other direction. This causes the electricity to flow back and forth along the power lines and through the electrical circuits in your home. Electricity travels at the incredible speed of 186,000 miles per second, so there is not the slightest chance of the alternations bunching up or overlapping. In the one thirtieth of a second during which the electricity is flowing in one direction, the energy in the electrical wave travels over 1500 miles. The electrons do not travel that fast. The wave of energy runs along the wire with the motions of electrons contributing to the wave, in the same way that air molecules permit sound waves to travel through them.

HARMONIC VIBRATIONS

(3-11) Tie a piece of string to a small stone or to a bolt and then let the object dangle. Now swing it back and forth like a pendulum, as shown in Figure A-1. If you look closely you will see that at each end of its swing, the weight on the string comes to a dead stop before it moves again. Then it moves in the other direction, slowly at first, gradually gaining speed until it is at the bottom of its swing. After it swings past the bottom low point, it starts to slow down because it is

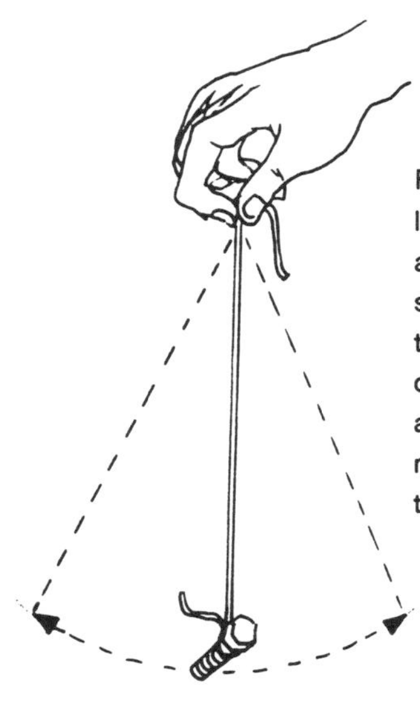

Figure A-1: Make a pendulum out of a piece of string and a weight. When you let it swing by itself, it demonstrates the nature of all vibrations, called "harmonic motion". The alternations in alternating current move with harmonic motion.

going against gravity. Finally it comes to a dead stop again at the other end of its swing.

This kind of motion, like a pendulum, is called *harmonic motion.* Everything that vibrates or oscillates goes through exactly the same kind of motion, slowing down, coming to a dead stop, then moving in the opposite direction, the same as a pendulum.

The alternations of alternating current are exactly the same. The electricity moves in one direction and then in the other direction, coming to a dead stop at each end of its run.

ALTERNATING CURRENT DIAGRAMS

(4-11) To demonstrate alternating currents in a form that could be easily visualized, a straight level horizontal line was chosen to represent time. It is actually a vector, so that it properly should have an arrow head, like Figure A-2. The flow of time is steady, without variation, so the vector for time is a straight line going from left to right.

Figure A-2: Time is a horizontal vector.

A current of electricity that is steady and direct is also diagrammed as a straight level line. The strength of the current is shown by its height above the time line. A steady current of ten amperes will be drawn 10 units above the line for time. This is shown in Figure A-3.

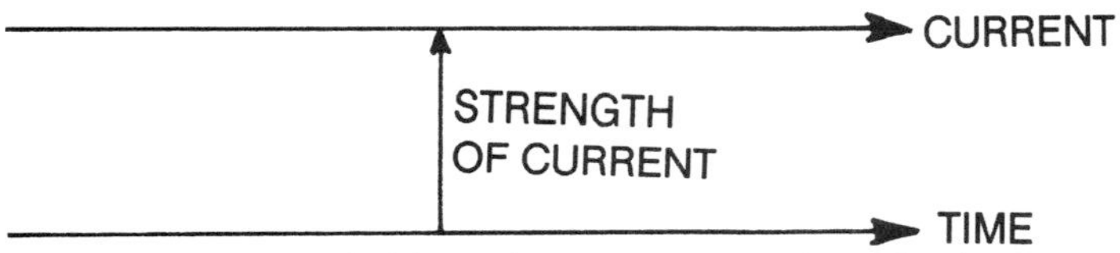

Figure A-3: The height of the current vector above the time vector shows the strength of the current.

To show a current going in the opposite direction, it cannot be turned around because time flows only in one direction. So to show a current going backwards it is drawn in below the time line. A steady current of five amperes going in the opposite direction is shown 5 units below the time line. Its direction is the same direction as time flow, as shown in Figure A-4.

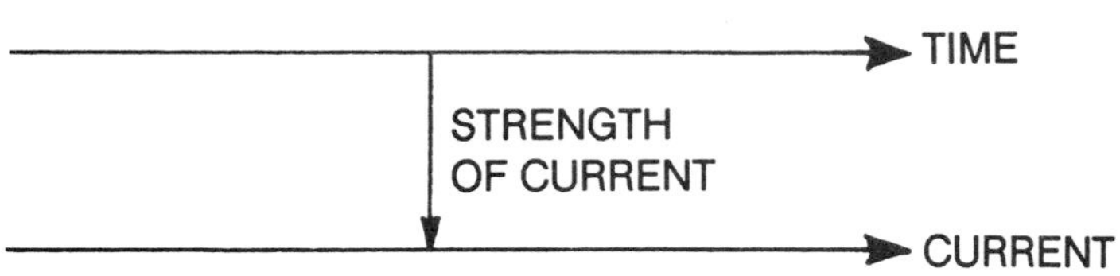

Figure A-4: A current in the opposite direction is a vector below the time line. Again the strength of the current is shown by the distance between the time vector and the current vector.

When a current changes direction, it crosses over the time line. At the point of the crossover, the current is zero. A current changing direction is shown in Figure A-5.

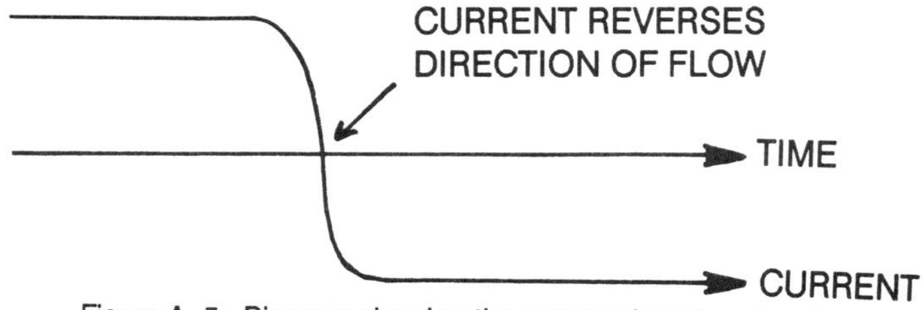

Figure A-5: Diagram showing the current changing direction.

SINE WAVE DIAGRAM

When an alternating current is plotted on the time graph, it has to cross the time line every time it changes direction, so it looks like Figure A-6. The upper half wave shows the current going in one direction and the lower half wave shows the current going in the other direction.

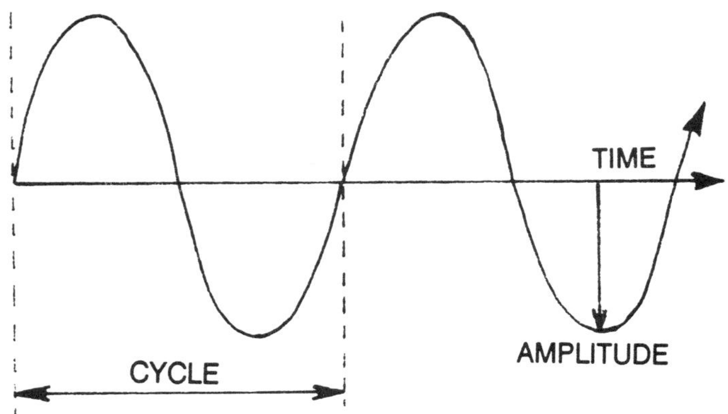

Figure A-6: The diagram for alternating current. The current changes direction twice every cycle.

Some of the terms that are used in talking about alternating currents are:

Sine wave; the shape of alternating current on the diagram.

Hertz; Cycles per second.

Amplitude; the highest point in a sine wave.

Cycle; One upper half wave and one lower half wave.

TRIGONOMETRIC DIAGRAM

The sine wave is named for the trigonometric diagram from which it originates. The word trigonometry just means "the study of triangles". Angles are measured in degrees. There are 360 degrees in a circle, as in Figure A-7. The circle is cut into quadrants of 90 degrees each, as shown.

On a circular diagram, the vector that represents the current spins around exactly like the second hand of a watch or clock. The only differences are that the vector goes anti-clockwise and zero is not at the top, but at the right side of the circle on the time line.

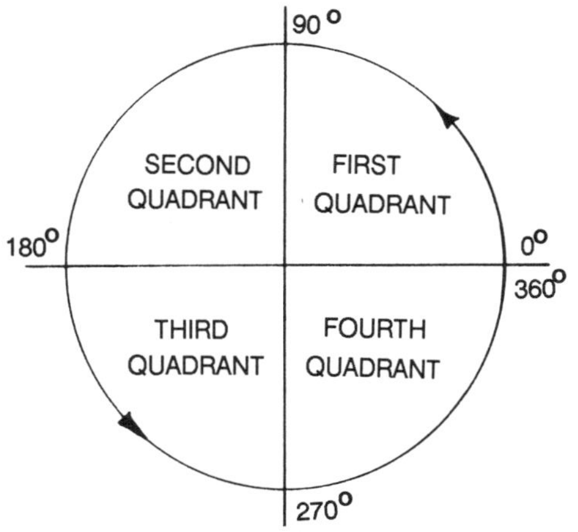

Figure A-7: Four ninety degree angles make a full circle.

The amplitude, or vertical height, of the current vector can be shown at any given moment by the dotted line in the diagram, Figure A-8. When the amplitude is divided by the vector, the result is the sine of the angle that the vector is making in the circle (sine is spelled sin for short). The steady rotation of the vector produces the sin wave form shown in Figure A-6. For this reason the shape of the wave in the a-c dia gram is called a sin wave.

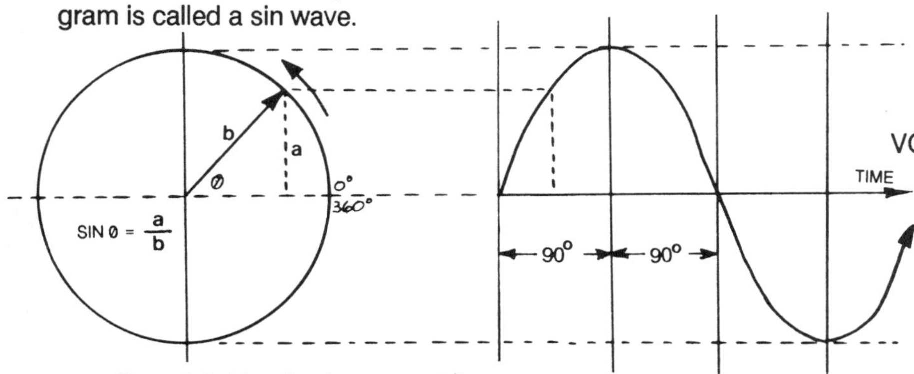

Figure A-8: How the sine wave got its name.

GENERATING ALTERNATING CURRENTS

(5-11) Wind a coil of circuit wire on an iron bolt and then arrange to spin a magnet near the coil, as shown in Figure A-9. An alternating current will register on your meter. When the magnetic flux of the N pole of the magnet cuts through the conductors of the coil, a voltage and a current are generated in the circuit. When the S pole comes around, the magnetic flux has the opposite polarity and so the current goes in the opposite direction.

In Figure A-10 the coil of wire and the bolt are mounted on a shaft with bearings. The two ends of the wire in the coil are soldered to brass rings (called slip rings or collector rings). As the slip rings rotate, two metal contacts rub on them and transfer the electrical current through the meter. The magnet is mounted on a support nearby. When the coil is rotated an alternating current registers on the meter.

These examples show that the magnet can be rotated or the coil can be rotated. The result is the same. In electrical terms, the part that rotates is called the rotor. The part that stays stationary is called the stator. The rotor and the stator are usually made of thin steel laminations, described in the following paragraph.

EDDY CURRENTS

(6-11) When alternating currents are used anywhere near iron, the magnetic flux causes currents to be generated inside the iron. These currents bounce back and forth in time with the alternations of the magnetic flux. Little eddies or vortices of current are activated in the iron and the iron gradually becomes red hot from electric heating. This represents a substantial loss of energy and power.

To minimize and to prevent losses of power from eddy-currents, the iron-cores of a-c devices are specially designed. The core is made out of thin laminations of steel alloyed with cobalt. The natural coating or platina of oxide on the laminations prevents current from going from one lamination to the next. This minimizes the size of eddy-currents, and the special alloy of the steel also minimizes eddy-currents. In practice, the iron core still gets hot, but the heat is the result of the normal operation of the coils. These losses are due to resistance in the conductors and are called R I^2losses.

VOLTAGE AND CURRENT IN A-C CIRCUITS

(7-11) In direct current devices (d-c) the effective voltage or current in a circuit is easy to measure because it does not change in value from moment to moment. However, in a-c circuits the voltage and current flucuate regularly from zero

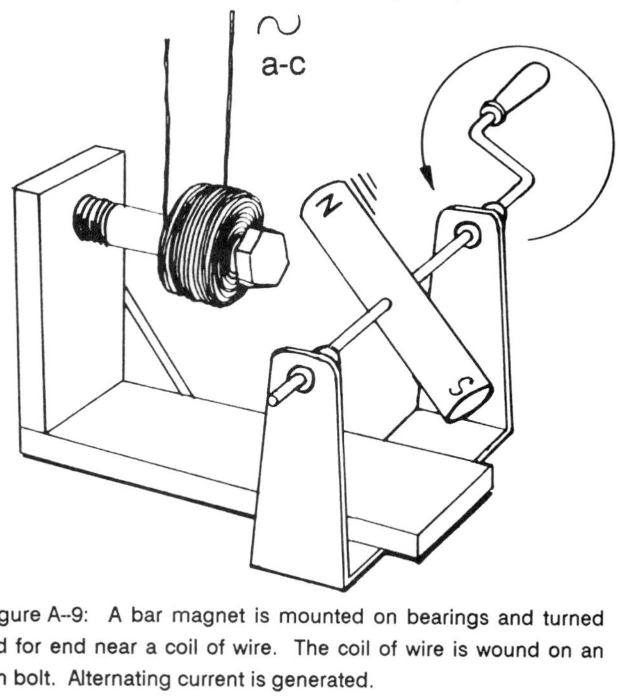

Figure A-9: A bar magnet is mounted on bearings and turned end for end near a coil of wire. The coil of wire is wound on an iron bolt. Alternating current is generated.

to a maximum value. Therefore, measuring the effective value of voltage and current in a-c circuits requires a different approach.

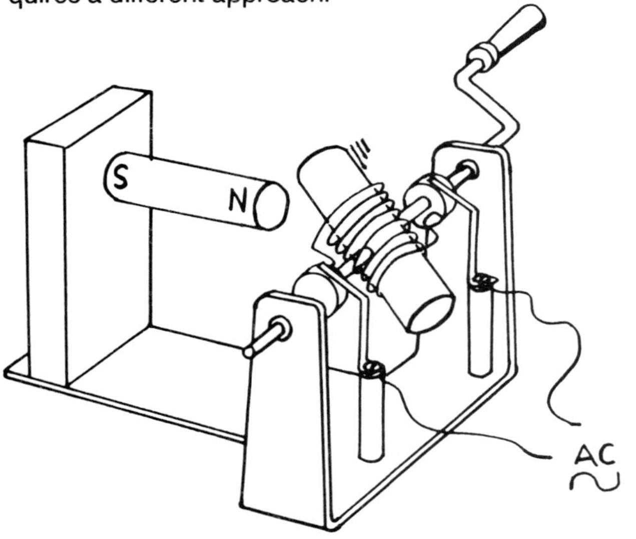

Figure A-10: Rotating the coil and bolt while the magnet is stationary. Alternating current is generated.

There are two ways of measuring alternating current. One way is to carefully measure the amount of heat produced by an a-c circuit in a measured amount of time. When that is established, the same circuit is then placed on direct current. When the same amount of heat is generated during the same time interval, the voltage is measured. The voltage of the d-c circuit will be the voltage for the a-c circuit. To summarize, one ampere of alternating current will generate the same amount of heat in a given time interval as will one ampere of direct current.

The other way to determine the effective voltage and current in an a-c circuit is to measure the maximum voltage in the circuit. It is known mathematically that the average value of a sinusoidal wave form is equal to one half the square root of two, multiplied by the peak value. The effective current in an a-c circuit is:

$$I = \frac{\sqrt{2}}{2}\, I\ (\text{max}). \qquad (34)$$

The average voltage in the same circuit will be:

$$V = \frac{\sqrt{2}}{2}\, E\ (\text{max}). \qquad (35)$$

When the current or voltage of an alternating current circuit is mentioned, it is understood that the effective value is the value intended. The effective value of the a-c circuits in the home is 120 volts (115 volts in some localities).

PHASE RELATIONSHIPS IN ALTERNATING CURRENTS

During one complete cycle of alternating current the current reverses itself once and returns to the starting point. At the same time, during one cycle, the vector on the circle diagram (Figure A-7, page 130) rotates once, turning through 360 degrees of angle.

When the alternating voltage is applied to a circuit, the current that is produced may not be in phase with the voltage. If by chance the circuit contains only resistance, then the current will be in phase with the voltage. However, almost all electric circuits do have some inductive elements and some capacitive elements. These elements act to delay or speed up the current so that there is a tendency for the current to be out of phase with the voltage to some degree.

INDUCTIVE REACTANCE

(9-11) In most circuits there is usually some

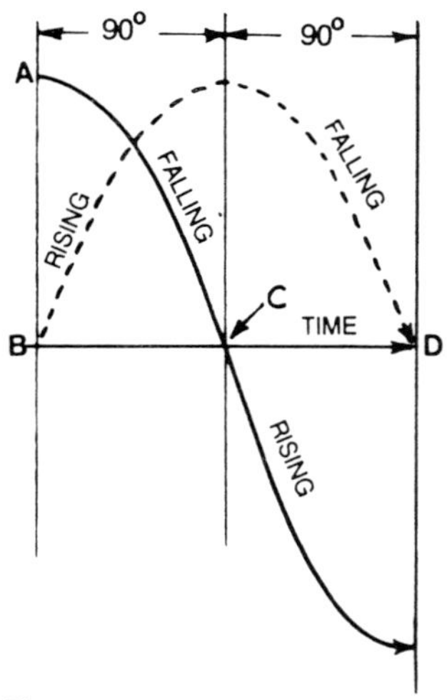

Figure A-11: When Lenz's law is considered, the induced current will oppose the in-phase current. When the applied voltage is falling, the induced current is rising. When the applied voltage is rising, the induced current is falling.

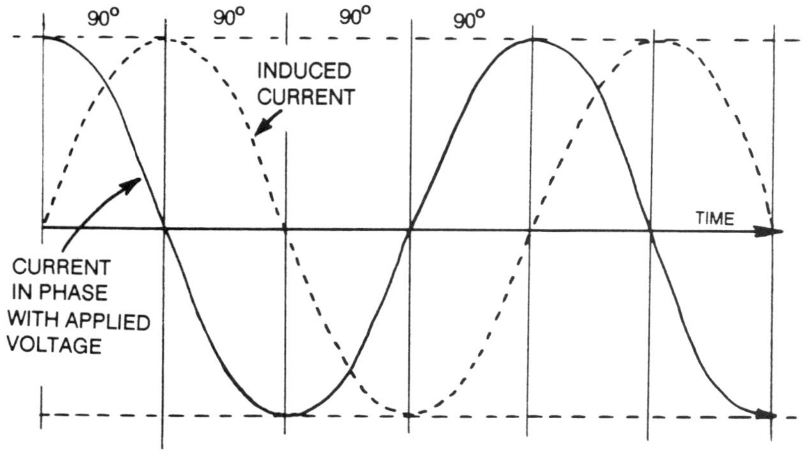

Figure A-12: The inductive current continues to respond as predicted by Lenz's law. The undulations in the induced current form a sin wave that is exactly 90 degrees out-of-phase with the applied voltage. The induced current lags behind the applied voltage and current.

inductance caused by the self-induction of the conductors in the circuit. If a coil is present then there will be an inductive current in the circuit. When Lenz's law is considered, the induced emf acts to oppose the applied emf.

Lenz's law indicates that when the applied voltage and current are falling, the self induced current is rising. But when the applied voltage is rising in the opposite direction, the induced current begins to fall. These relationships are shown in the diagram, Figure A-11.

The induced current continues to respond in opposition to the applied voltage and current. This is shown in the diagram, Figure A--12. The rising and falling of the induced current is always exactly 90 degrees out of phase with the applied voltage.

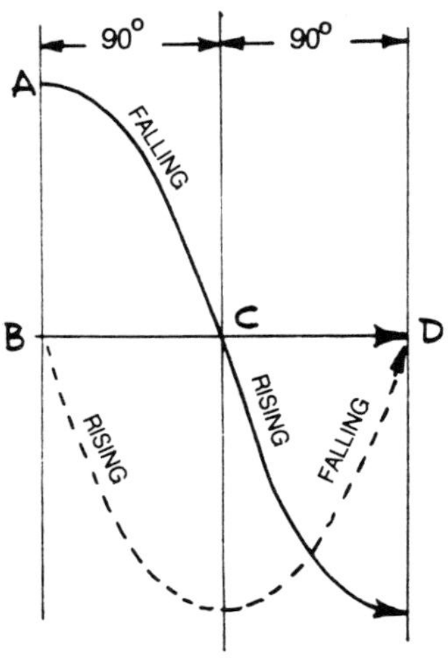

Figure A--13: A capacitor is fully charged when the voltage is at a peak. When it is fully charged, the current is momentarily stopped. Meanwhile, the current in phase with the voltage is at full flow in the circuit. When the voltage drops, the charge in the capacitor is released into the circuit, running ahead of the voltage.

The inductive current in an a-c circuit lags behind the applied voltage. This lagging current mixes in with the current that is naturally in phase with the voltage and throws it out of phase. The out-of-phase current represents a loss of power, which will be described in a following paragraph.

The current in an inductive circuit can be expressed mathematically as:

$$I = \frac{E}{2\pi f L} \qquad (36)$$

In this equation, I is the current, E is the voltage and L is the inductance in henrys. The symbol is the Greek letter Pi, and has the value 3.14159. (Pi is the circumference of a circle divided by the diameter of the same circle).

The quantity $2\pi f L$ has a name. It is called the inductive reactance, and is shortened to the symbol X_L. It behaves exactly like a resistance in a circuit, and so it can be evaluated in ohms, when the inductance is in henrys. The inductive reactance is

$$X_L = 2\pi f L \qquad (37)$$

X_L = 2 x 3.14159 x frequency x henrys.

CAPACITIVE REACTANCE

(10-11) A capacitior responds instantaneously to the applied voltage. It reaches its highest charge exactly when the voltage is highest. The capacitor is also a storage device. When it is fully charged, it is full of energy. No current is moving in the capacitor at

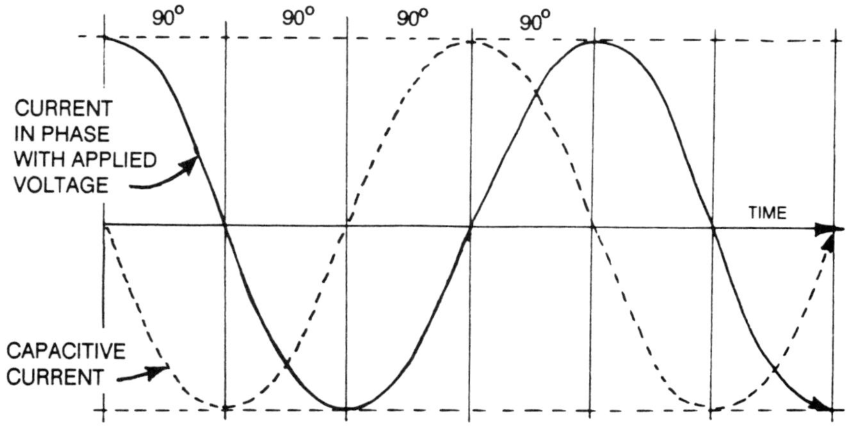

Figure A-14: As the capacitor continues to respond to the voltage, the resulting capacitor current continues to run ahead of the voltage, leading by ninety degrees. At peak voltage, there is no current flow in the capacitor. But at zero voltage, the capacitor current is at a maximum. The rise and fall of the capacitive current is always exactly 90 degrees ahead of the voltage and the in-phase current.

the moment that it is fully charged. A capacitor is also like a fence which prevents a charge from going any further. It is the end of the line. From these considerations it is evident that in the pendulum-like flow of electricity back and forth, the capacitor current is at the dead end of its run when the capacitor is fully charged. This means that when the voltage is at a maximum and the in-phase current is at maximum flow, the capacitive current is standing still.

These relationships are detailed in the diagram in Figure A-13. When the voltage and in-phase current are at maximum flow at (A), the current in the capacitor is standing still at (B). But when the voltage falls, the charge is released from the capacitor and flows back in the reverse direction (it has nowhere else to go). When the voltage reaches zero at (C) the capacitive current has reached a maximum. When the applied voltage rises in the opposite direction, the capacitor becomes charged again. It is fully charged with stored energy at point (D).

As detailed in Figure A-14, as the capacitor continues to respond to the changing voltage, the currents that are running in the capacitive circuit form a sine wave that is exactly 90 degrees out-of-phase with the voltage. The capacitive current leads the applied voltage by 90 degrees.

The capacitive current in a circuit with a capacitor runs out in front of the applied voltage. The leading capacitive current mixes with the current that is naturally in phase with the applied voltage. The mixture throws the current out-of-phase with the voltage and has the effect of lowering the efficiency of the circuit. This effect is described in a following paragraph.

The effect of capacitance is like a resistance in an alternating current circuit. It is called capacitive reactance, and can be expressed in ohms, like a resistance, when the capacitance is measured in farads. Capacitive reactance is expressed by the equation

$$X_C = \frac{1}{2\pi f C} \qquad (38)$$

VECTOR DIAGRAMS

(11-11) Vector diagrams for inductive current and capacitive current are shown in Figure A-15 and Figure A-16. In Figure A-15, the voltage vector V coincides with the time vector. The vector is moving

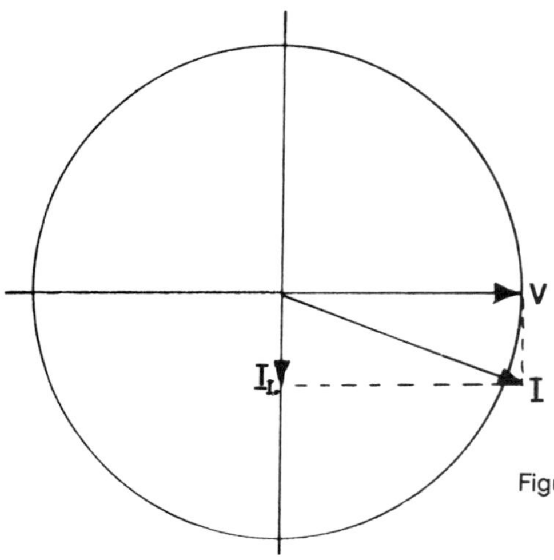

Figure A-15.

anti-clockwise, so the inductive current vector I_L is pointing straight down, lagging the voltage by 90 degrees. By adding the vectors for voltage and inductive current, the actual current is shown at I.

In Figure A-16 the capacitive vector I_C points straight up because it leads the voltage by 90 degrees. When the voltage vector and the capacitive current vector are added together, the result is the current I. It is out of phase by several degrees.

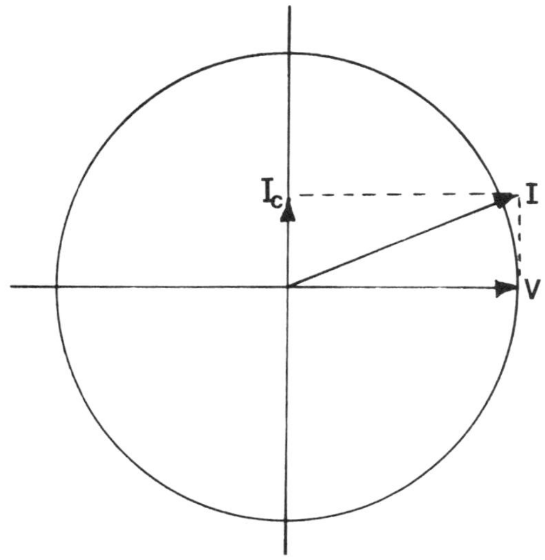

Figure A-16: When the voltage vector coincides with the time vector, the capacitive current points straight up and the inductive current points straight down. One leads the voltage by 90 degrees and the other lags the voltage by 90 degrees.

COMBINING RESISTANCE, INDUCTANCE AND CAPACITANCE.

(12-11) Currents associated with the applied voltage are always in phase with the applied voltage. If there is a resistance in the circuit, the in-phase current is not affected. It remains in-phase with the voltage. In current phase relationships, the effect of resistance can be ignored.

Inductive current and capacitive current are something else. The inductive current lags by 90 degrees and the capacitive current leads by 90 degrees. This means that they are 180 degrees apart and stand in direct opposition to each other. In the special case where the inductive current is equal to the capacitive current, they exactly wipe each other out and there is no current at all. In wave terminology, this is called total interference. One set of waves totally interferes with the other set of waves.

In a-c circuits containing both capacitance and inductance, the respective vectors are subtracted from each other. This is shown in the diagram of Figure A-17. If one of the pair is slightly larger than the other, there will be a residue of out-of-phase current left over. Any residue is called impedance. The impedance of a circuit can also be expressed in ohms, just the same as if it was a resistance. It is expressed by the equation

$$Z = \frac{E}{I} \qquad (39)$$

The impedance Z can be reduced to a mathematical quantity by applying Pythagorus' theorum for right triangles, which is that the sum of the squares of two sides is equal to the square of the long side. The impedance Z is then expressed:

$$Z = \sqrt{R^2 + (X_L - X_c)^2} \quad (40)$$

Then the current in the same circuit can be shown with the equation:

$$I = \frac{E}{\sqrt{R^2 + (X_L - X_c)^2}} \quad (41)$$

TUNED CIRCUITS

(13-11) If the impedance is totally eliminated from a a-c circuit, the circuit achieves a condition of resonance. The power and efficiency of a circuit are then at a maximum. This happens when the inductive reactance X_L is equal to the capacitive reactance Xc. That is, when

$$2\pi f L = \frac{1}{2\pi f C} \quad (42)$$

When these terms are rearranged, the natural frequency of any a-c circuit turns out to be

$$f = \frac{1}{2\pi\sqrt{LC}} \quad (43)$$

In a tuned circuit the resistance current is in phase with the voltage and the reactance currents have been eliminated by total interference.

The diagram, Figure A-17, shows a circuit approaching resonance. When the capacitance vector C is equal to the inductance vector L, the circuit is tuned.

POWER IN ALTERNATING CIRCUITS

(14-11) The power of an a-c circuit is determined by multiplying the current value and the voltage value together at any given moment in time. When the current and voltage are at a maximum, the power of the circuit is at a maximum also. However, the power and efficiency of the circuit can only reach a maximum value when the current and the voltage are in phase.

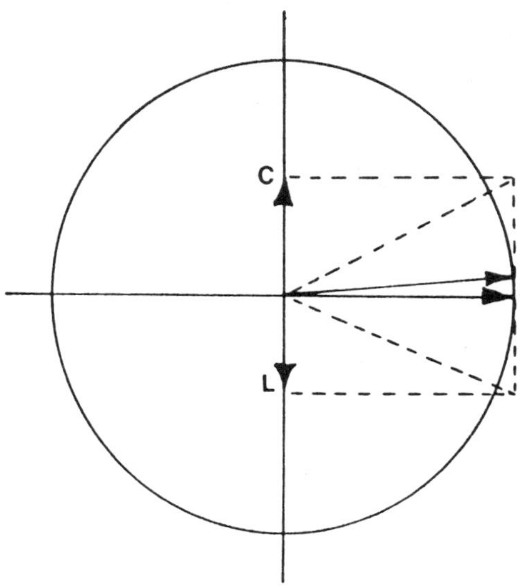

Figure A-17,

When the voltage vector and the current vector exactly overlap, a maximum condition exists and the power of the circuit can be expressed

$$P = EI \quad (44)$$

But if the current is slightly out of phase as in Figure 11-17, the power must be found by the equation

$$P = EI\cos\theta \quad (45)$$

In this equation, E is the voltage, I is the current, and the cos is the cosine of the small angle between the current and the voltage. Because the cosine is always less than one, the effect is to make the circuit less efficient.

The loss of power that can occur in an alternating current circuit is shown in the diagram of Figure A-18. In this diagram, V is the voltage, I is the out-of-phase current, and P is the power. The large dotted lobe is the amount of power that is lost due to the current being out-of-phase with the voltage.

The power is measured in watts, and that is the instantaneous current multiplied by the instantaneous voltage. Figure A-18 emphasises the importance of the power factor in a circuit.

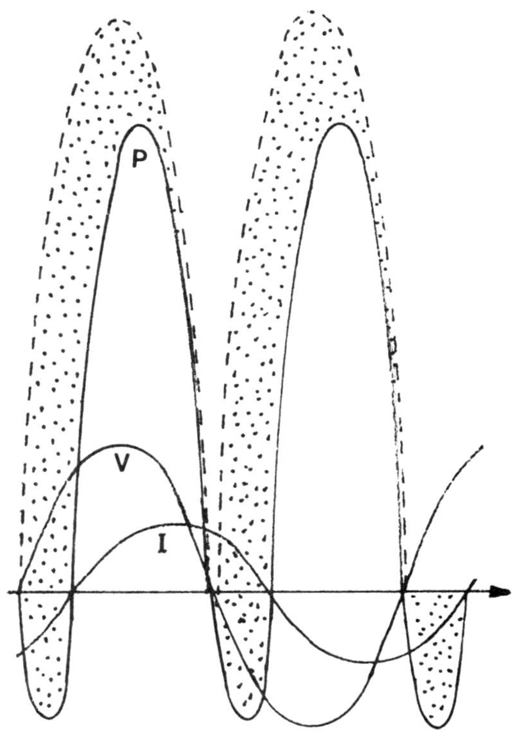

Figure A-18: The power of a circuit is reduced when the voltage and the current are not in phase, mainly because the actual power is the instantaneous value of the current multiplied by the voltage at the same moment. The large dotted lobe in the diagram shows the amount of power that is lost when the voltage and current are out-of-phase.

POWER FACTOR

(15-11) The *power factor* of a circuit is a quick method of expressing the efficiency of a circuit. The power factor compares the actual conditions in an alternating current circuit with the optimum conditions that would be possible. The power factor is expressed

$$\text{Power Factor} = \frac{E\,I\cos\theta}{E\,I} \qquad (45)$$

The maximum power factor is one. The power factor can often be improved by installing a suitable capacitor unit to reduce the inductive lag. In the terminology used when referring to the power factor, the term volt-amps may be used instead of watts. The reason is that except when the circuit is actually resonant, volt-amps is a greater value than watts. A wattmeter measures the voltage and the amperage at a given moment, and they may not be in phase. Therefore they would not be at a maximum value.

TESTING FOR POWER FACTOR

(1-15)The power factor of an induction motor (or the power factor of the motor circuit) can be measured by determining the power with a wattmeter, the current with an ammeter and the voltage with a voltmeter. The wattmeter has two coils, as shown in the diagram of Figure 15-1. One coil has a very high resistance in the circuit which prevents the current from entering, but measures the potential drop across the circuit at the moment the unit is connected. The other coil is of thick wire which permits the current to flow without hindrance. It measures the magnetic flux generated by a free flow of current in the circuit.

The connections for the ammeter, the voltmeter and the wattmeter are shown in Figure 15-1. The power factor is then derived by dividing the result from the wattmeter by the product of multiplying the amperage by the voltage. The power factor should be slightly less than 1 (one) if it has been done correctly.

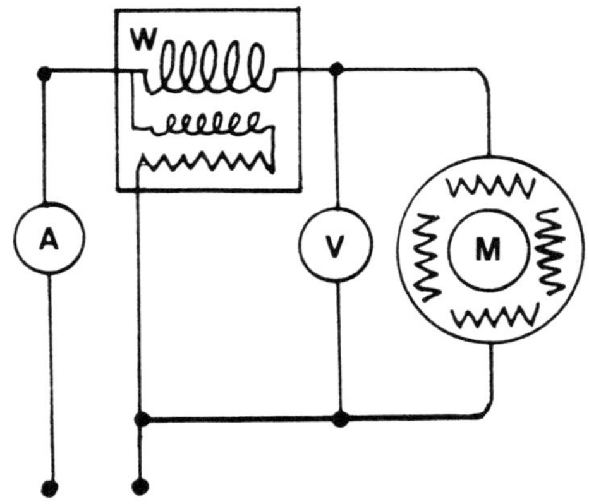

Figure 15-1: Connections for measuring the power factor of the induction motor (M). A is an ammeter, W is the wattmeter and V is the voltmeter.

THE CATHODE RAY TUBE

(2-15) The cathode ray tube is a useful device for measuring the conditions within an alternating circuit. When used for this purpose, it is usually called an oscilloscope or an oscillograph. A diagram of a cathode ray tube is shown in Figure 15-2.

The large end of the tube is coated with a phosphor on the inside surface of the glass. It is usually marked off with grid lines so that voltage relationships can be measured by eye. The narrow end of the tube has a heated element designed to permit electrons to boil off from its surface. Down the tube a short distance is a plate which carries a positive charge. The electrons are accelerated towards the positive plate. A tiny hole in the center of the plate lets a narrow stream of the electrons pass through. The narrow stream of electrons is the cathode ray.

Two horizontal plates with a small space between them are used for controlling the vertical motion of the cathode ray. Downstream a bit further, two vertical plates with a small space between them can control the sideways deflection of the cathode ray. By placing the plates in the circuit that you want to test, the shape of the actual wave can be made to appear

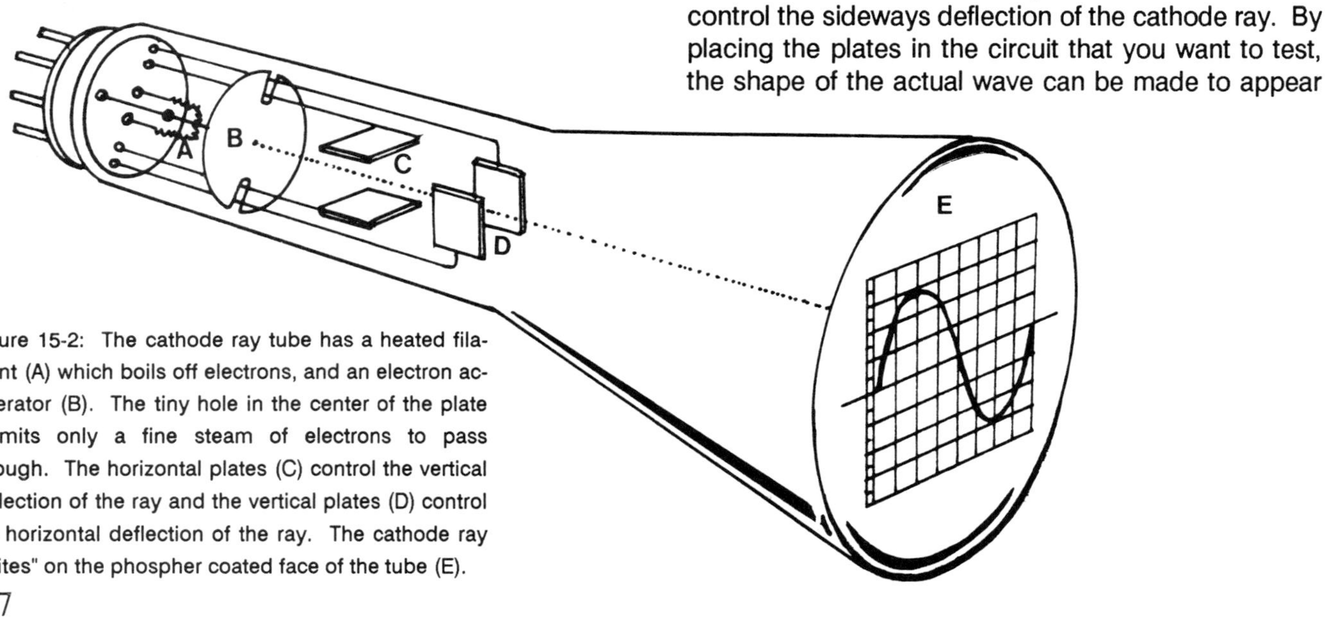

Figure 15-2: The cathode ray tube has a heated filament (A) which boils off electrons, and an electron accelerator (B). The tiny hole in the center of the plate permits only a fine steam of electrons to pass through. The horizontal plates (C) control the vertical deflection of the ray and the vertical plates (D) control the horizontal deflection of the ray. The cathode ray "writes" on the phospher coated face of the tube (E).

on the screen as a thin bright line.

To make the cathode ray trace out a shape on the screen, the vertical plates are gradually charged, and then the charge is suddenly released. This causes the dot on the screen to sweep rapidly from right to left and sweep slowly from left to right. During the slow sweep, the shape of the test wave is traced on the screen. The frequency of this action can be set on a dial. When set at 60 hertz, the shape of the wave in regular a-c power can be seen.

If the shape of the wave on the oscilloscope screen is not a perfect sin wave, there is a defective element in the circuit that is generating transient waves. By changing the main elements in the circuit temporarily, the defect can be located and corrected.

TESTLIGHTS

(3-15) Test lights are useful for determining if a circuit or element is live and operating. It is often important to know if a circuit is live before touching it, as a safety precaution. For standard a-c circuits a 10 watt or 15 watt lamp can be used. The leads can be three feet long, which will give six feet when stretched out. Probes or alligator clips are used on the ends of the leads. The type of small screw-base lamps used for Christmas tree lights are suitable.

Typical test lights are shown in Figure 15-3. For testing batteries, a small flashlight lamp can be used, and for testing circuits around your automobile, a socket and bulb from a headlight or dash instrument can be used. Test lights are also available at your automotive retailer and from electrical wholesale concerns.

SUBSTITUTION BOXES

(4-15) A substitution box has two lead wires with alligator clips or probes. On the box there are one or more dials which can be set at a substitute resistance to connect into a circuit while an element from the circuit is removed.

One substitution box is for resistances of any amount from zero to infinity. The other substitution box is for capacitance. It also has a wide range of choice so that a capacitive device which has been temporarily removed from the circuit can be replaced quickly and accurately. A typical substitution box is shown in Figure 15-4.

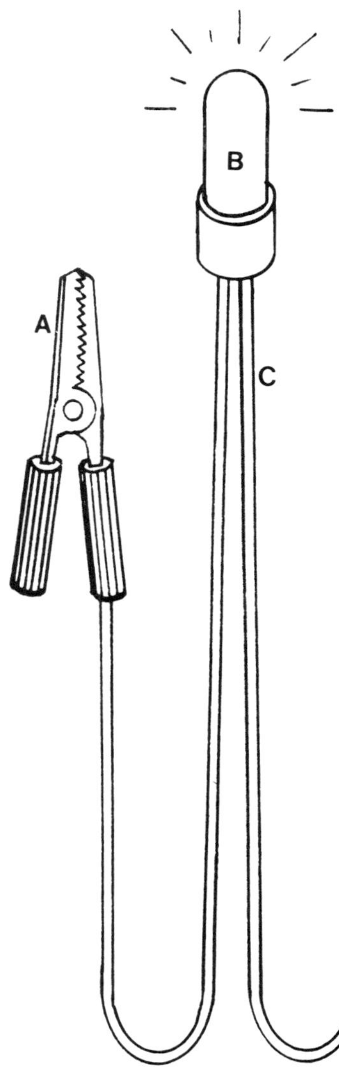

Figure 15-3: Test lights are inexpensive but are useful for testing circuits and batteries. In the drawing, (A) is an alligator clip, (B) is the bulb, (C) is the wire leads and (D) is a metal probe, usually chrome plated steel.

EXAMPLE:

What is the power factor when the voltage is 120 volts and the effective currrent is 4 amps. The current is lagging the voltage by 3.5 degrees.

ANSWER: Turn to the cosine chart on page 173 and write down the cos for 3.5 degrees (0.99813). Then the equation is:

$$\text{Power Factor} = \frac{E(120) \times I(4) \times \cos(0.99813)}{E(120) \times I(4)}$$

$$= 0.99813$$

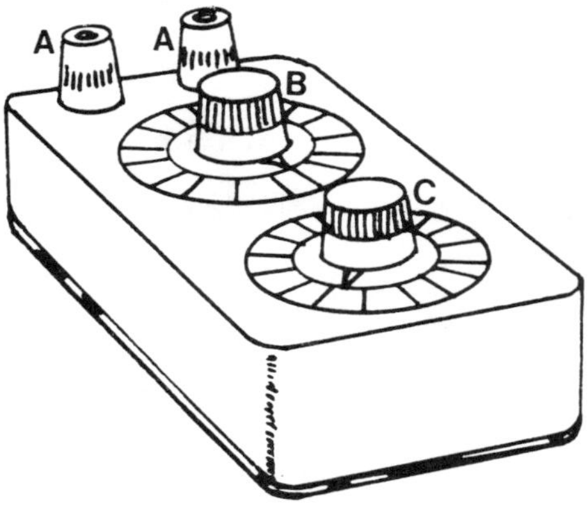

Figure 15-4: A substitution box has a pair of sockets (A-A) for inserting the the wire leads. There is a dial (B) for setting small to medium resistances, and a dial (C) for setting medium to large resistances. The numbers for the resistances or capacitances are engraved all around the dial. (The markings are not shown on the drawing).

EVENTS FROM THE PAST

The artists who rendered these drawings of locomotives and trains worked as illustrators for newspapers and the popular newsmagazines of the day. The realism and proportions of the locomotives in their drawings can be judged by comparing them with the real locomotive at upper left.

ELECTRICAL MACHINERY

THE SINGLE PHASE ELECTRIC MOTOR

Electric motors may look the same on the outside but many of them are manufactured to fit into special applications. Motors that have a wide use are built by the thousands. A popular motor for constant run jobs like furnace fans is the squirrel cage induction motor. Another motor that is popular for all-purpose small power requirements is the universal motor. It will turn an electric drill or start your car as well as hundreds of other jobs. For light torque the shaded-coil motor is often used. The eddy-disc motor is used occasionally, and for clocks, the miniature synchronous motor is used. All of the motors mentioned run on *single-phase power*. Large industrial motors are usually wired for *three-phase* power, discussed in a later paragraph.

THE UNIVERSAL MOTOR

(1-12) The universal motor is the workhorse of electric motors. It will run on a-c or d-c. It is sturdy and reliable. It is used just about everywhere when variable torque is needed. It will open your car windows, drive a truck hoist, run a vacuum, shave you or open a can of beans. No job is too big or too small.

How a universal motor works is shown in schematic form in Figure G-1. The rotor windings and the stator coils are all connected in series so that each coil gets the full *current* in the circuit. While only one rotor coil is shown in the diagram, an actual motor has many identical windings spaced around the rotor. The rotor is made up of laminations of electric steel (steel alloyed with cobalt). Slots around the circumference of the rotor permit the windings to be pressed into place so the windings are imbedded in the rotor. A typical rotor lamination for a universal motor is shown in Figure G-2.

The schematic in Figure G-1 shows the current being fed to the rotor coil through a split ring arrangement. The split in the ring is placed so that when the rotor conductor is energized, it is immersed in the magnetic flux of the stator poles. The stator poles then push the live rotor conductor sideways, as shown in the diagram of Figure G-3. The diagram shows the shape of the magnetic field in a universal motor.

When the rotor conductors enter the magnetic fields of the stator poles, they are energized one after another by the split ring arrangement. As the conductors enter the fields of the stator poles, they are forced sideways across the poles. As soon as their work is done, the split ring cuts off the current until they approach the magnetic poles again.

When d-c is used for a universal motor, as in an automobile, the magnetic action is continuous. When a-c is used, the polarity of all conductors and

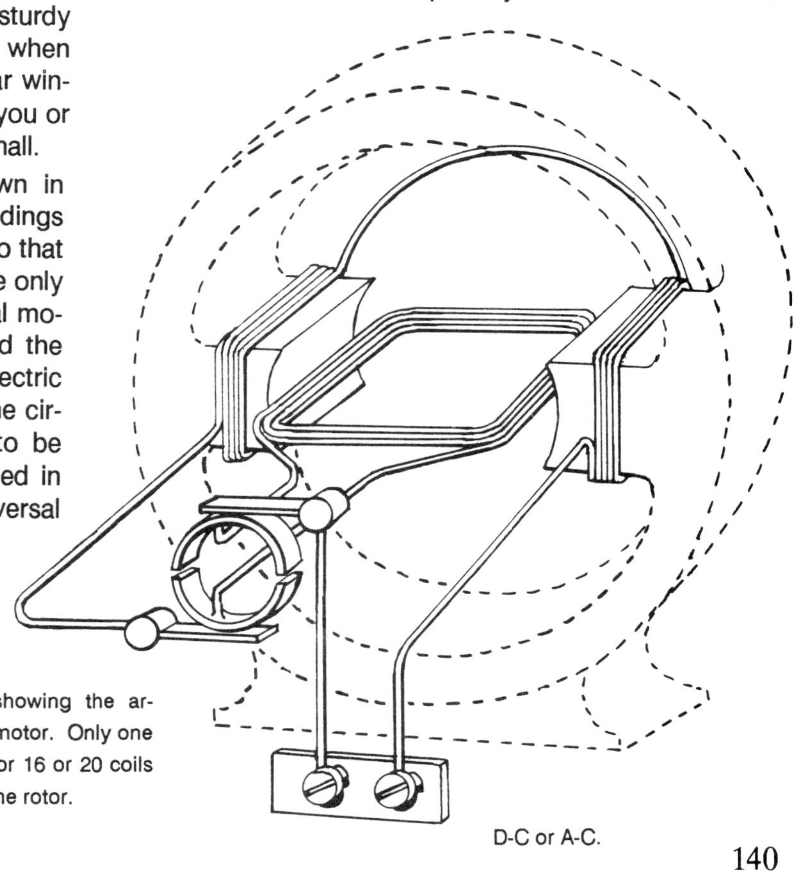

Figure G-1: Schematic diagram showing the arrangement of the coils in a universal motor. Only one rotor coil is shown. In an actual motor 16 or 20 coils would be spaced at intervals around the rotor.

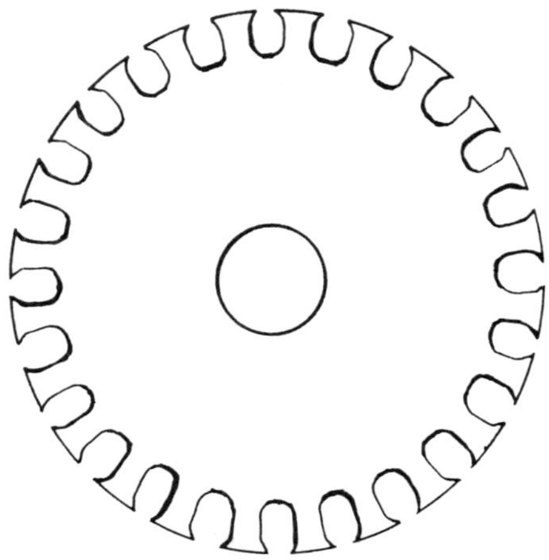

Figure G-2: Typical rotor lamination.

magnetic poles change 60 times per second. Since they all change at the same time, the motor operates efficiently and the rotor conductors are always pulled around in the same direction.

In a universal motor, the split ring has one split in it for each rotor coil. It is called a commutator. The commutator is drum shaped with many small wedges of brass alloy arranged in a circle. The metal blocks are insulated from each other. The ends of the wires from each rotor coil are soldered to the inside of the metal blocks. Each rotor coil is soldered to two of the brass alloy blocks which are exactly opposite each other, on opposite sides of the commutator.

Two tempered carbon blocks, called brushes, wipe against the commutator as it rotates. The electric current is transferred from the carbon blocks to the rotating brass wedges in the commutator. The blocks of tempered carbon are held firmly in spring loaded brackets. The carbon does not wear quickly as would be expected. It is extremely hard and wears for many years before it needs to be replaced. The brushes and the commutator are shown in the diagram of Figure G-4.

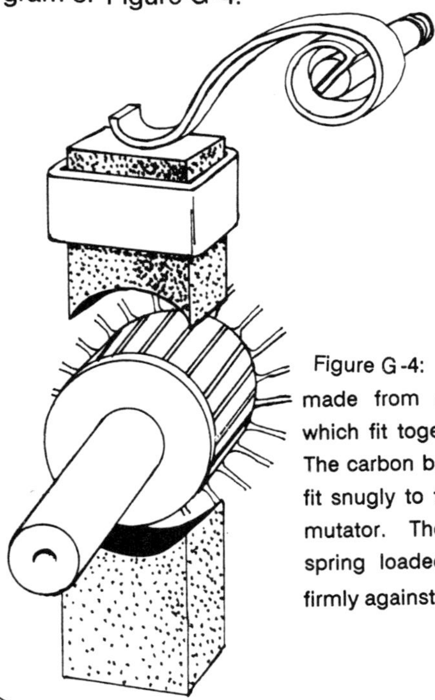

Figure G-4: The commutator is made from many brass wedges which fit together to form a circle. The carbon brushes are curved and fit snugly to the curve of the commutator. The carbon brushes are spring loaded so that they press firmly against the commutator.

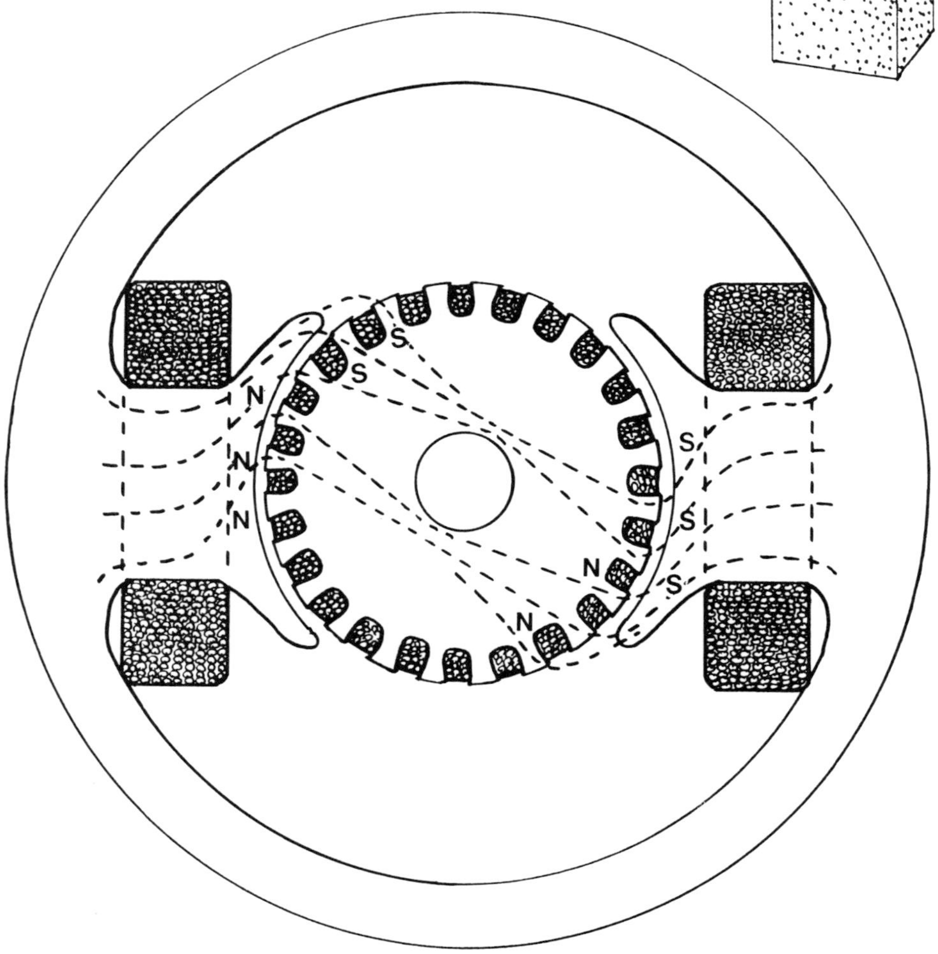

Figure G-3: The magnetic flux of the live conductors in the rotor blends with the magnetic field of the poles. The warped magnetic field pulls the conductors sideways as it responds to the attraction between unlike poles.

THE AMAZING MOTOR

The most unique feature about the universal motor is that theoretically it has no limiting RPM (revolutions per minute). As the motor runs faster it automatically controls the amount of amperage in the coils. More and more voltage can be added until the motor turns so fast that it begins to fly apart from centrifugal force. Of course there is no advantage to destroying a perfectly good motor. WARNING: Do not try this as an experiment. You could lose a finger or even an eye.

COUNTER EMF IN ELECTRIC MOTORS

(2-12) All electric motors have the same characteristic; they begin to act like generators as soon as they start up. Because of Lenz's law, every electro-magnetic effect generates an opposing effect. For example, the applied voltage in a motor has to overcome a counter voltage.

What happens is that the same magnetic field that drives the rotor conductors around also generates a back voltage in the same conductors. When the motor is just starting to turn, the counter emf is insignificant. But as the motor gains speed, the counter emf grows larger and stronger. The counter emf acts to place a limit on the speed (RPM) of the motor.

The counter emf or back voltage also reduces the amperage that is moving through the coils. The value of the current is reduced because the effective voltage is the difference between the applied voltage and the counter voltage.

THE SQUIRREL CAGE INDUCTION MOTOR

(3-12) The famous inventor Tesla originated the idea of making a magnetic field jump from one place to another. He demonstrated that if you place two coils near each other, and then energize them in quick succession (one after the other) the magnetic field will jump from the first coil to the second one. He also demonstrated a rotating magnetic field and used it to rotate a rudimentary induction motor.

Many inventors improved on Tesla's first motor design. Now the modern induction motor is about as simple and as efficient as can be. When in operation, the rotor is the only moving part.

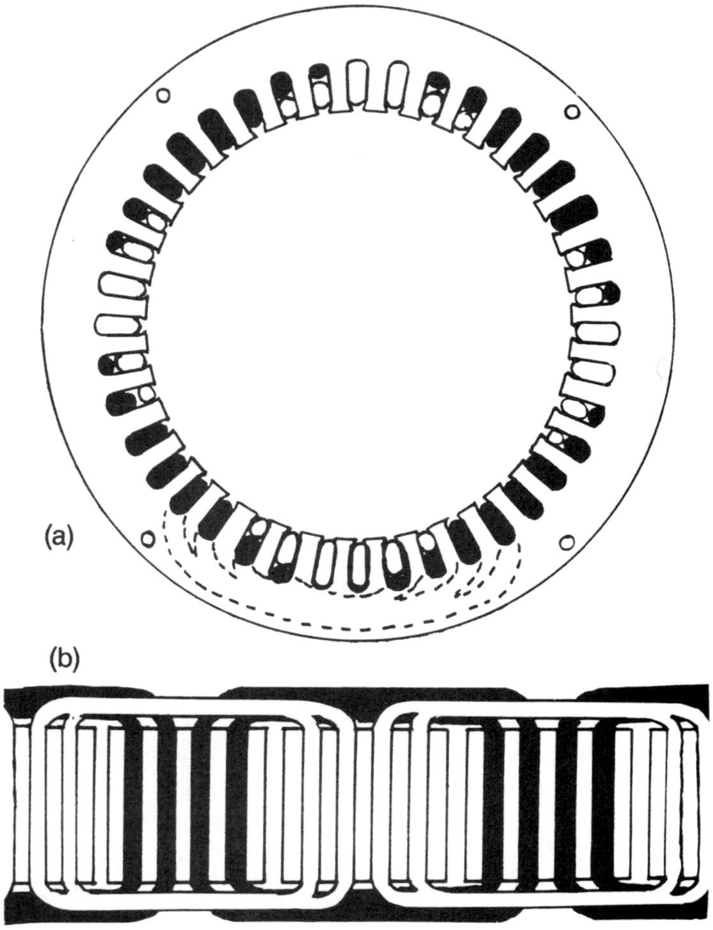

Figure 12-5: Two views of the coils in the stator of an induction motor. The running coils are shown in black and the starting coils are shown in white. View (a) is a sectional view. View (b) is a view from inside the stator.

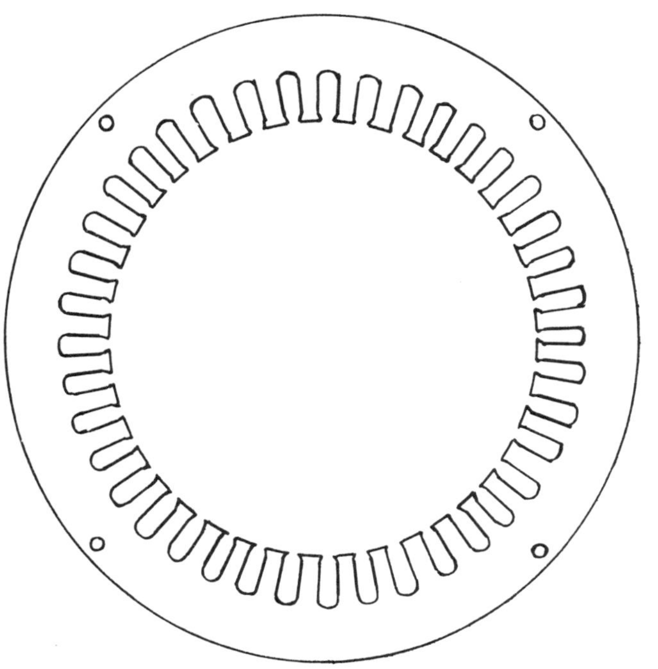

Figure 12-4: A stator lamination for an induction motor. The coils fit into the slots. The laminations are steel alloyed with cobalt.

When alternating current is connected through a wound coil, the magnetic flux in the coil also alternates. The polarity of the magnetic flux faces one direction for half a cycle and then collapses and reappears with the polarity facing the other direction for half a cycle. If a second coil is placed nearby as part of a separate loop of circuit, an alternating current and magnetic flux will be induced in the coil. The difference between the two coils is that in the induced current the cycle is 90 degrees or one quarter cycle late.

Tesla demonstrated that the magnetic field in the first coil would blend with and jump into the second coil. He succeeded in making the magnetic field jump through several coils, one after the other. That demonstration was the origin of the induction motor, because that is exactly how an induction motor makes a magnetic field rotate.

THE INDUCTION MOTOR

In an induction motor the stator is made of many thin laminations like the example in Figure G-4. The windings or coils are made square or rectangular so that they will fit into the slots of the stator as shown in Figure G-5, (a). In most single phase induction motors, four poles are used, two N and two S. Similar poles are opposite each other on each side of the stator.

Half way between the first set of coils, several other coils are inserted into the slots. They exactly overlap the first set of coils. In Figure G-5 (b), the first set of coils is shown in black. The second overlapping set is shown in white. Where the coils fit into the slots, they are separated into small bundles so that the coil conductors can be distributed into the slots. In Figure G-5 (a) the dotted outlines at the bottom of the drawing show how the coils for one pole are distributed into several slots.

THE SQUIRREL CAGE

The rotor for the induction motor is shown in Figure G-6. It is like a ladder curved into a circle. Perhaps the exercise wheel sometimes seen in the cages of a pet squirrel or pet hamster will remind you of the rotor in the induction motor. If it does, then you know why it is often called the squirrel cage motor. It is a misnomer, but it is too late to change it.

The rotor shown in Figure G-6 is from a small shaded coil motor. The actual rotor in an induction motor looks like the drawing in Figure G-7. The solid bars are very close together and the sides of the rotor are fitted with cast-on impellers which blow air against the coils to keep them cool. In the actual motor, you cannot see the center bars because they are imbedded inside the rotor core, which is made up of many steel laminations like the one shown in Figure G--8. When the rotor is manufactured, the "cage" and impellers are die cast right onto the laminated core, so that it is actually one piece when it is completed.

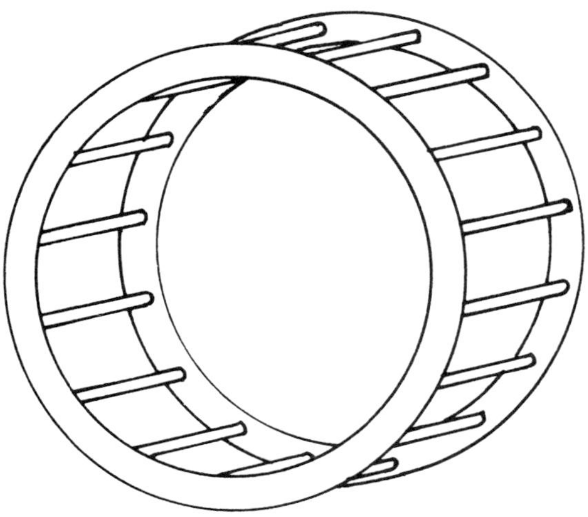

Figure G--6: The shape of the conductors in the rotor of an induction motor might remind you of a hamsters or pet squirrels exercise wheel. That is how the misnomer "squirrel cage motor" originated.

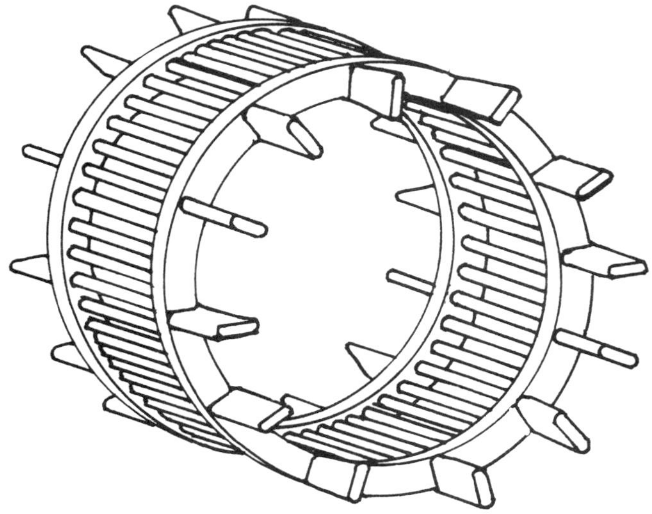

Figure G--7: This drawing illustrates what the actual conductors in the rotor of an induction motor look like.

HOW IT WORKS

The induction motor operates as follows. The black coils in Figure G--5(a) and (b) are heavier than the white coils. The black coils are the working coils. The white coils are only used momentarily to start the motor. The black coils are energized first and the magnetic flux rises through them. Then the current is induced in the white coils a quarter of a cycle later. The magnetic fields jump into the poles formed by the white coils. At that time, the black coils have changed polarity so the magnetic field jumps into the poles of the black coils. Again the white coils are energized, the magnetic field jumps again, and so on. The result is that four magnetic fields rotate around the stator at the rate of 1800 revolutions per minute. The speed of revolution is dictated by the 60 hertz alternations in the current.

The illustrations in Figure G--9 show that when unlike magnets are brought close together, the magnetic fields form circles or loops of magnetism, like elastic bands.

In Figure G-10 the four magnetic fields are shown as white loops like elastic bands. As indicated in the diagram, the loops of magnetism drag through the ladder bars of the rotor. That action is the same

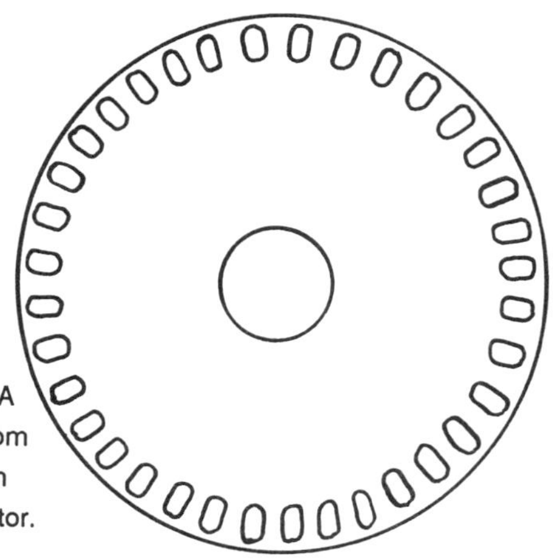

Figure G--8: A lamination from the rotor of an induction motor.

as a magnetic field cutting through conductors, so currents are generated in the rotor bars. That creates poles in the rotor so that it quickly starts to move, following the rotating magnetic fields.

Attached to one side of the rotor there is a centrifugal switch. As soon as the rotor starts to turn rapidly, the centrifugal switch disconnects the white coils from the induction circuit so that they are not functioning. They are not needed until the motor has to be started again. The starting coils are wired in a separate circuit that is in parallel with the motor coils. The current in the starting coils should be 90 degrees (or nearly) later than the main coils. This 90 degree lag is usually done by induction. However, there are several ways of doing it. In expensive motors it is common to use a capacitor in the starting circuit which develops a large current 90 degrees out-of-phase. In some motors a pure resistance is used in the starting windings and the resistance current then leads the inductive currents in the main coils by 90 degrees.

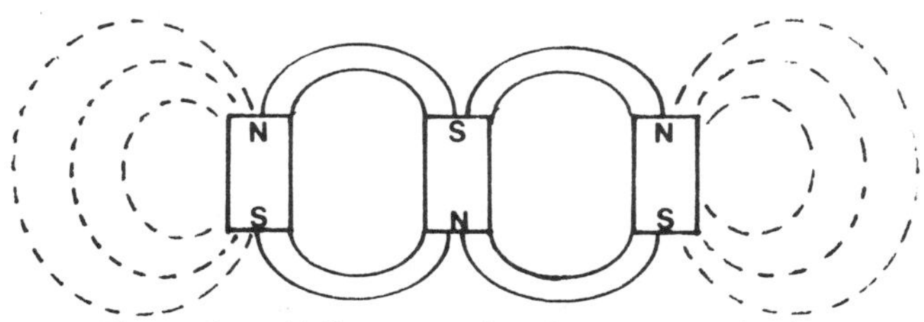

Figure G-9: Unlike magnetic poles attract each other and the lines of magnetism are concentrated into loops or bands when unlike poles are brought close together.

Figure G-10: Although the voltage in pairs of coils is opposed, the currents flow in the same direction where the coils meet. The concentration of current forms tight loops or bands of magnetism between the conductors. These loops rotate around the stator at 1800 RPM.

RPM IN INDUCTION MOTORS

(4-12) The RPM of a universal motor is limited only by the amount of voltage that is applied. On the other hand, the RPM of induction motors is locked to the 60 hertz alternations of the current. At 60 cycles per second, the number of cycles in a minute is 3600. When four poles are used in the rotor of an induction motor, the RPM of the rotating field is 1800.

The action that applies torque to the rotor is the magnetic field cutting through the conductors of the rotor. Therefore, the RPM of the rotor must be less than 1800 for the above criteria to be effective. The top speed of a four pole induction motor when idling (not working) is 1750 to 1775 RPM. When it is working, the RPM may be from 1600 to 1700, depending on the job that is being done.

The difference in speed between the rotating magnetic field and the rotor is called the "slip" of the motor. It has to have some slip in order to perform work.

For some special jobs (like vacuum cleaning) a very high RPM is desirable. For high speeds only two poles are used instead of four. The RPM of the rotating magnetic field is then 3600 and the top RPM of the motor is about 3550 revolutions per minute.

When a load is placed on a motor, the rotor turns slower. This makes the magnetic fields cut through the rotor conductors faster. The higher speed generates more voltage and current, so there is a greater torque on the rotor. The torque of the rotor automatically adjusts itself to the job at hand. If the load is too great, the motor will stall. It will not start again until the fuse has been replaced or the circuit breaker has been reset.

REVERSING MOTOR SWITCH

(5-12) There are many motor applications where it is an advantage if the motor can be switched into reverse. A screw cutting lathe is an example; it is usually equipped with a reversing motor and switch. If the motor is of the universal type, it can be put into reverse by changing the direction of the current in the stator windings. If the motor is of the induction type, it is put into reverse by changing the direction of the current in the starter windings.

When a motor is switched off, the residue magnetic flux in the windings collapses through the coil, generating a sudden pulse of reverse voltage. This will cause a poorly made switch to arc and carbonize. For this reason it a common practice to switch a motor off before reversing it.

The switch that is used for reversing a motor is shown in Figure G-11. It is called a two pole, double throw switch. It has six terminals. The connections that the terminals make when the switch is used are shown in the diagram, Figure G-12. The method of wiring a reversing switch is called a cross-over connection, shown in Figure G-13. On the bottom of the switch, wires are soldered on connecting the diagonal terminals. The two wires are insulated from each other where they cross.

To reverse a motor, the switch is connected to the terminals of the coils as shown. If the motor is of the universal type, the switch is wired into the circuit for stator coils. When the switch is used for reversing a motor, it is first switched to the "off" position for a moment before it is switched over to reverse.

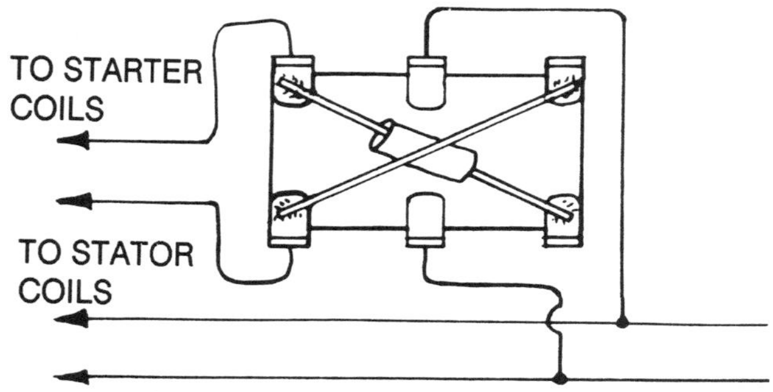

Figure G-12 (a): The cross-over connection on the bottom of the switch. This switch diagram is for reversing an induction motor.

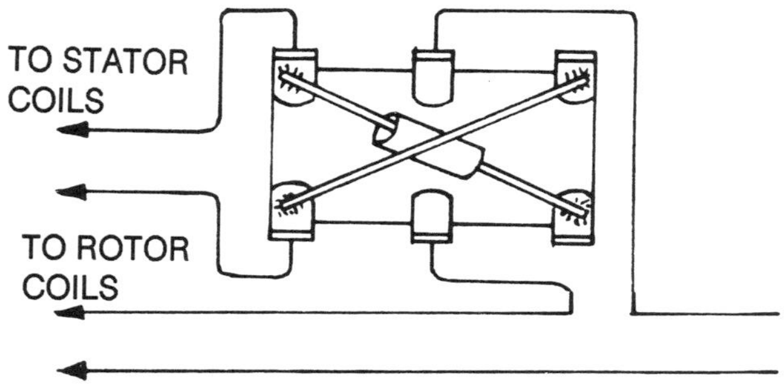

Figure G-12 (b): This switch diagram shows the connections to make for reversing a universal motor.

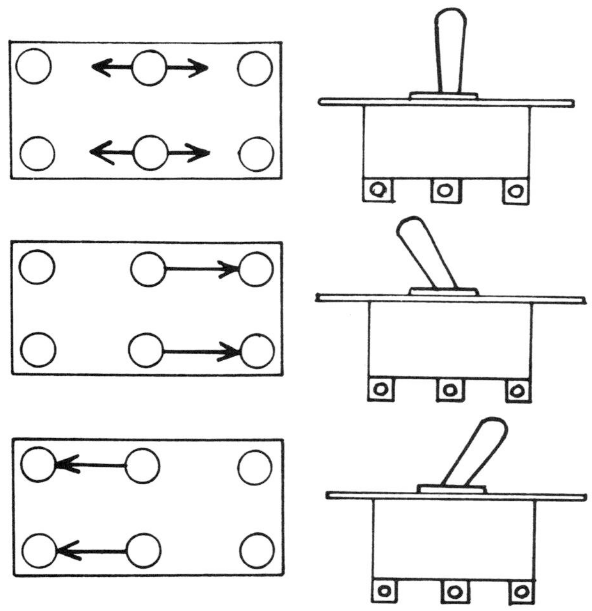

Figure G-11: A two-pole, double throw switch. The diagrams show the three positions of the switch.

THE SHADED COIL MOTOR

(6-12) The shaded coil motor operates on a-c power only. It is usually manufactured with a built-in gear box so that the small amount of torque that is developed can be stepped up to do a useful job.

Two views of the motor are shown in Figure G-13. There is only one wound coil (A) and the shaded coils (B). The shaded coils are made of four pieces of copper wire imbedded in the stator laminations (C). The rotor (D) is made like the rotor of an induction motor. A typical shaded coil rotor is shown in Figure G-6.

When working, the shaded coil motor operates very much like a transformer. It is sometimes called a transformer motor. The conductors in the rotor act like the secondary winding in a transformer. The energizing coil is off to one side of the motor. When the magnetic flux rises through the laminations, waves of flux flow through the rotor like the spreading waves on a pond when you throw a stone into the water.

Each conductor that the flux flows through develops a voltage and current. The first conductors that are affected are the shaded coils nearest to the main coil. The current in the shaded coils make a reverse magnetic field and that prevents the main waves of flux from flowing into the rotor on that side. In other words, one side of the rotor is "shaded" from the flux by the shaded coils.

The magnetic flux flows through one side of the rotor and drives it around similar to an induction motor. The shading effect of the shaded coils is ef-

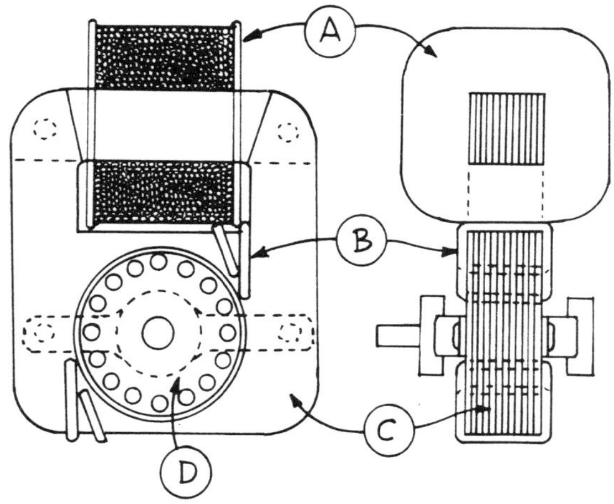

Figure 12-13: Two views of a shaded-coil motor. The shaded coils (B) are rings of copper wire. They "grab" the magnetic flux and shade one side of the induction rotor, creating a bias that makes the rotor turn.

fective for a split second and the flux pushes through all of the laminations. Now the magnetic flux begins to collapse back as the alternation completes one cycle. This time the shaded coils on the lower side hold back the magnetism and again the rotor is forced around. This rotor turning action happens every one-half cycle. That means the rotor is given a small push 120 times per second. Anything repeating that fast is like a steady push so the rotor keeps spinning as long as it is turned on.

THE EDDY CURRENT MOTOR

(7-12) The eddy current motor is used mainly in electric meters, but it is sometimes used in small mechanical devices like store window turn-tables and small advertising displays. A typical eddy current motor is shown in Figure G-14. It has a disc rotor (A) which is usually copper or aluminum, and a driving coil (B) with a laminated steel core (C). The coil is mounted on a bracket with a thumb screw so that it can be moved to alter the speed of the motor. The

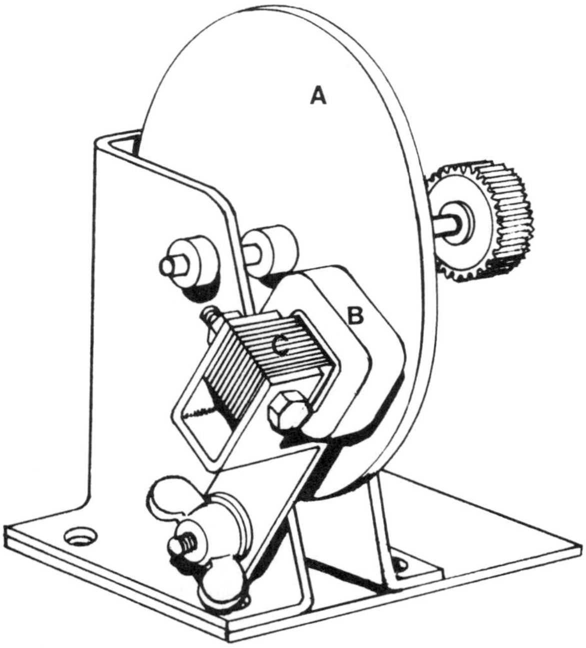

Figure G--14: The eddy current motor is the lowest form of electric motor. The example is used for animating a window display

end of the core is not quite touching the disc.

It operates by generating an alternating magnetic flux through one side of the copper disc. This developes a circular eddy current around the location of the core. The magnetic flux from the eddy current then opposes the magnetic flux of the coil (by Lenz's law). That places two north or two south poles next to each other, and they promptly move away from each other, driving the disc around as they repel each other. This action happens 120 times per second, so it is like a steady push on one side of the rotor.

THE SYNCHRONOUS CLOCK MOTOR

(8-12) Small synchronous motors are made mainly for operating electric clocks. They are also used in other mechanisms where a very small amount of accurate power is needed. There are at least three kinds that are popular, but all of them operate on the principle of lock-step action.

In one motor the rotor has twelve (or more) very tiny magnets spaced around the diameter at intervals. One of the tiny magnets is in line with a little iron keeper which keeps the rotor from turning. A coil is placed so that the magnetic flux from the coil counters the effect of the one tiny magnet for a split second. That releases the rotor and it begins to turn like an induction motor but is immediately locked again when the flux collapses. This action happens 60 times a second so it is almost like steady motion.

ALTERNATORS AND GENERATORS

(9-12) Inventors and engineers have a goal or ambition to reduce every device to the fewest moving parts. The induction motor is a classic example: it only has one moving part. The alternator is another machine that meets the criteria of an ideal machine, because it has only one moving part as well.

The plan for most small alternators is to place a set of permanent magnetic poles in the rotor. The rotor is then set spinning and the magnetic poles move past a series of stationary coils. The moving magnets generate a voltage in the coils and the output voltage rises and falls with each alternation as the magnetic fields energize each coil in sequence. The coils are imbedded in laminations of electric steel. A plan for an alternator is shown in FIgure G-15.

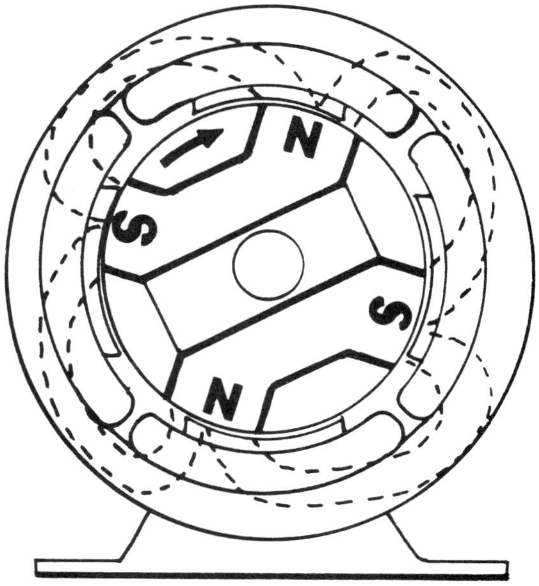

Figure G--15: The only difference between an electric motor and an alternator is that the alternator needs permanent magnets or an exterior source of current. The dotted lines show where the loops of magnetism spin through the stator coils.

INDUSTRIAL ALTERNATOR

(10-12) The larger industrial alternator is usually wired for three-phase output. Industrial alternators often use a commutator as well, like a universal electric motor. The commutator makes it possible to control the strength of the magnetic poles in the rotor. The output from an alternator cannot have variations of phase or power fluctuations. The controlled rotor poles permit a steady, uniform output.

Like a three-phase motor, the three-phase alternator can be wired with the coils connected in a "Y" or in a "Delta". As the magnetic poles in the rotor cut across the conductors of the coils, the output voltage from the three sets of coils are 120 degrees out of phase with each other.

The voltage between any two of the three output conductors is $\sqrt{3}$ multiplied by the voltage in one of the wires. A diagram of an industrial alternator is shown in Figure G-16. The diagram showing the connections is in Figure G-17.

The rotor of an alternator is usually driven by a steam turbine or by a diesel engine. The rotor must turn at 1800 RPM to generate 60 hertz output voltage. That means the driving engine has to be carefully controlled. As the rotor spins the flux from the rotor poles cuts through the coil conductors in the stator. The flux from the stator coils then induces a back voltage or counter emf in the rotor windings. This has the effect of making the rotor more difficult to turn.

Every increase in the torque necessary to spin the rotor is matched by an increase in the power output. The controls permit the alternator output to vary over a narrow control range without changing the RPM of the rotor. The amount of rotational energy converted to electrical power is measured by the wattage in the output. The power developed by a three-phase alternator is

$$\text{Power} = \sqrt{3}\ E\,I\cos\theta \quad (47)$$

In this equation, E is the emf measured between any two of the three-phase wires, and I is the line current. The cosine is applied to any small angle of lag that may exist between the output voltage and the output current.

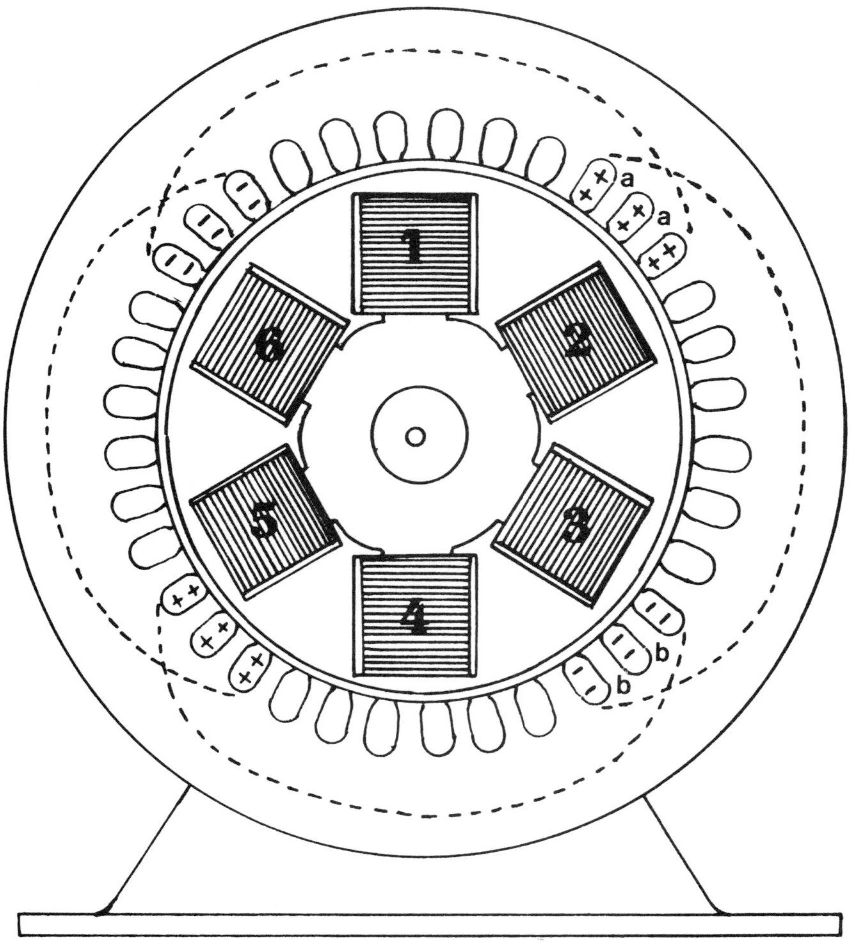

Figure G-16: This alternator has three sets of windings, one of which is shown in section, marked (a) where the positive conductors enter and marked (b) where the negative conductors come out. As shown by the positive and negative signs, six bundles of conductors fill up three slots. The rotor has 6 imbedded electro-magnets. These are energized by an exterior source of current, usually from an exciter generator using permanent magnets.

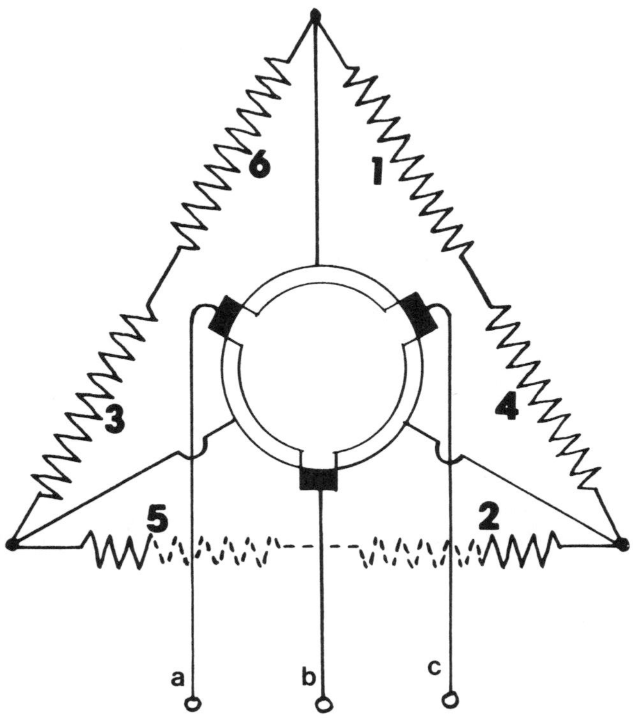

Figure G-17: The rotor of the alternator (Figure G-16) has a commutator with three segments as shown. The three-phase input from the exciter enters through the conductors marked (a), (b) and (c). The connections through the coils are made so that coils number 3 and 6 are energized in series. Similarly, as the commutator turns, the coils 1 and 4 are energized together, and the coils 2 and 5 are energized together.

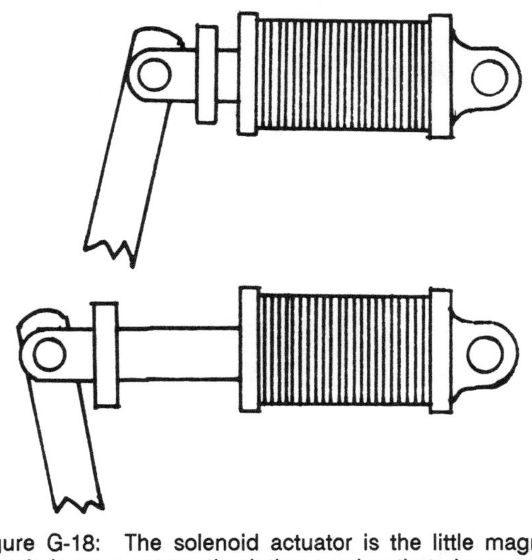

Figure G-18: The solenoid actuator is the little magnetic muscle in your automatic clothes washer that always opens and closes the proper valves at the exact moment. The sliding core of the solenoid is spring loaded so that it will return to its original position after it has completed its work. It is then ready to function again for thousands of operations without a failure.

As in all electrical devices, the alternator is highly efficient. It is basically a method for converting heat energy or rotational energy into electricity and power. The following list of power losses also applies to motors and to other electro-mechanical devices:

(1) Friction in the bearings absorbs some energy usually in the form of wasted heat energy.

(2) Air resistance around the spinning rotor also converts some energy to wasted heat energy.

(3) Core losses in the iron core due to electrical heating and hysteresis. Hysteresis is a word that means the natural drag of magnetism against being changed in pollarity.

(4) The loss of heat energy in all of the conductors due to resistance.

THE EXCITER

For efficient generation of three-phase power, it is essential that the magnets used are saturated with magnetism. While successful alternators are made with permanent magnets, a much stronger magnetic field can be obtained by suitable coils and laminated cores.

(11-12) The exciter is a small and compact generator adapted for delivering d-c current at a suitable voltage. The d-c voltage is then fed through slip rings to the coils in the rotor of the large alternator. The d-c current is used so that the polarity of the rotor magnets can remain unchanged like permanent magnets.

The exciter can be arranged to be mounted on an extension of the alternator shaft. Usually it is a separate motor or engine. Sometimes a bank of batteries is used for the initial start up and to energize the rotor magnets. Once started, the alternator then supplies the power for any accessories that need power.

THE SOLENOID ACTUATOR

(12-12) The solenoid actuator is like a little muscle that is available for instant action. It will perform any little push or pull job like closing a switch or opening a valve.

The solenoid is a coil with a movable core. The core is usually laminated. Both the core and the coil have a hinge or eye for connecting them up to a machine. There are greased grooves in the coil and the core part can slide back and forth easily. Often the actuator is spring loaded so that after the work is done, the core will return to the closed position. There is a diagram showing the solenoid actuator open and closed in Figure G-18.

EVENTS FROM THE PAST

Our unit of power was named horsepower by James Watt and other early inventors. With a glance over these antique illustrations, it is easy to see that it was inevitable. The power and speed of horses was the only thing that could be compared to the power of the steam engine.

POWER SUPPLIES

SINGLE PHASE POWER

The power that is supplied to a residential block is called single phase power. The power that is supplied to a commercial or industrial region is called three phase or *polyphase* power. The single phase power is adapted to all the requirements of a home. Polyphase power is adapted to operate any kind of industrial or commercial device or lighting.

RESIDENTIAL POWER

(1-13) The power that is delivered into your home comes in on three wires. There are two live wires and a neutral or rerturn wire. Each of the two live wires carries single phase alternating current, but the difference in the phases between the two wires is 180 degrees. The voltage in each of the two live wires is 120 volts (115 volts in some localities).

When the power first enters your home it goes to the electrical panel for control and distribution. A diagram of a typical electrical panel is shown in Figure P--1. The neutral wire is the one in the center of the box. The two live wires go down each side. The layout of the wires may be different in boxes by different manufacturers, but the general idea is the same.

Showing through the front lid of the box are the *circuit breakers,* usually in two vertical rows. In the diagram, some of the circuit breakers are marked 15 amps and 20 amps. The circuits that operate through these circuit breakers go to lighting circuits and wall outlets, as indicated. The 20 amp breakers may go to the outlets for fridges, freezer, washer, furnace motor, etc. For appliances that need greater voltage, like the clothes dryer and the electric stove, the two live wires are connected through the appli-

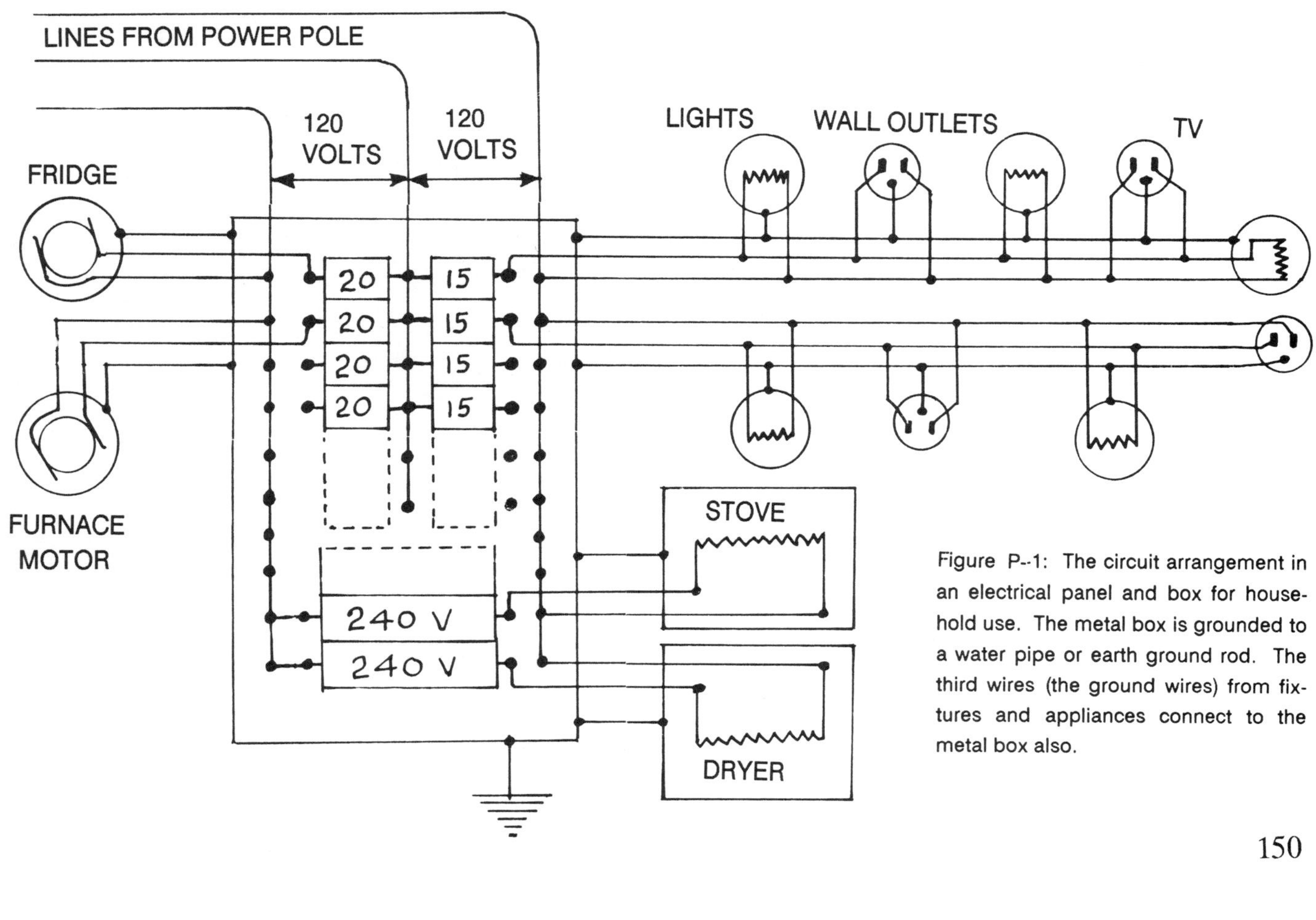

Figure P--1: The circuit arrangement in an electrical panel and box for household use. The metal box is grounded to a water pipe or earth ground rod. The third wires (the ground wires) from fixtures and appliances connect to the metal box also.

ance and through a special 240 volt circuit breaker, as shown.

The metal electrical panel box is grounded to a water pipe or to an earth ground by a heavy wire or cable. Then all of the lights, appliances, outlets, etc., have a ground wire that connects to the metal box, as shown. Don't confuse the ground wire with the neutral wire coming into the panel box. They are both a "third" wire, but the jobs they do are entirely different. The ground wire assures that any short or accidental crossing of live wires in any appliance will quickly run into the earth before it can cause a fire or other damage. The ground wire itself is not a live wire. The neutral wire is a live wire. As you can see by the electrical panel diagram, the neutral wire forms a part of the live circuits throughout the house.

CONVERTING SINGLE PHASE TO TWO PHASE

(2-13) Two phase current is sometimes needed in the home for an appliance or motor. Some brands of garage door openers use two phase current. Single phase current is converted to two-phase current by placing a large capacitor unit or a special transformer in a parallel circuit. The second phase is 90 degrees out of phase with the house current. The connections for a two phase garage door motor (and some furnace motors) are shown in Figure P-2.

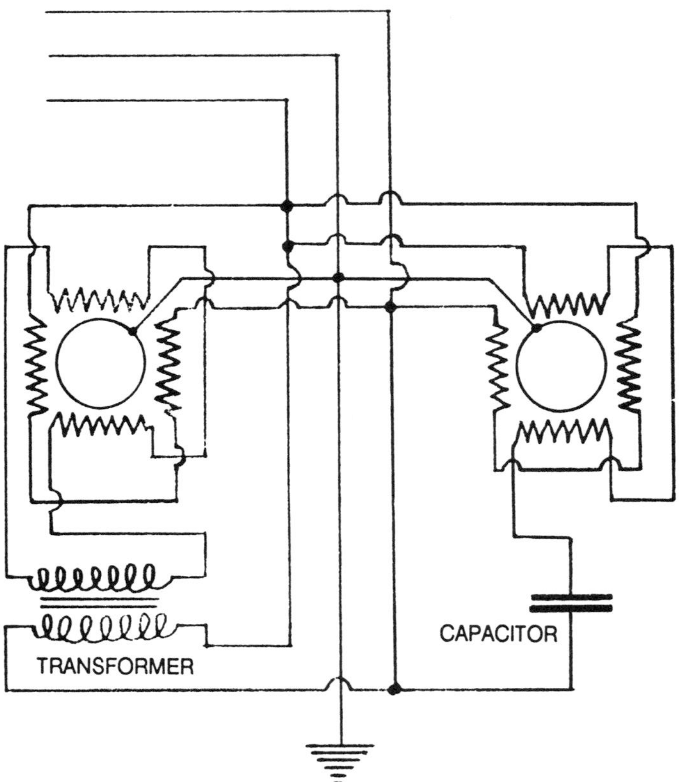

Figure P--2: A transformer or a capacitor unit can be used to convert single-phase current to two-phase. Both are shown in the diagram. They are connected in parallel with the motor coils, as indicated.

Two-phase electric motors may have four magnetic poles as shown in the diagram, or the larger motors may have eight magnetic poles. Motors with four coils have an RPM of 3550 and motors with eight coils around the rotor have an RPM of 1775.

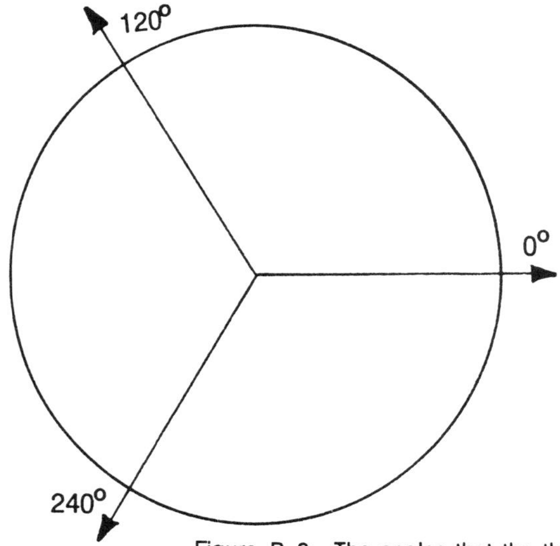

Figure P--3: The angles that the three vectors make in three-phase power. The vectors are 120 degrees apart.

THREE PHASE POWER

(3-13) Industrial plants and factories and some commercial areas are supplied with three-phase current. The 60 hertz frequency is standard. Each phase of the current is 120 degrees different. The positions of the three vectors on the circular-angle diagram are shown in Figure P--3. The vectors are equally spaced.

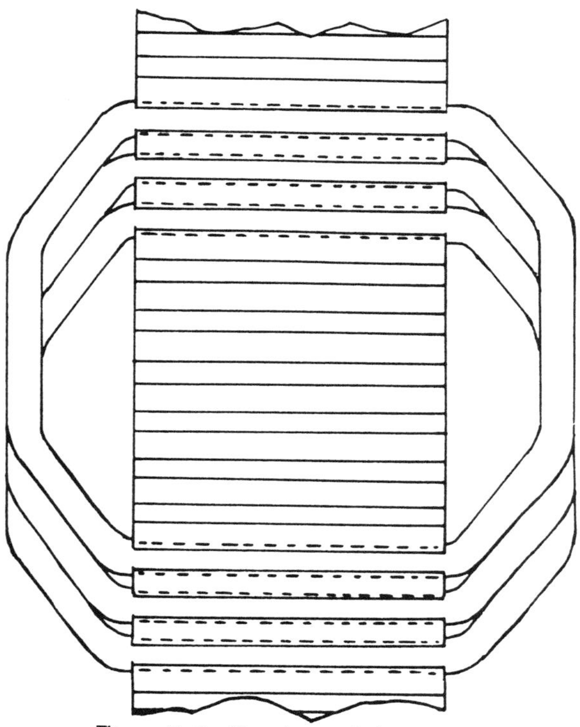

Figure P--5: The shape of the nesting coils when seen from inside the rotor.

A three phase motor may have six or twelve windings. The stator for a large industrial motor has 36 slots. There is a separate coil in every slot. The coils are shaped in a jig during manufacture so that they will nest together, one after another. The diagram in Figure P--4 shows a 36 slot stator with some of the coils in place. Figure P--5 shows the shape of the coils when seen from inside the rotor.

Three-phase current is brought into a factory building with four wires. Three of the wires carry the the three phases of the current. The fourth wire is a neutral or return wire for 120 volt lighting circuits and outlets.

When the three-phase system enters the building, the conductors go directly to an electrical panel and box. The diagram for the layout of the circuit breakers and connections in the box is shown in Figure P-6.

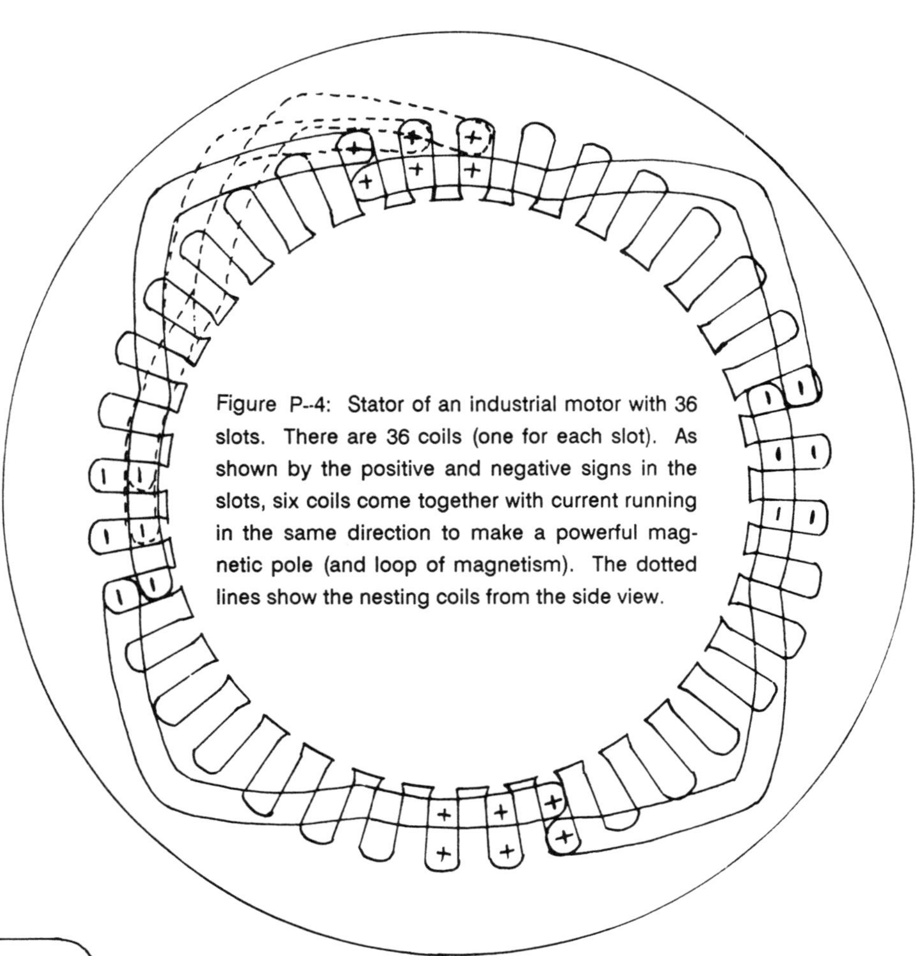

Figure P--4: Stator of an industrial motor with 36 slots. There are 36 coils (one for each slot). As shown by the positive and negative signs in the slots, six coils come together with current running in the same direction to make a powerful magnetic pole (and loop of magnetism). The dotted lines show the nesting coils from the side view.

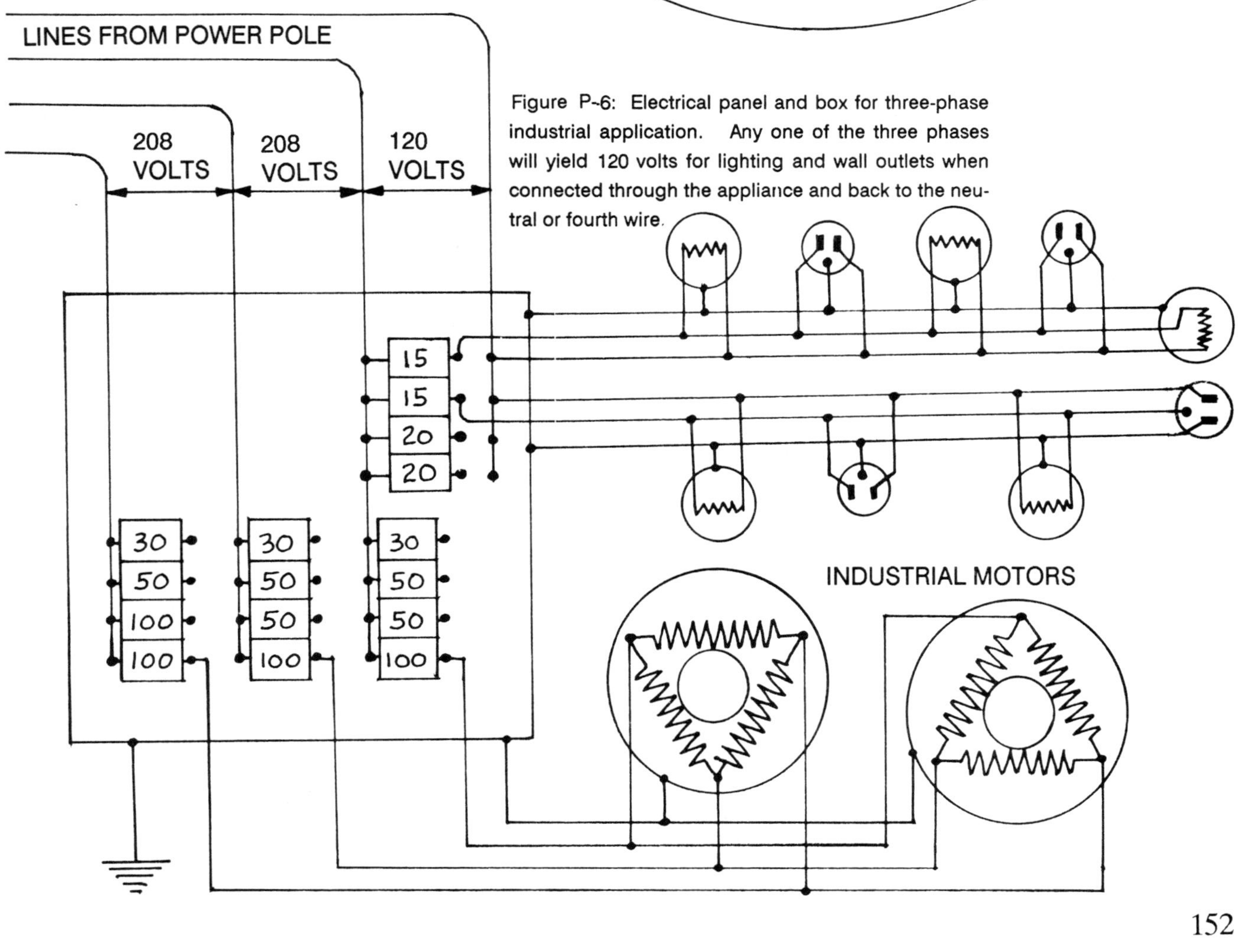

Figure P-6: Electrical panel and box for three-phase industrial application. Any one of the three phases will yield 120 volts for lighting and wall outlets when connected through the appliance and back to the neutral or fourth wire.

Each of the three phases is 120 volts. However, when two of the phases are connected through a circuit, the maximum voltage is not 240, but rather 208. The difference is due to the phase relationship. This is shown in the diagram, Figure P--7. However, any one of the three phases can be connected through a circuit to the neutral wire and produce 120 volts as shown in Figure P-6.

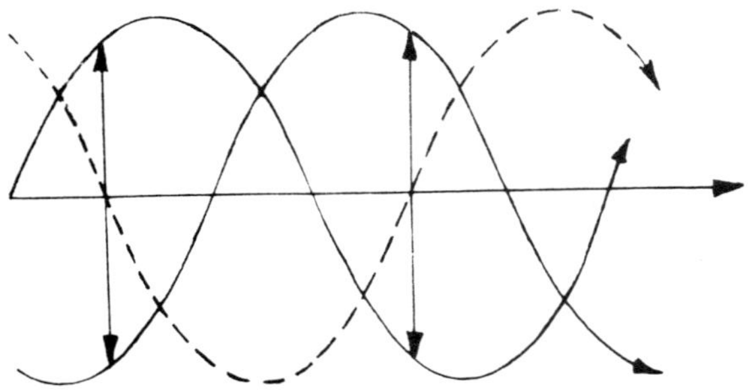

Figure P--7: The maximum voltage between the three phases occurs between two of the phases when the third phase is at zero voltage. The vertical arrows in the diagram show the locations of maximum voltage.

THREE PHASE CONNECTIONS

(4-13) The connections for three-phase motors and other three phase electrical devices can be connected in the form of a "Y" or in the form of a triangle. The two methods are called respectively the "Y" connection and the "Delta" connection (a triangle is the Greek letter Delta). The connections are shown in the diagram of Figure P-8.

Some of the large induction motors have internal slip rings, one for each phase. Connected to these rings are external resistances in sets of three, which can be automatically controlled with variable adjustment mechanisms. This permits a load to be under constant control during start up and during running. The load can be brought up to speed slowly. When the operating RPM of all of the machinery has been attained, the exterior control can be switched off until needed again. A diagram showing the slip-ring arrangement and the variable resistances is shown in Figure P-9.

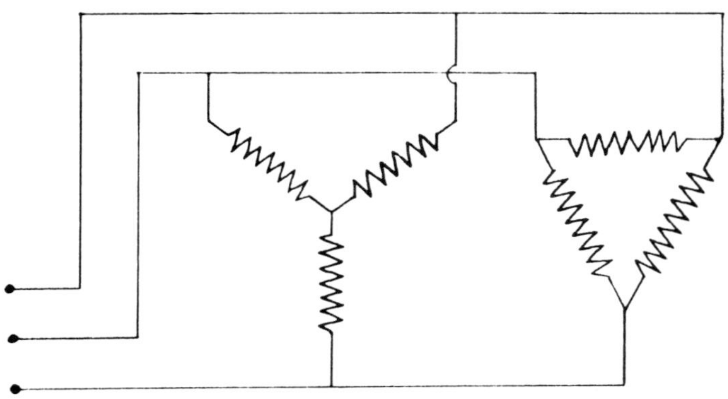

Figure P-8: Three-phase power can be connected in a "delta" (triangle) or in a "Y". Both methods are equally effective for operating three-phase motors or other three-phase devices.

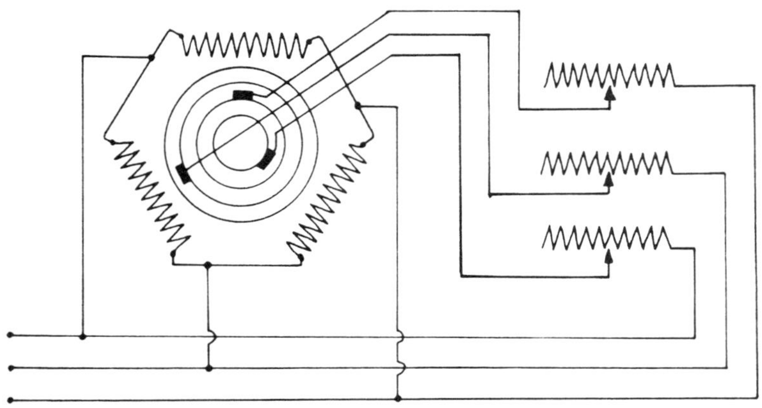

Figure P--9: Induction motors are primarily "constant speed" motors. They lack the high torque of the universal motor at low RPM. To give industrial induction motors more control and more of the characteristics of a universal motor, slip rings are used on the rotors to control the rotor poles. External resistances are adapted for variable control as indicated in the diagram.

EMERGENCY AND REMOTE POWER SUPPLIES

There will always be a need for emergency power for hospitals. A surgeon cannot lay aside his scalpel in the middle of a delicate and critical operation because the power lines have been blown down or disrupted by lightning. The emergency power system is designed to cut-in within seconds of a shutdown and replace the regular power supply with exactly the same quality of power.

The initial flow of current from an emergency supply is delivered from an array of batteries. The direct current from the batteries goes through a converter that changes it to alternating current of the same frequency and voltage as the regular supply. As soon as the control system in the emergency unit senses a drain on the batteries, a switch closes that starts up a gasoline or diesel powered generator or alternator. In some localities, arrays of solar cells are used to keep batteries fully charged up.

The supply of emergency power and remote power is a field that is open to new ideas and inventions. One example is the portable generator that can be carried about like a suit case or back-pack. After you set up your camp and it starts to get dark, it is simply a matter of starting up the gasoline-powered unit. The small ones deliver 1200 watts at 120 volts with a 60 hertz alternator. Larger units provide two 1500 watt circuits, which is enough to light up the average small home. These units make it convenient to take a plug-in radio along on your camping and fishing trips. A small plug-in television set is also nice for those who don't like to get too far away from civilization.

The mechanical conversion of d-c power to a-c power in power supplies is accomplished with an electric vibrator in the small units. This method is used for lighting up a neon sign when only battery power is available. The vibrator causes the d-c current to be delivered in a series of pulses with 60 hertz frequency. Suitable transformers then convert the train of pulses to a regular sine wave current alternation.

The large units, as used in emergency power supplies, comprise a d-c electric motor driven by the batteries. The d-c electric motor then drives an alternator on the same shaft. A synchronous control device governs the frequency of the alternations. The control device is a timer, like a quartz crystal, that has accurate timed vibrations.

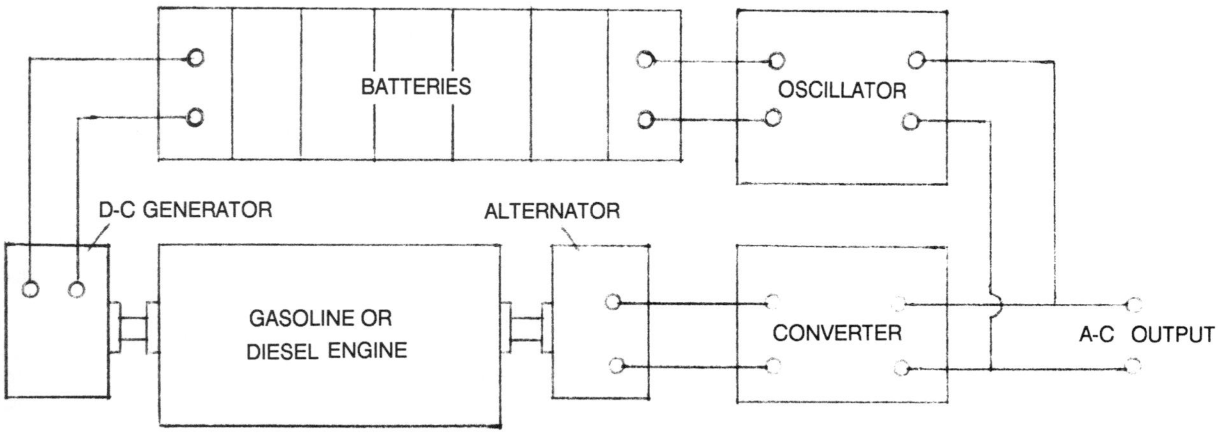

Figure P-10: Block diagram of an emergency power supply. The batteries are first to respond. The d-c current from the batteries powers an oscillator circuit with a 60 cycle frequency control unit. The drain on the batteries signals the engine to start up automatically. Power from the engine drives a generator to recharge the batteries and an alternator to take over the job of running the operating room or hospital. The control circuits are timed by quartz crytals to assure accuracy in the 60 hertz output.

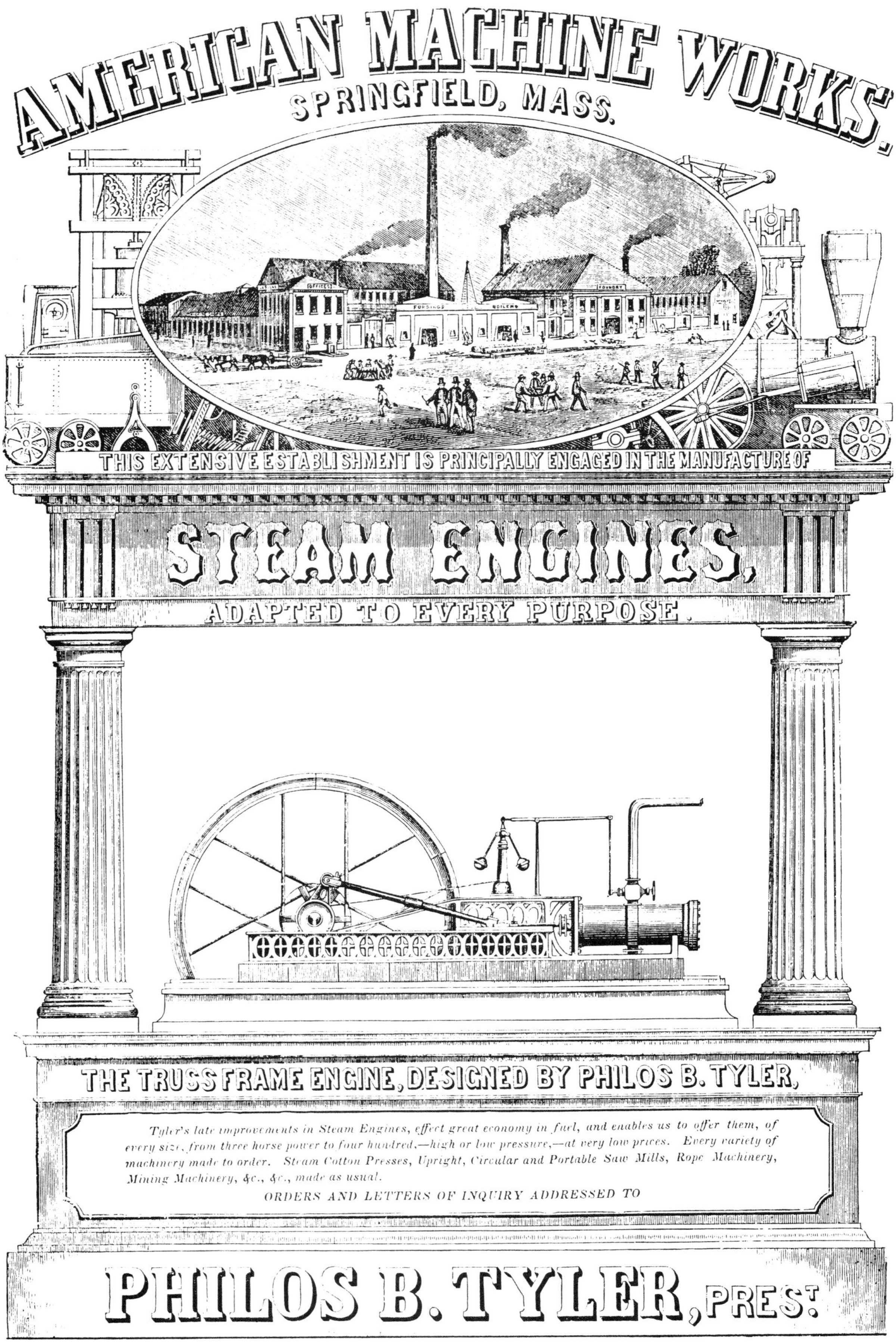
AMERICAN MACHINE WORKS,
SPRINGFIELD, MASS.
THIS EXTENSIVE ESTABLISHMENT IS PRINCIPALLY ENGAGED IN THE MANUFACTURE OF
STEAM ENGINES,
ADAPTED TO EVERY PURPOSE.
THE TRUSS FRAME ENGINE, DESIGNED BY PHILOS B. TYLER,
Tyler's late improvements in Steam Engines, effect great economy in fuel, and enables us to offer them, of every size, from three horse power to four hundred,—high or low pressure,—at very low prices. Every variety of machinery made to order. Steam Cotton Presses, Upright, Circular and Portable Saw Mills, Rope Machinery, Mining Machinery, &c., &c., made as usual.
ORDERS AND LETTERS OF INQUIRY ADDRESSED TO
PHILOS B. TYLER, PRES^T

ELECTRICAL DEVICES

THE TRANSFORMER

(1-14) It was the special attributes of the transformer that made a-c power the universal method of electrification. In the early days of electricity the advocates of d-c power had very strong arguments against the use of a-c. Direct current is devoid of all of the problems associated with self-inductance and lagging phase relationships. The fact that the transformer can do so many things at such little cost made the use of a-c inevitable. The transformer can control loads, change voltage up or down, change the phase, divide up available current, divide up the available voltage, and still operate with efficiency as high as 98 percent.

A diagram in cross section of a typical neon transformer is shown in Figure H-1. The core is made of thin laminations of high grade electrical steel (steel alloyed with cobalt and rolled thin). The core is built to go all the way around each coil separately so the magnetic pathways in the cores are complete without air gaps.

There are two coils. The 120 volt coil is called the primary winding. The output coil may be from 2000 to 15,000 volts, and it is called the secondary winding. The coils are carefully made and wrapped tightly with bands of insulation to prevent them from vibrating. A wire that can move even slightly due to the magnetic vibrations of the system will eventually crystalize and break, rendering the transformer defective.

After it has been assembled, the completed transformer is fastened into a leak-proof container of enamelled steel. The container is then filled with a tarry composition that sets and hardens. The tar filler has three jobs to do:

(1) It is a fire retardent in case of fire.

(2) It conducts heat away from the conductors in the windings.

(3) It prevents the wires of the transformer coils from moving or vibrating.

PRIMARY AND SECONDARY WINDINGS

The number of turns in a transformer winding is important for two reasons. One reason is that the

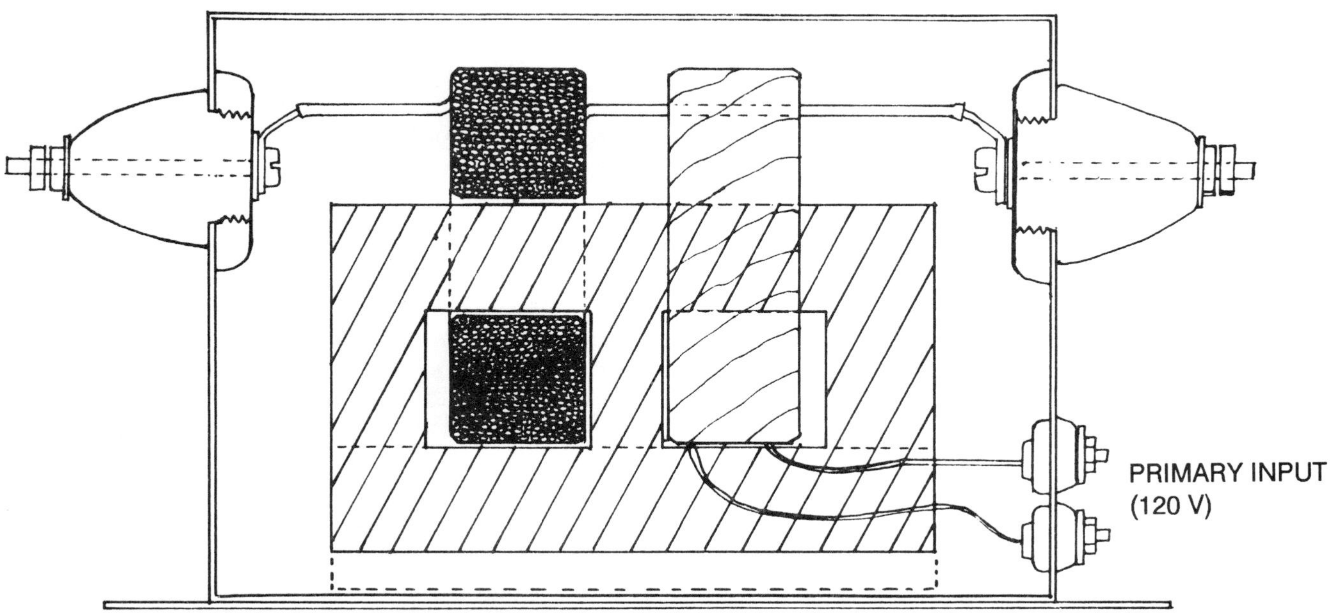

Figure H-1: A transformer has two coils. The number of turns in the secondary can be more than 100 times the number of turns in the primary winding. The coils are bound with linen and shellacked to prevent individual conductors from vibrating. After assembly the transformer container is filled with a composition that sets hard.

voltage output of the secondary is exactly related to the number of turns of wire it contains compared to the primary. If the secondary has twice as many turns of wire, it will have twice the voltage output. If it has ten times the number of turns, it will have ten times the voltage output. The relationship is shown in the equation:

$$\text{Output voltage} = \frac{\text{Input voltage} \times \text{Secondary turns}}{\text{Primary turns}}$$

You cannot get something for nothing, so when the voltage is increased, the current is reduced. The output voltage multiplied by the output current will always nearly equal the input watts. Remember that watts is the voltage multiplied by the current.

$$\text{Watts (input)} = \text{Output voltage} \times \text{output current.}$$

The other reason the number of windings in the secondary is important is because the secondary winding comprises many turns of fine wire. That means that there will be a considerable electrical resistance in the coil. When the transformer is in the designing stage, before it is manufactured, the amount of voltage drop in the secondary due to resistance alone is calculated. Then the excess voltage drop is expressed in terms of turns of wire in the coil. Since each turn of wire in the secondary increases the output voltage by a known amount, a few extra turns of wire on the coil will bring the voltage up to the proper rating for the transformer.

CHOKE COIL EFFECT

Another reason why the number of turns in transformer coils is a critical factor in its operation, is the *choke coil* effect. The primary winding must have enough turns of wire so that when the secondary is open circuit (not connected to anything) the current in the primary will be choked off. The primary is made so that it will have almost total self-induction by itself. With the normal amount of voltage applied to the primary, the choke effect cuts off the current in the primary except for a small "trickle current" to keep the transformer viable.

However, when the secondary coil is connected to a load, the current in the secondary is exactly 90 degrees out of phase with the voltage in the primary. The magnetic flux of the secondary is in phase with the self-induction of the primary. The opposing magnetic flux from the secondary enters the core of the primary and counteracts the choke coil effect. Then current flows freely in the primary exactly up to the requirements of the secondary. In this way a transformer controls its own power factor with an efficiency as high as 80 percent. Large power transformers have an efficiency rating as high as 99 percent.

STEP UP AND STEP DOWN

(2-14) The loss of power that affects a circuit is because of the resistance in the conductors. The formula for heat energy losses is

$$\text{Loss} = R\,I^2$$

The loss is proportional to the current squared, so evidently if the current is reduced, the losses will be greatly reduced. When a transformer acts to yield a high voltage, the current is proportionately reduced. If you increase the output voltage by ten times, the output current is reduced to one tenth.

The great advantage of transmitting power over great distances at a very high voltage is that the current is reduced to a trickle. The heat losses due to resistance in the conductors is largely eliminated.

When the power has reached its destination then step-down transformers are used to make the power suitable again for household use and for factories and stores. The line transformers on the power pole near your residence not only reduce the voltage to a useful value, they also convert the phase of the alternations. Single phase is produced for households and light commercial. Three-phase or polyphase voltages are produced for industrial use.

THE AUTO TRANSFORMER

The name "auto-transformer" might give the impresssion that it came off of an automobile. However, the use of the term auto is short for automatic, and it describes the way the transformer works. There is only one coil instead of two. The lead-in wires of

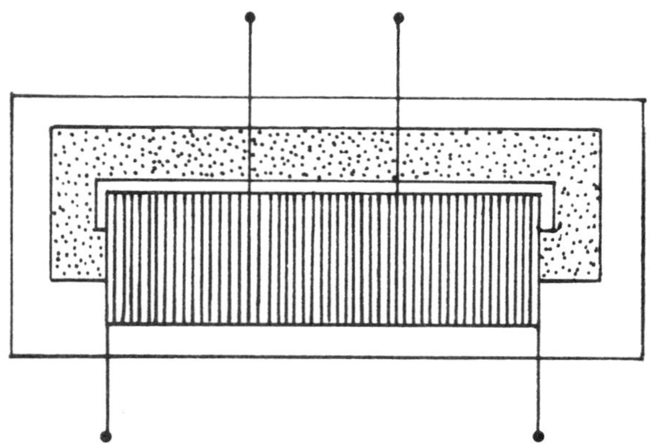

Figure H--2: The main feature of an auto transformer is that it has only one coil. It is used as an inexpensive step-down transformer.

the secondary are attached to the coil conductors with a metal to metal contact. The output voltage is equal to whatever proportion of the turns in the winding that is bracketted by the lead-in contacts. If the voltage output is to be one half of the primary voltage, then one half of the turns in the coil are included in the secondary portion of the coil.

The auto-transformer is used for lowering the high voltages in transmission lines down to a manageable voltage. It is also used as a step down transformer in electronic toys and devices.

A diagram of an auto transformer is shown in Figure H-2. The main difference between it and a regular transformer is that it has only one coil instead of two. Either the secondary or the primary is tapped into the coil at an appropriate place depending on the output voltage desired. The output voltage cannot be kept entirely separate from the input voltage because of the direct connection inside the coil. However, the choke effect of the primary controls the input current to the minimum needed by the output. A good core design is essential so that the magnetic field has a complete path without air-gaps.

THE TAPPING TRANSFORMER

(4-14) A tapping transformer is used in a situation where several voltages are needed in a piece of electrical equipment. As can be seen in the diagram of Figure H-3 the output or secondary coil has several wires connected to the coil at intervals along the coil. Any two of the output wires can be chosen to yield a particular voltage that is needed. Some tapping transformers have a sliding contact which changes the output voltage as the sliding contact is moved from one position to the next.

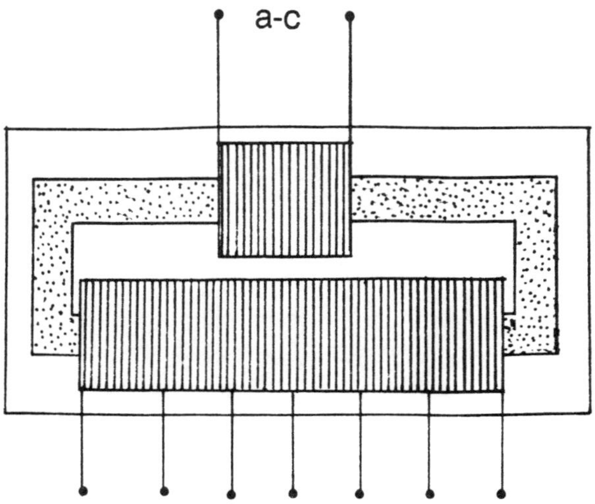

Figure H-3: The tapping transformer is tapped at several places in the secondary coil. Different voltages are obtained from different pairs of tap wires.

THE RECTIFIER

(5-14) Regardless of the great versaltility of a-c power, the use of d-c is essential for some electrical devices. The application of d-c in communications equipment is necessary for the functioning of vital parts. One example of device that operates on d-c is a battery charger. If only a-c power is available, then it is necessary to have a method of converting a-c to d-c that is cost-efficient and effective.

The device used for converting a-c to d-c is called a rectifier. For low voltage and small current requirements, the solid state rectifier is used. When large amounts of current are needed, a diode or triode rectifier is used. A diode is the name given to all vacuum tubes with two electrodes. A triode has three electrodes.

A rectifier functions by letting the a-c current through in one direction but stopping it from getting through in the opposite direction. Only the half wave of the a-c alternations get past the rectifier. As a result, the output from a single acting rectifier looks like Figure H-4.

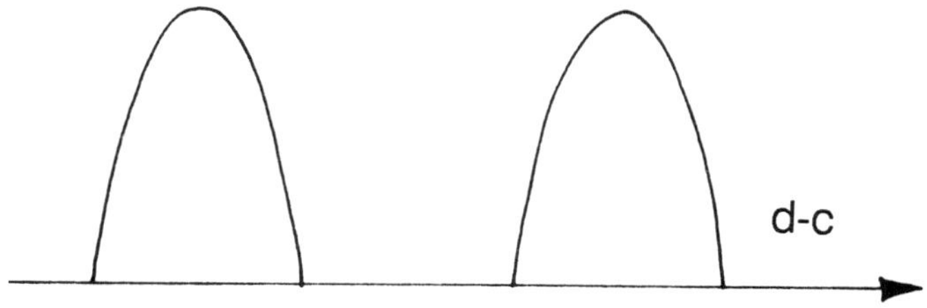

Figure H-4: When a-c is converted to d-c, using a single rectifier, the output from the rectifier will have the shape shown in the diagram. Only half of the wave of the a-c is used.

FULL WAVE RECTIFIER

(8-14) The symbol for a rectifier is an arrow head pointing to a fence. The arrow head points in the direction the current is permitted to pass through. When four selenium rectifiers are connected together as in the diagram of Figure H-5, the output will be in the form of a "fullwave". The output is shown in the diagram of Figure H-6. To remove the worst of the "bumps" in the output, a filter can be used, comprising an inductance and a capacitor connected as shown in Figure H-7. The filtered output then looks like Figure H-8.

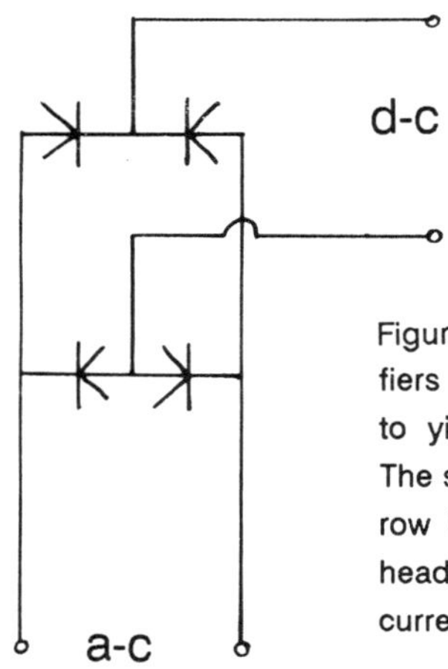

Figure H-5: Four solid-state rectifiers can be connected as shown to yield a full-wave d-c output. The symbol for a rectifier is an arrow head or a "K". The arrowhead points in the direction of the current.

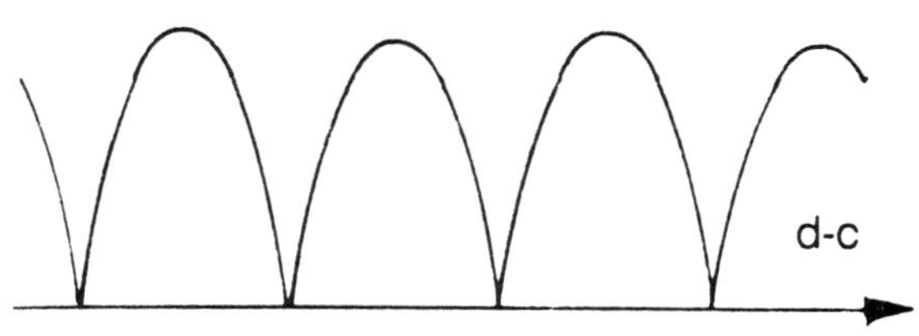

Figure H-6: The output from a full-wave rectifier system is a series of lobes of current touching each other at the base.

THE SELENIUM RECTIFIER

(6-14) The selenium rectifier is a small steel plate with a thin layer or coating of selenium on one side of it. Both sides of the steel plate are then coated with a selenium alloy which acts as the electrodes for placing the unit in a circuit. When the unit is connected to an a-c supply, the alloy layer becomes negative where it is touching the layer of selenium. The negative field repels the a-c current on that side. This permits the current to pass through freely in one direction, but is blocked from the other direction.

CUPROUS OXIDE RECTIFIER

(7-14) The copper-oxide rectifier works on exactly the same principle as the selenium rectifier already discussed. In the copper-oxide rectifier, one of the electrodes is a thin layer of copper plated onto a small steel plate. The electrode on the opposite side is a layer of cuprous-oxide.

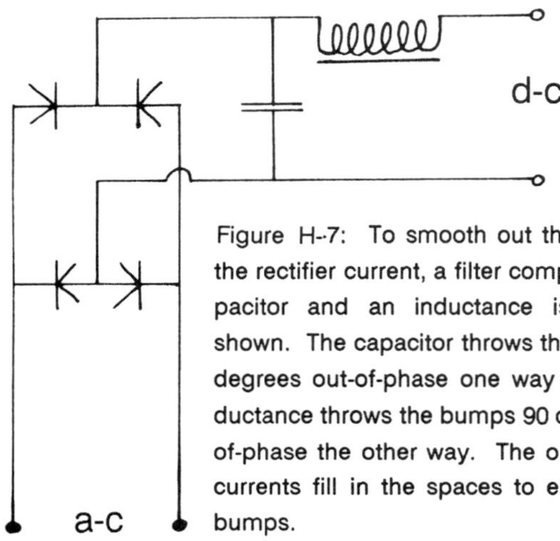

Figure H-7: To smooth out the bumps in the rectifier current, a filter comprising a capacitor and an inductance is used, as shown. The capacitor throws the bumps 90 degrees out-of-phase one way and the inductance throws the bumps 90 degrees out-of-phase the other way. The out-of- phase currents fill in the spaces to eliminate the bumps.

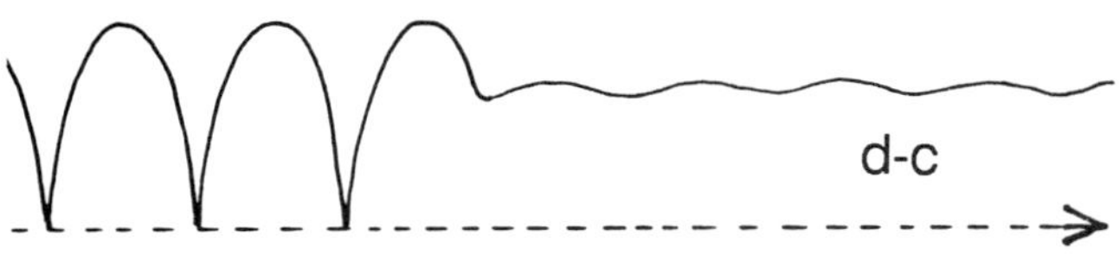

Figure H-8: Before and after the rectifier output current goes through the filter.

ELECTRON TUBE RECTIFIERS

(10-14) A vacuum tube called a *diode* (two electrodes) will permit the a-c to pass through the tube in one direction only. As shown in Figure H-10, a heated element in the tube releases electrons into the evacuated space within the tube. When the alternating current is applied, the electrons are drawn to the plate when it is positive and the current flows through. When the plate goes negative, the electrons are driven away. They cannot go back into the heated filament so they accumulate until the plate goes positive again. The action is one way only. The output is a half-wave form like Figure H-4.

CHOICE OF VOLTAGE

(9-14) The diagram shown in Figure H-9 uses two solid-state rectifiers and a transformer to yield full-wave rectification as well as bumping up the voltage to a predetermined amount. If a tapping transformer is substituted, several d-c voltages are then available from the a-c input.

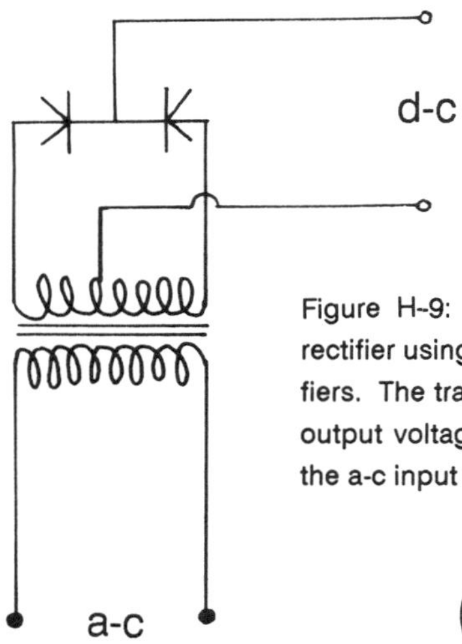

Figure H-9: Circuit for a full-wave rectifier using two solid state rectifiers. The transformer permits the output voltage to be greater than the a-c input voltage.

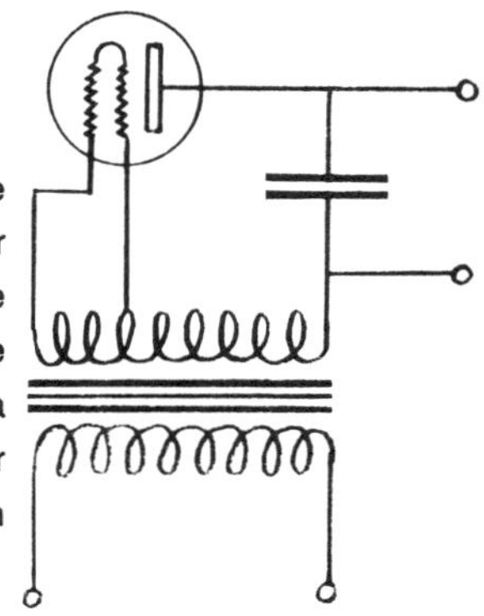

Figure H-10: A vacuum tube used as a rectifier permits a larger flow of current than a solid state device. The secondary coil of the transformer is tapped to provide a current for the heater filament, or a separate small transformer can be used.

Using a *triode* with the heated element between two plates, as shown in Figure H-11, full-wave rectification is the result. The plates are driven alternately negative and positive in time with alternations of the a-c. A separate small transformer winding provides the current for heating the filament, as shown. To complete the d-c circuit, it is taken off the filament coil and the plate coil.

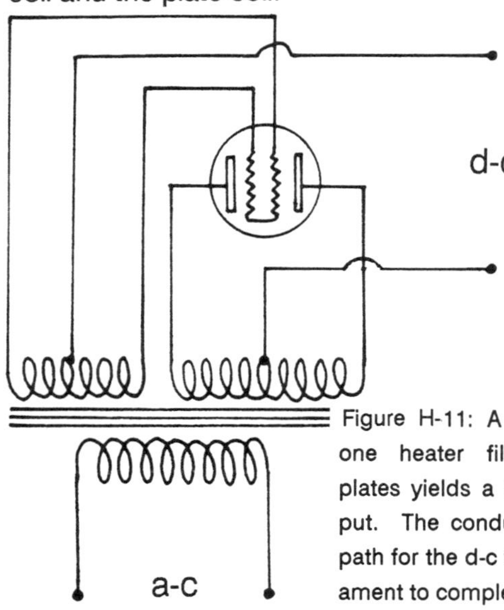

Figure H-11: A vacuum tube with one heater filament and two plates yields a full-wave d-c output. The conductors provide a path for the d-c to return to the filament to complete the circuit.

TUNGAR DIODE RECTIFIER

(11-14) A device called a tungar rectifier is a vacuum tube containing argon at a very low pressure. The cathode is a spiral-shaped winding of tungsten. There is a large graphite annode centered within the spiral cathode.

The argon gas becomes ionized when bombarded by electrons from the hot cathode. The argon becomes a conductor and permits a large flow of current during the negative half wave on the cathode. The output from one such tube is a half-wave. To produce full-wave rectification, two identical tubes are used side by side.

MERCURY ARC RECTIFIER

(12-14) The mercury arc rectifier operates on exactly the same principle as the tungar rectifier. The difference is that for the cathode a pool of hot mercury is used. It has a graphite annode, as before. During the negative half cycle of the a-c, the mercury vapor in the tube conducts a large current to the annode. For full-wave rectification, two such tubes are used side-by-side.

THE THERMOCOUPLE

A useful device for controlling circuits where temperature is a factor, is the thermocouple. The way it works is based on the fact that when two different metals are brought into contact, there is a natural potential difference between them. For iron and copper, for example, the potential difference at the freezing point (0 degrees centigrade) is 3.7 microvolts per degree.

When a circuit is made from a copper wire and an iron wire, there is no current unless there is a temperature difference between one junction and the other junction of the two metals. When one junction is placed in the furnace pilot flame and the other junction is outside of the furnace, a sufficient emf is generated to control a gas valve. If the pilot light goes out, the thermocouple circuit turns the gas off by closing a valve.

BI-METALIC STRIP

The bi-metalic strip is a useful device for opening and closing switches where temperature is a factor. The bi-metalic strip consists of a strip of flat copper and a co-extensive strip of sheet iron. The two strips are fastened together by solder-sweating or by special adhesives. When the temperature changes around the bi-metalic strip, the iron strip has a different co-efficient of expansion than the copper strip. This makes the bi-metalic strip bend into a curve.

To save space, and to give the strip more effective power, the bi-metalic strip is usually wound into a spiral. It is used in thermostats which keep your house heated to a steady temperature, and it is used in many electrical circuits where temperature is a factor. A typical bi-metallic strip is shown in Figure H-12. When the bi-metallic strip moves, it tips the little sealed vial of mercury, causing the mercury to flow and make contact. This is a sensitive kind of electric switch.

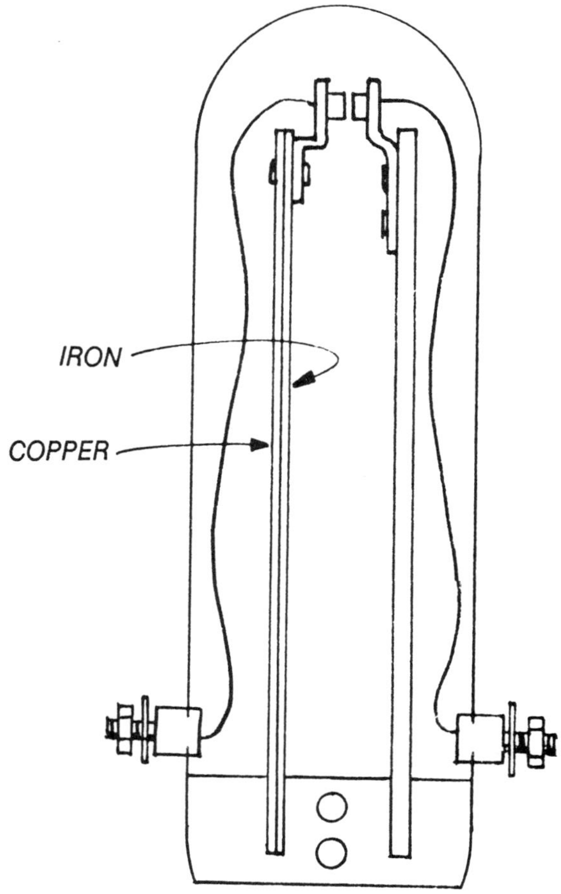

Figure H-12: When heated, copper changes in dimension more than iron. When an iron and a copper strip of metal are adhered together, a change in temperature makes the strip bend into a curve. The resulting motion can be used to open or close a switch. The bi-metalic strip in your household thermostat is wound into a spiral to make it more efficient. When it moves, it tips a little bottle of mercury so that the mercury runs to the end of the bottle and makes an electrical contact. This action turns on your furnace or fridge.

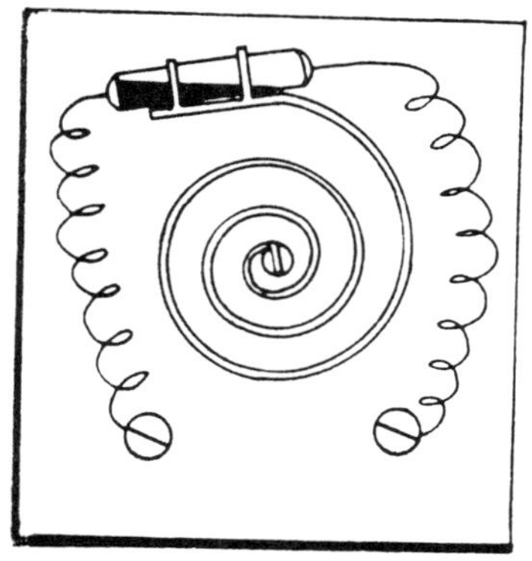

Figure H-12.

EVENTS FROM THE PAST

When the competition for sales between automobile manufacturers began to be a factor in the market place, then cars became more elegant and beautiful.

BASIC ELECTRONICS

ELECTRONICS STARTED WITH EDISON

When Edison was working on the electric light bulb, one of the things that attracted his attention was the layer of luminescence that surrounded the hot filament. Edison surmised that molecules of air remaining in the evacuated bulb might account for the effect. However, when he placed a magnet near the bulb, the luminescence reacted strongly to the magnet. Then he knew that he was observing a new phenomenon. He recorded the results of these experiments and called it the "Edison Effect". The same effect is called a "space charge" today, but Edison gets credit as the first one to observe and record the phenomenon.

ACCELERATED ELECTRONS

Electrons are usually tied to materials by the attraction of unlike charges. When the material is heated, the individual particles bounce around within the confines of their normal space positions in the material. However, when a material is heated to incandescence, the electrons are bounced around so rapidly that some of them are accelerated right out of the material. To get out, they must have enough energy to overcome the natural electric attraction of the material. Although the electrons are attracted back to the material, there are more electrons flying out to take their places. The result is a cloud of energetic electrons in the space around an incandescent filament.

THE THERMIONIC VALVE

In England, John A. Fleming experimented with the Edison effect. He found that a magnet would control the position of the electron cloud. Then he wondered what an electric charge would do to it. He arranged to place an electrode in the tube with wires to it so that he could control the electric charge on the electrode. When he had accomplished that, he had invented the first diode. Amplification of a signal by voltage control became possible.

LEE DEFOREST

In United States, Lee DeForest was also experimenting. He got the idea that a grid to control the flow of electrons to the plate would be much more sensitive, and he was right. DeForest's triode made radio communication as we know it a reality. DeForest began to manufacture triodes in small quantities and put them on the market. That was the beginning of the DeForest Vacuum Tube company.

Meanwhile, G W Pickard and H Dunwoody were using a piece of galena crystal to detect weak radio signals. The galena acted like a rectifier which captured weak alternating currents and converrted them to d-c currents. The crystal converted the natural

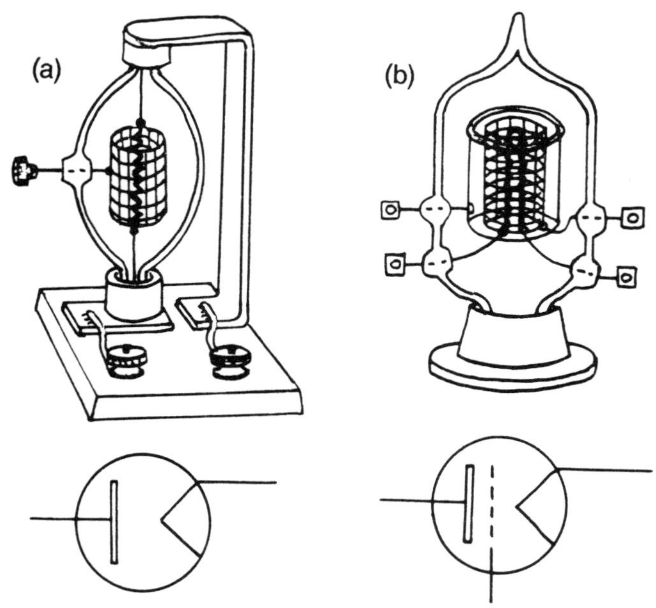

Figure 17: Typical of the early vacuum tubes, these drawings illustrate (a) a diode and (b) a triode. The symbols for these valves (as they were called) are also shown. The diode was an evacuated glass bulb with a vertical wire through it which could be heated to incandescence by an exterior battery. A metal screen around the hot filament could be positively charged by a battery so that it would draw off the electrons as needed.

The triode had an additional metal grid positioned between the filament and the plate. The grid could be activated by a signal to generate a sensitive varying current. A steady voltage (called the bias) was applied to the plate to draw off whatever electrons the grid permitted to pass.

amplification of a tuned circuit to make earphones audible.

EDWIN ARMSTRONG

About the same time as DeForest was making his triode valve, another inventor, Edwin Armstrong was applying for a patent on his triode. Armstrong independently invented superheterodyne tuning methods for radio. He was also the sole inventor of frequency modulation, as well as many other inventions for radio.

All of these events happened just after the turn of the century. By 1912, radio technology was a booming industry. Plucking music out of the air seemed like magic to most people, so that radio became a kind of magic act that everyone wanted to be a part of.

Figure 18: The crystal was held in a little cup with a set screw to assure contact. The surface of the galena was then tickled with a "cat's whisker" until a sensitive spot was found on the surface. The loudness of the music that was heard was the criteria for sensitivity.

CRYSTAL SET

The typical circuit for a crystal set is shown in Figure 20. The coil was often variable instead of the condenser. As the components were mainly made by hand by small lot manufacturers, it was easier to make a variable coil than to make a variable condenser. To make the coil variable, it was carefully wound on a form. A metal finger could be slid along a brass bar to make contact with any portion of the coil, similar to the potentiometer invented by Fleming.

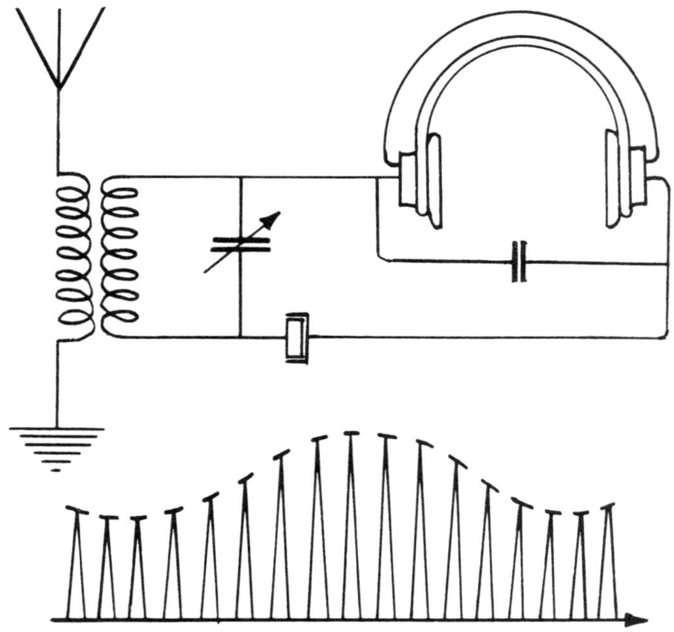

Figure 20: Diagram for a crystal set is shown at (a). The magnet coils in the earphones could not respond to an individual rectified d-c pulse, but they did respond to the audio "envelope" created by a train of pulses, as shown at (b).

RADIO WAVES IN SPACE

Early experimenters with electricity were at first surprised by the effects at a distance. Both magnetism and electrostatic potential could be made to show a response in nearby circuits. Hertz theorized that there was no limit to the distance over which these effects could be noticed. He made a receiver with a loop of wire and two metal knobs on the ends of the wire with the knobs almost touching. He made a sender using a coil with a spark gap. His idea was that when a spark was formed in the sender, the same spark should show up in the receiver. The very first try was successful.

Maxwell's mathematical considerations had shown that a light ray, which travels across the universe, is an electro-magnetic propagation. Hertz surmised that the waves he was sending with his apparatus were the same kind as light waves, but with a much lower frequency. Subsequent experiments proved that Hertz was correct.

Experimenters tried sending waves by rapidly charging and discharging a metal ball. That worked, but a

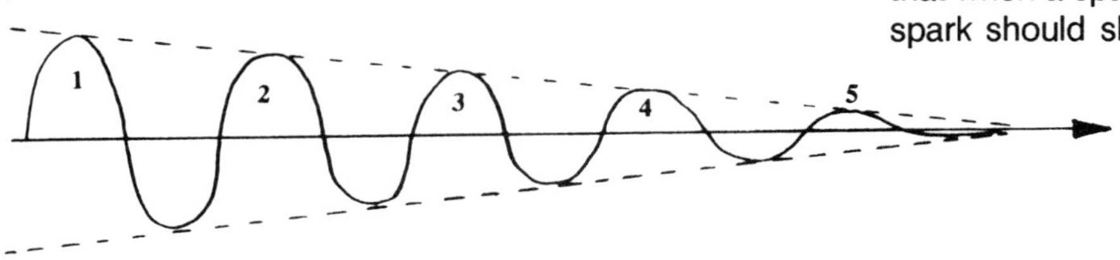

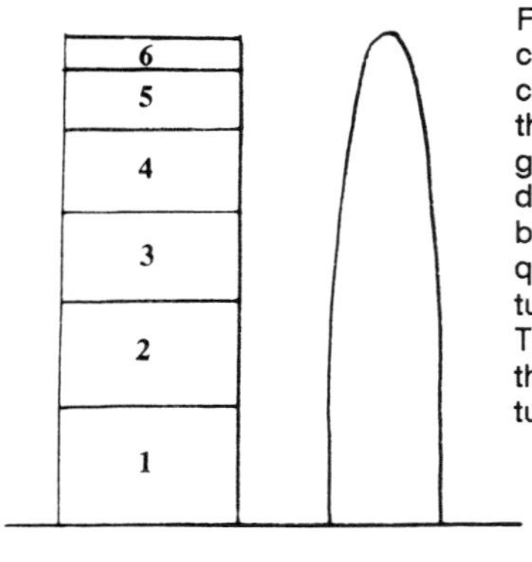

Figure 19: The amplification factor of a tuned circuit is caused by the overlapping or piling up of many signal increments. If only one pulse is introduced to a tuned circuit, the pulse will flow back and forth like a pendulum and gradually die out (called attenuation). However, like a pendulum, the pulse retains its frequency. Multiply this action by providing a steady series of pulses at the same frequency as the tuned circuit, and the pulses will pile up in the tuned circuit like the example in the diagram at the left. This gives the tuned circuit an amplification factor equal to the total of all of the attenuation pulses. See the section on tuned circuits in the chapter on Alternating Currents.

metal ball did not work as a receiving antenna. A metal rod was tried and that worked. Some configurations that they tried worked better than others, and gradually the best shapes for sending and receiving radio signals were discovered.

THE RADIO FREQUENCY OSCILATOR

Radio frequencies range from 30,000 hertz, which is the top range of human hearing, upwards towards the infrared. The abbreviation for radio frequencies is r-f. In the early days of communications, pulses of radio frequencies were used for sending the dots and dashes of the telegraph by wireless. The r-f oscilator evolved from the need for a steady reliable source of radio frequency waves. It was soon realized that all types of radio communications could be accomplished by modifying and modulating r-f frequencies.

The radio frequency oscilator is the basic unit for sending and transmitting signals. A typical r-f oscilator is shown in Figure 21. This circuit is powered by batteries or by d-c voltages. It will deliver a steady output of controllable frequencies.

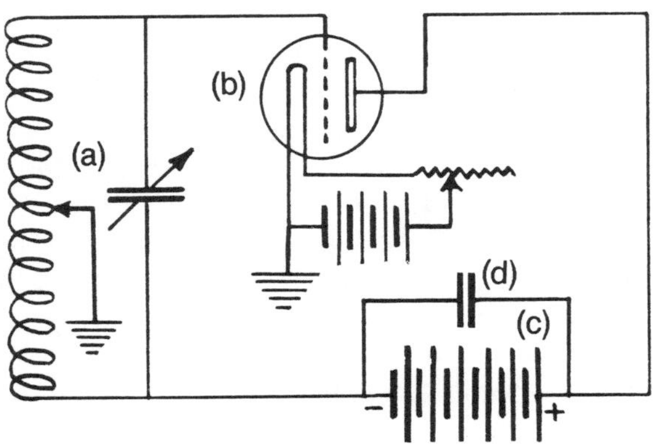

Figuire 21: Oscilator circuit is the electrical equivalent of a pendulum which has to be given a little shove to keep it swinging. The tuned circuit provides the steady frequency to time the swings of the pendulum. The plate circuit provides a train of pulses at the proper frequency to keep the tuned circuit functioning.

The diagram is simple if you consider the elements separately. The section marked (a) is a tuned circuit, comprising a coil and a variable capacitor. Tuned circuit (a) has a ground tapped into the center of the coil. A tuned circuit swings from maximum positive to maximum negative. Since the ground is neutral, this allows the maximum swing to be generated.

The filament circuit is marked (b). It has its own battery which heats up the filament and makes it produce electrons. It is adjustable for maximum effect. A supply of electrons has to come from somewhere to replace the ones that form the main current. Therefore the filament circuit has a ground wire attached to permit electrons to enter.

The main battery (c) supplies the voltage bias to the grid and plate. In order that the high frequency pulses can bypass the slow moving battery, there is capacitor (d) which forms a path for the high frequency pulses.

The way that the oscilator works is easy to understand if you think of the tuned circuit (a) as a pendulum which has been given a shove to make it swing. Unless the pendulum gets another shove from time to time, it will slow down and stop swinging. As the current oscillates through the tuned circuit, the grid is driven alternately positive and negative. Because of the battery bias (bias just means voltage) the plate is always positive with respect to ground. When the grid goes negative, the electrons from the filament are repelled and there is no current. When the grid is maximum positive, electrons are drawn from the filament. During part of the cycle the plate is more positive than the grid, and the electrons then pass through the grid to the plate. This means that there is a surge or pulse of plate current every time the tuned circuit causes the grid to go positive. The capacitor (c) permits these pulses to charge right through the circuit and energize the tuned circuit again. The tuned circuit is given a little shove with every pulse, and so it keeps oscilating. The oscilator circuit will continue to generate a train of pulses at the tuned frequency until the battery runs down.

The electrons that enter the plate circuit would soon build up and stop the action, so the ground wire that is tapped into the coil provides a path for the surplus electrons to escape.

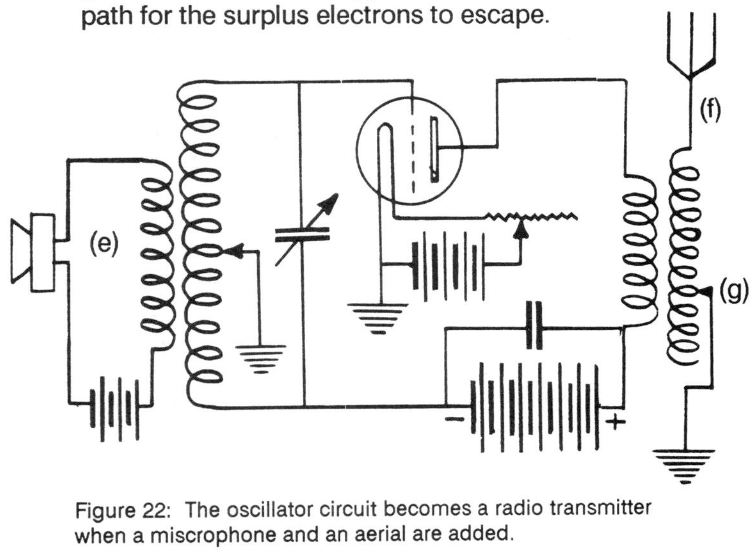

Figure 22: The oscillator circuit becomes a radio transmitter when a miscrophone and an aerial are added.

THE RADIO TRANSMITTER

The circuit diagram for the radio transmitter, Figure 22, is the same as the diagram for the r-f oscilator in Figure 21. The only differences are that a microphone (e) has been added, coupled to the main coil as shown. Also, a broadcast antenna (f) has been added. The antenna can be tuned to the same frequency as the oscillator by adjusting the sliding coil tap (g).

THE CARRIER WAVE

The output from the oscillator is called the carrier wave. The signal from the oscillator looks something like the diagram (a) in Figure 23. The audio signal from the microphone looks like diagram (b). When the voice and audio frequencies are imposed on the oscillator frequency, the output then looks like diagram (c). This is the final shape of the wave that is broadcast for radio reception.

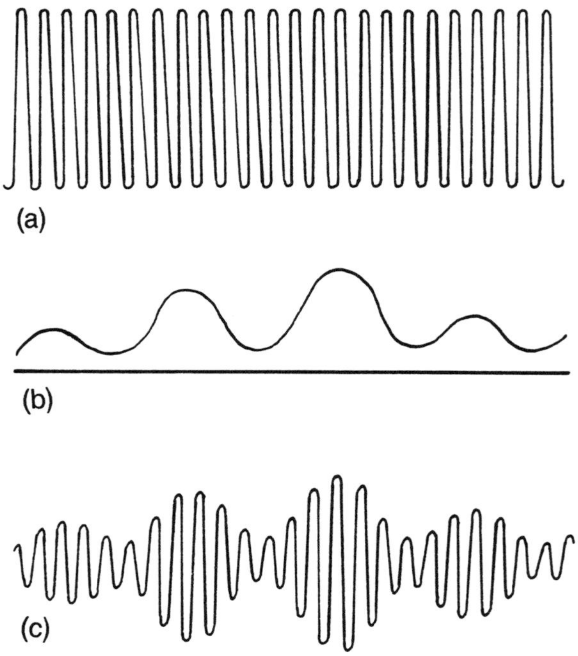

Figure 23: The oscillator generates a steady, uniform radio frequency wave as shown at (a). The signal produced by the microphone is represented schematically by diagram (b). When the audio frequency is used to modulate the radio frequency, the combination looks something like diagram (c). The diagrams fail to show that there are usually hundreds, and sometimes thousands, of r-f waves in the carrier frequency for each audio frequency wave.

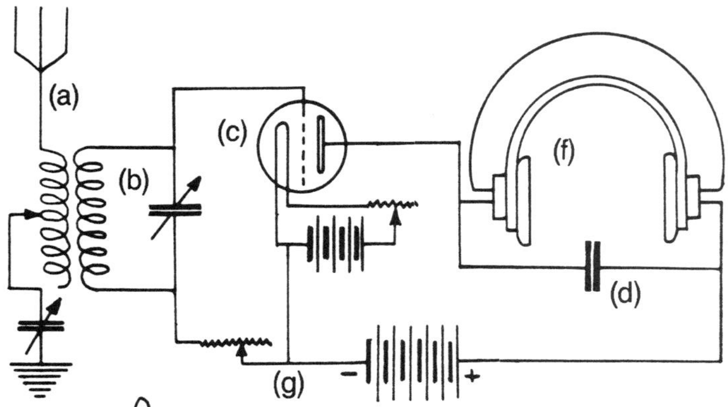

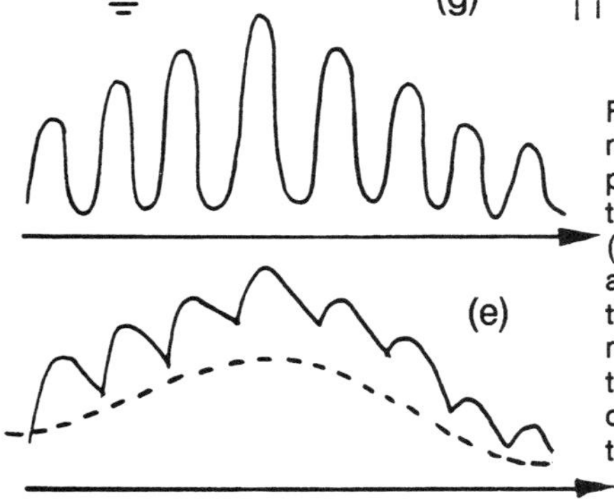

Figure 24: Simplified circuit for a radio receiver. The individual d-c pulses are too rapid to be felt by the earphones, but the capacitor (d) responds to a lower frequency and tends to fill in the spaces between the pulses, as at (e). The resulting lumps of d-c current retain the original audio frequencies. The earphones respond to these lower frequencies.

THE RADIO RECEIVER

The circuit diagram for a simple radio receiver is shown in Figure 24. The parts are called (a) the antenna, and coupled to the antenna coil, the tuned circuit (b), comprising a coil and a variable capacitor. The vacuum tube triode (c) provides detection and amplification of the signal. The condenser (d) responds to the voltage average of the train of d-c pulses from the plate of the triode. As shown in diagram (e), the average voltage is the shape of the original voice signal. This voice signal then activates the earphones (f) to produce audible sound.

The batteries in the circuit diagram power the filament of the triode, and the plate and the grid of the triode. Note that the complete circuit is arranged to generate a series of d-c pulses in the plate circuit. The d-c voltage from the batteries delivers a steady current of electrons to the grid of the triode. Electrons that are released into the circuit by the hot filament are replaced through the connection at (g).

VACUUM TUBE CHARACTERISTICS

Before the transistor and solid state circuits were widely used, everything was done with vacuum tubes. The engineering of vacuum tubes and the related technology was an enormous industry. The vacuum tube was the subject of thousands of inventions. Some basic criteria became standard in that the success of a radio circuit depended on vacuum tubes that were of uniform high quality. The characteristics of vacuum tubes and how they compared to each other were of great importance.

The same characteristics show up now in solid state devices, so that the same type of diagrams are still in use today to describe circuit elements and how they function. The most important attribute of a vacuum tube is the response when a voltage is applied to the grid. How much current is supplied at the output? The other important factor is the "linearity" of the response. That means, does the current increase at the same rate as the voltage?

The diagrams shown in Figure 26 demonstrate the two most important tube characteristics. In the diagram (a) the arrows pointing straight up represent current strength in the plate circuit. The arrows pointing to the right represent the amount of voltage impressed on the grid of the tube. In three different positions, marked (1), (2) and (3), an increase in grid voltage is accompanied by an increase in plate current. The dotted lines show how the current and voltage values are plotted across and up until they meet. The dot where they meet is taken as the characteristic of the tube.

The three dots show the results of only three voltages. If all possible voltage variations are included in the diagram, then it looks like diagram (b). The dots have joined to become a line. Where the line is straight, that means that the plate current increased in exact proportion to the grid voltage. That part of the tube shows a "linear" response. Where the line is curved, it takes more or less voltage to achieve the same result. Most standard tubes have a characteristic curve like a long "S", similar to the curve shown in diagram (b).

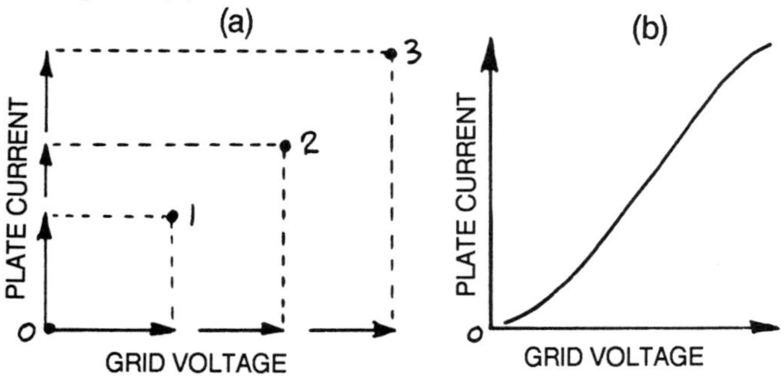

Figure 26: The most important attributes of a vacuum tube are linearity and current to grid-voltage ratio. The sloped line in diagram (b) is made from many dots like (1), (2) and (3) in diagram (a).

THE RADIO FREQUENCY AMPLIFIER

The amount of energy that it takes to power a radio speaker so that audible sound is produced is many times greater than the energy of the r-f signal. To build up the strength of the r-f signal one or more stages of amplification are used. As many as four amplifier stages may be designed into a radio receiver to give it fidelity and power.

A typical r-f amplifier is shown in Figure 27. It may have an air core transformer and a tuned circuit, as shown at (a). In many radio receivers, the output from the previous stage is fed directly to the grid of the next stage with only a capacitor between the stages, as shown at (b). The capacitor separation permits the r-f signal to proceed through the stages while the d-c components of up stream stages are blocked. In a multistage amplifier, the battery circuit is arranged to power all of the stages.

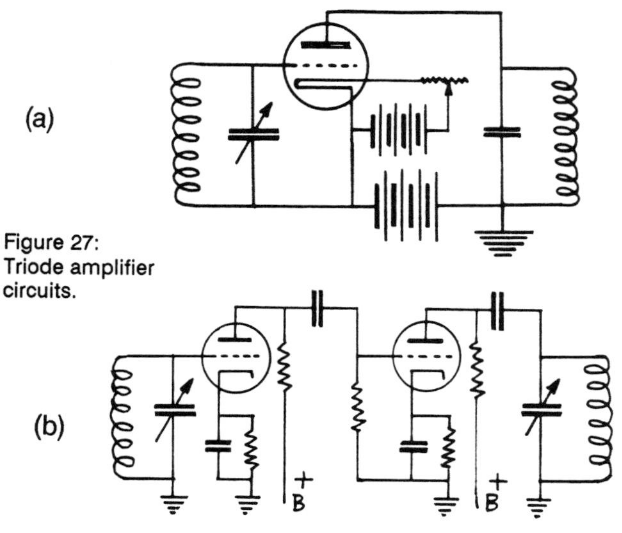

Figure 27: Triode amplifier circuits.

SOLID STATE COMPONENTS

Solid state devices are as old as the galena crystal in the crystal set of 1910. When the transistor was invented, the vacuum tube became out-of-date. The transistor did not replace the vacuum tube immediately, but the trend was set. There are very few vacuum tubes in use today.

Solid state components are usually made from silicon or germanium. Both of these elements are insulators, but when they are mixed with certain other atoms they become semi-conducting. The added atoms may be the natural impurities found in silicon and germanium, or special impurities can be designed into the material. The crystals are then said to be "doped" with the impurity. Some "doping" atoms give the crystal an ability to conduct negative charges and some produce a reverse effect, permitting positive charges to move through the substance.

POSITIVE HOLES

When electrons are removed from a region in a crystal by doping, the excess positive charges in the region are permitted to drift through the substance under the influence of an exterior charge. In this condition, the positive charges are called "holes" and they act like positive charges. They respond to electric or magnetic fields and they attract electrons. The holes can behave like a current of positive electricity.

The base crystal can also be doped in a way that causes an electron excess in a region. Bits of crystal that are doped in this fashion are called N regions, because of the negative charge they retain. Bits of crystal with an excess of positive charges, or holes, are called P regions.

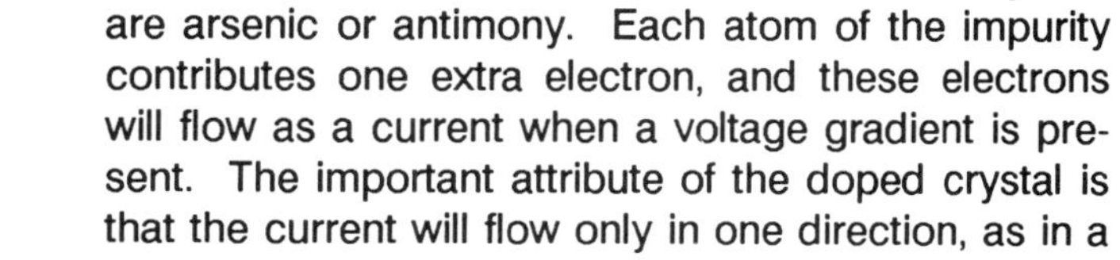

Figure 28: Schematic diagrams of diode rectifier and two kinds of solid state triodes.

The doping substances used for N crystals are arsenic or antimony. Each atom of the impurity contributes one extra electron, and these electrons will flow as a current when a voltage gradient is present. The important attribute of the doped crystal is that the current will flow only in one direction, as in a rectifier.

The doping substances used for P regions are aluminum or indium. Each atom of the impurity takes an electron out of circulation in the crystal, leaving an excess of positive holes. The doped material conducts electricity better in one direction than the other, and in the opposite way to that of an N region.

A diode can be made using a PN junction of the two kinds of doped crystal. Or a triode can be made using a PNP combination or an NPN combination.

FROM CRYSTALS TO MICRO-ELECTRONICS

The crystals for solid state circuit components are made from natural crystals or they are grown in a special process with the degree of impurity carefully controlled. The growth of crystals in a saturated solution was one of the first methods used. This is similar to lowering the temperature of water to slightly below the freezing point. When a "seed" is introduced into the ice cold water, ice crystals form around the seed.

The techniques progressed towards the ultimate invention of micro-electronics, as later methods included depositing thin layers of crystal material on a substrate bed of inert material such as glass. This is done in a vacuum chamber. Layers of different elements are fused together by the process of deposition to provide a combination with the desired characteristics.

A vapor of the desired layer is created in the chamber by a combination of chemical and thermal action. The vapor is then condensed onto the substrate, forming a thin film of the desired crystal.

Another method involves the printing of the desired component on a ceramic base. The component can then be the exact shape desired. The printed ceramic is then fired in an oven causing the special ink to fuse to the ceramic. When cooled, another layer of doped ink can be printed over the first, and again fused in an oven. Silk-screen printing is used to control the shapes of the elements.

SOLID STATE CIRCUITS

The circuits used with transistors and solid state diode rectifiers are similar to the circuits designed for vacuum tubes. The objectives of a transistor circuit is to detect or amplify an r-f signal. The final output is the high fidelity sound and video we are accustomed to.

Some typical circuits using solid state components is shown in Figure 29. Note the symbol for a transistor is similar to the symbol for a vacuum-tube. One of the connections in the solid-state triode is called the emitter. It is equivalent to the filament in the vacuum-tube. The second connection is called the collector. It does the same function as the plate in a vacuum-tube. The other connection is called the base. A bias voltage makes the base function like the grid in a vacuum-tube. The analogy is not an exact one because of the unique attributes of crystals.

The collector current divided by the emitter current is the amplification factor of the triode. A factor of 3 is average for a transistorized amplification stage. Three is a small number but if you multiply it by itself three times you are up to 81.

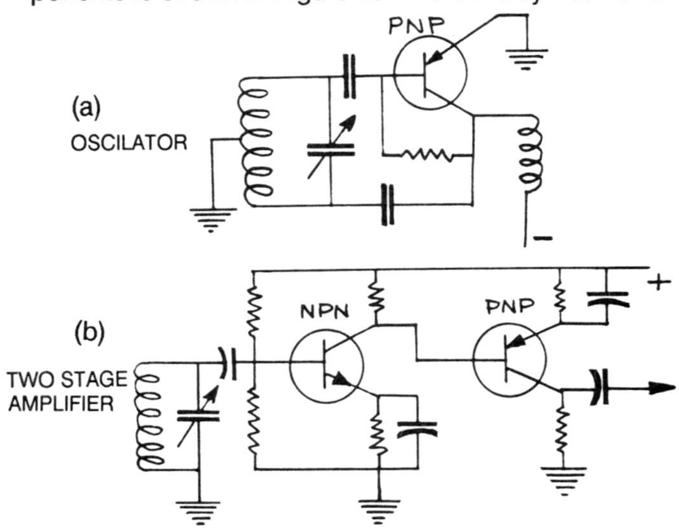

Figure 29: Typical solid state circuits.

THERMAL EFFECTS OF ELECTRICITY

The conductors in an electric circuit are heated by the passage of the electric current. The heating is due to the resistance of the conductor. Conductors that have no resistance, as when they are subjected to low temperature techniques, do not produce heat. The science of low temperature applications is called Cryogenics.

Thermal and thermionic effects occur in electric circuits in several ways. In a good conductor, the electrons are free to move around to a certain degree within the spaces between and around the atoms and molecules. The conductor material behaves like a container within which the electrons are like a superfine gas. Each different metal conductor has its own characteristic electrical pressure. So when two different metals are touched together, there is a small pressure difference which can be measured as voltage. When the temperature at the contact point is increased or lowered, the contact pressure is increased or lowered.

Three different effects were noted by different early scientists, all caused by the same phenomenon. These are (1) The Seebeck effect, (2) The Peletier effect, and (3) The Thompson effect.

THE SEEBECK EFFECT

When two different conductors, such as iron and copper, are formed into a loop, a current will flow when one of the contact junctions is heated or cooled.

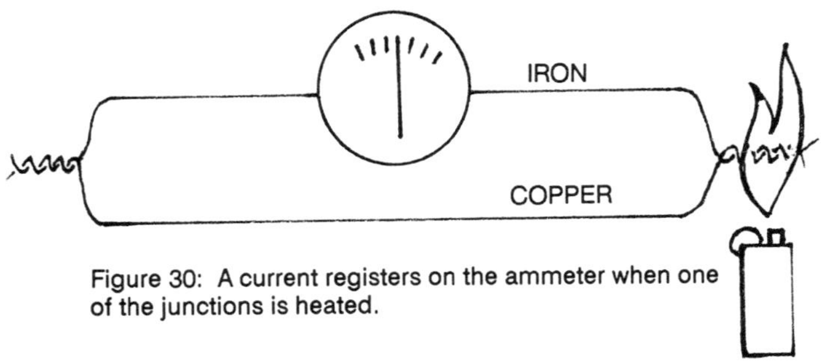

Figure 30: A current registers on the ammeter when one of the junctions is heated.

THE PELETIER EFFECT

When two different conductors are formed into a loop, one of the contact junctions will become heated and the other junction will become cool when a current of electricity is passed through the loop.

THE THOMPSON EFFECT

When a conductor made of one kind of metal is formed into a loop, thermal gradients in the loop will behave like voltage gradients. That is, if one side of the loop is heated, an emf is created along the sides of the heated region where the temperature is dropping off.

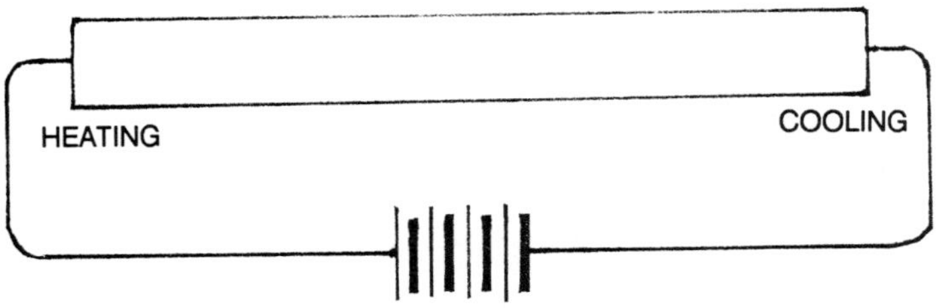

Figure 31: When a current of electrons passes from a wide conductor into a fine one, the electrons are accelerated. The kinetic energy needed to accelerate the electrons is absorbed from the surrounding metal at the junction, leaving the junction cooler.

PHOTOELECTRIC EMISSION

Early experimenters discovered that the electrostatic charge on metal objects could be made to disappear by flooding the objects with light of different colors. They found that ultraviolet light was the most effective. It was later discovered that a charge could be created by light rays impinging on metal surfaces.

THE PHOTOELECTRIC CELL

When light rays shine on certain metals, and particularly cesium, the cesium behaves like an emitter or like a heated filament. The released electrons can then be adapted to control a current.

CASCADE CELL

It was also discovered that certain metals will release several electrons when the surface of the metal is struck by a single electron. In a vacuum tube a flow of electrons is directed onto the first of several small metal panels arranged in a series of steps. Each of the steps adds to the cascade of electrons, increasing the output current by a large factor.

IONIZATION CELLS

When a small stream of electrons is directed into a gas at a very low pressure inside a vacuum tube, the gas is ionized and becomes a conductor. By this method a current of a few electrons from a cesium spot can be made to control a large current through such a tube. The cesium is activated by a varying ray of light to produce a large current with the same variations.

X-RAY TUBE

Within a vacuum tube a hot tungsten filament produces electrons. The electrons are accelerated to about half the speed of light by a large potential difference between the hot cathode and the cold tungsten anode. When the high velocity electrons strike the surface of the cold tungsten, a spray of X-rays are produced. The direction of the x-ray spray is controlled by lead shielding. The remaining x-ray spray is allowed to pass through an object placed in front of a photographic film. The film reveals the various density configurations of the object through its depth. Very dense regions show up dark and less dense regions show up in lighter shades. X-rays are dangerous to health.

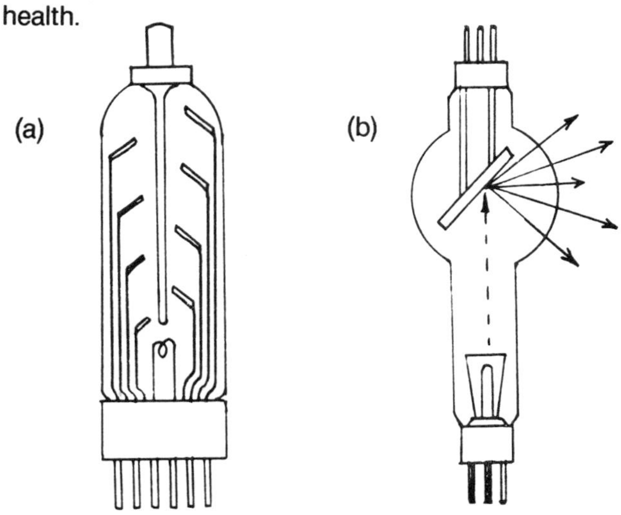

Figure 32: Schematic diagram (a) shows a cascade or multiplier tube. Electrons striking the metal steps knock extra electrons out of the metal, which add to the current. Diagram (b) shows an X-ray tube. Accelerated electrons strike an angled plate of tungsten, causing a spray of X-rays, shown by the arrows.

LIQUID CRYSTALS

Some liquids instantly convert to crystalline films when immersed in an electric field. This is something like water suddenly freezing into ice. Also, a clear liquid can be made to turn opague under the influence of a magnetic or electric field. The technology of digital read-out and digital monitors of the type found in instrumentation depend on these phenomenae. Watch and clock faces, oven timers and indicators are examples of the many applications.

EVENTS FROM THE PAST

The first typewriter to be mass produced by the Remington Corporation.

USEFUL MATHEMATICS

BASIC ARITHMETIC

Everything in our technical world runs on mathematics. In order to be a survivor in the modern world, everyone must know the basics of arithmetic, and geometry. Just to play poker effectively, you have to be a mathematician of sorts. Every time you ante up a quarter, you are adding to the pot. Handling money in our day to day existence, buying groceries, filling the car with gas, ordering a pair of jeans.....some basic arithmetic is essential.

To get along in life you need a surprisingly little knowledge of mathematics. But you do need to have some understanding of how numbers work. To be a competent and knowledgable inventor you need some basic ideas about arithmetic. There is also some specialized knowledge you need, which is explained in detail. Mathematics is not difficult if you let your handy pocket calculator to do all the adding, multiplying, subtracting and dividing. A scientific calculator will also do some amazing things, and they cost only a few dollars.

There are two reasons why you need to have some basic arithmetic. These are, (1) so that you can estimate costs and selling prices when you start manufacturing your invention, and (2) so that you can easily follow the courses on basic electricity and electronics in this book. You should know the rudiments of geometry, algebra and trigonometry. But don't let those big words fool you. They are just fancy names for the kinds of things that you do with numbers and proportions every day.

Decimal notation is something you do every day when you handle money. There is a dot between the dollars and the cents. Everything to the right of the dot is cents and everything to left of the dot is dollars. The dot is called a *decimal point.* When you are dealing with numbers, everything to the left of the dot is a whole number. Everything to the right of the dot is part of a number, just like ten cents is part of a dollar.

Some examples are shown of simple arithmetic, decimal notation and algebra. For algebra you have to know what an *equals sign* is for. When you use an equals sign, everything on one side of the

+ PLUS SIGN means ADD TOGETHER $6 + 2 = 8$ $\quad \begin{array}{r} 6 \\ +2 \\ \hline 8 \end{array}$	**X** TIMES SIGN means MULTIPLY $6 \times 2 = 12$ $\quad \begin{array}{r} 6 \\ \times 2 \\ \hline 12 \end{array}$
— MINUS SIGN means SUBTRACT (Take away) $6 - 2 = 4$ $\quad \begin{array}{r} 6 \\ -2 \\ \hline 4 \end{array}$	**÷** DIVISION SIGN means DIVIDE BY (Goes into) $6 \div 2 = 3$ $\quad \frac{6}{2} = 3$
= EQUALS SIGN means THE SAME VALUE $6 + 3 = 5 + 4$ $9 = 9$	**.** DECIMAL SIGN means TO FENCE OFF WHOLE DOLLARS FROM PART DOLLARS AND TO FENCE OFF WHOLE NUMBERS FROM PART NUMBERS

Decimal Examples:

10.10 = Ten and One Tenth
100.01 = One Hundred and One Hundredth
0.001 = One Thousandth

EQUATION: $\frac{6+4}{2} = \frac{17-2}{3}$

THE UNKNOWN MR. X:

$$\frac{6+x}{2} = \frac{17-2}{3}$$

Change the side without the x to one number (this is called simplifying).

$$\frac{6+x}{2} = 5$$

$$\left(\frac{6+x}{2}\right) \times 2 = 5 \times 2$$

Multiply both sides by 2.

$$6 + x = 10$$

$$(6 + x) - 6 = 10 - 6$$

Take away 6 from both sides. (subtract).

The unknown Mr. X is equal to 4.

Figure J-1:

equals sign must add up to the same number as everything on the other side of the equals sign. In other words, the equals sign tells you that what goes in front of it is equal to what follows it. This equality on each side of the sign is called an *equation.*

ALGEBRA

Algebra is useful when one of the factors you are dealing with is unknown. Make an equation with the unknown Mr. X in it like the example in the diagram on page 35. First you do the arithmetic to simplify the numbers. You can use a pocket calculator for the arithmetic. Then you take one step at a time, always working towards the objective of getting Mr.X all by himself on one side of the equation. When that is done, the X is no longer an unknown. This process is called solving for X.

There are certain rules you must follow when solving for X. The first thing is to make your X's different than the X that is used for the multiplication sign. Usually a curvy or fancy X is used for the unknown and a plain X used for multiplication.

There are some other rules that are important. The operations listed below indicate the things that you are permitted to do once you have written down your equation. As you will see, these rules are made so that you will not do anything that would cause the equation to become unequal.

Here is a sample equation:

Example: $$\frac{6 \times 4}{4x - 8} = \frac{2 + 1 + 3}{x - 2}$$

PERMITTED OPERATIONS:

(1) Add the same number to both sides:

$$\left(\frac{6 \times 4}{4x - 8}\right) + 3 = \left(\frac{2 + 1 + 3}{x - 2}\right) + 3$$

(2) Subtract the same value from both sides:

$$\left(\frac{6 \times 4}{4x - 8}\right) - 3 = \left(\frac{2 + 1 + 3}{x - 2}\right) - 3$$

(3) Multiply both sides by the same quantity:

$$\left(\frac{6 \times 4}{4x - 8}\right) \times 8 = \left(\frac{2 + 1 + 3}{x - 2}\right) \times 8$$

(4) Divide both sides by the same number:

$$\frac{6 \times 4}{2(4x - 8)} = \frac{2 + 1 + 3}{2(x - 2)}$$

(5) Invert both sides (turn upside down):

$$\frac{4x - 8}{6 \times 4} = \frac{x - 2}{2 + 1 + 3}$$

(6) Multiply or divide both the numerator and the denominator (the top and bottom) by the same number: $\frac{3(x - 8)}{3(6 \times 4)} = \frac{\frac{1}{2}(x - 2)}{\frac{1}{2}(2 + 1 + 3)}$

The X represents an unknown number. There are certain rules for writing an X along with other numbers. Again, the rules are there to save you from making mistakes. The first rule is that the times sign or multiplication sign does not have to be used. Instead of writing 3 x X, you just write 3X, as shown. An X added to an X is two times X, and therefore is written as 2X.

When you want to show that two X's are to be multiplied together, you just write one X with a small two up in the right hand corner. If you want X to be multiplied 3 times, then write one X and place a small 3 up in the right hand corner. These are shown in the examples below:

PERMITTED METHODS:

$3 \times x = 3x \quad x \times x = x^2$

$x \times x \times x = x^3 \quad x + x = 2x \quad x + x + x = 3x$

$3(x - 4) = 3x - 12 \quad \frac{4}{2} = 2 \quad \frac{2x}{x} = 2$

$3(x + 4) = 3x + 12$

$-3(x - 4) = -3x + 12 \quad \frac{x}{x} = 1$

$4\left(\frac{1}{2} - \frac{2}{6} + \frac{3}{8}\right) = \frac{4}{2} - \frac{8}{6} + \frac{12}{8} \quad \frac{x^2}{x} = x$

One other rule that is important to know is that when two numbers that have minus signs in front of them are multiplied together, the answer is positive. This is shown in the example : (minus 3 times minus 4 equals 12).

UNDERSTANDING FORCES

Many types of inventions need brackets and stay-rods. When considering the structural parts of your inventions, a knowledge of the forces involved is an advantage. Whenever you include a bracket, a handle or a supporting part in one of your inventions, the amount of the force is a factor. Also, when you are studying electricity and magnetism, a knowledge of forces is indispensible.

A force can be shown in a diagram, and it is called a *vector.* The important thing about a vector is that the length of it shows how much force it represents. A vector that represents two pounds of force is twice as long as one representing one pound of force. Some examples are shown in Figure J-2. The other important thing to know about a vector is that it always points in the direction of the force that it represents.

ADDING VECTORS

Vectors must be added together in a special way. A vector always has an arrow head on one end to show what direction the force is going. If there are two or more vectors pointing in the same direction along the same line, they are added together in the ordinary way that measurements are added together. However, if the vectors are pointing in different directions, they must be added together utilizing the *parallelogram* method shown in the diagrams of Figure J-2. If two vectors are pointing in exactly opposite directions, then they are simply subtracted to leave a remainder.

Diagram (1) shows the addition of two vectors A and B. First mark a point on your drawing paper and mark it O, which stands for "origin". (Origin means where things start from). Then draw the two vectors with the tips of their tails on the point O. These two vectors happen to be at right angles to each other. Now draw in the dotted lines which are parallel to the vectors. Join the O to the point where the dotted lines cross. The new line across the middle is the addition of the two vectors. This line is shown as A + B.

Diagram (2) shows the addition of two vectors which are not at right angles to each other. The dotted lines which form the parallelogram are always *parallel* to the vectors. (The word parallel means that two lines which are side-by-side are going the same direction). The addition of the two vectors is shown by the arrow vector A + B.

Diagram (3) shows how several vectors are added together. Start by picking any pair of vectors and add them together by making a parallelogram for them. In this example, the vectors B and C were added together to make the vector D. Now make a new parallelogram using vector A and vector D. The final vector is the addition of A, B and C. It is shown with a heavy dotted line.

Diagram (4) shows how to determine the force exerted by a vector in any given direction. Suppose that you have a guy wire attached to a building and you want to make sure that the anchor bolt is not going to pull the bricks loose. First draw a vector to represent the force in the guy wire. Measure to the exact center of the vector and make a mark. Then use a compass to draw a circle around the vector, just touching its point and its tail, as shown. To find how much force the cable is exerting in a given direction C, draw a vector from O to C, touching the circle at C. Measure the length of the vector C to get the amount of the force in that direction.

In diagram (5) there are two vectors A and B. You want to know how much force the two vectors are exerting in the direction E. First add vectors A and B together, resulting in vector C. Then draw a circle around the vector C, as shown. Now draw in vector E so that it starts at point O and just touches the circle. Measure the length of the vector to get the force in that direction.

Figure J-2:

VECTORS

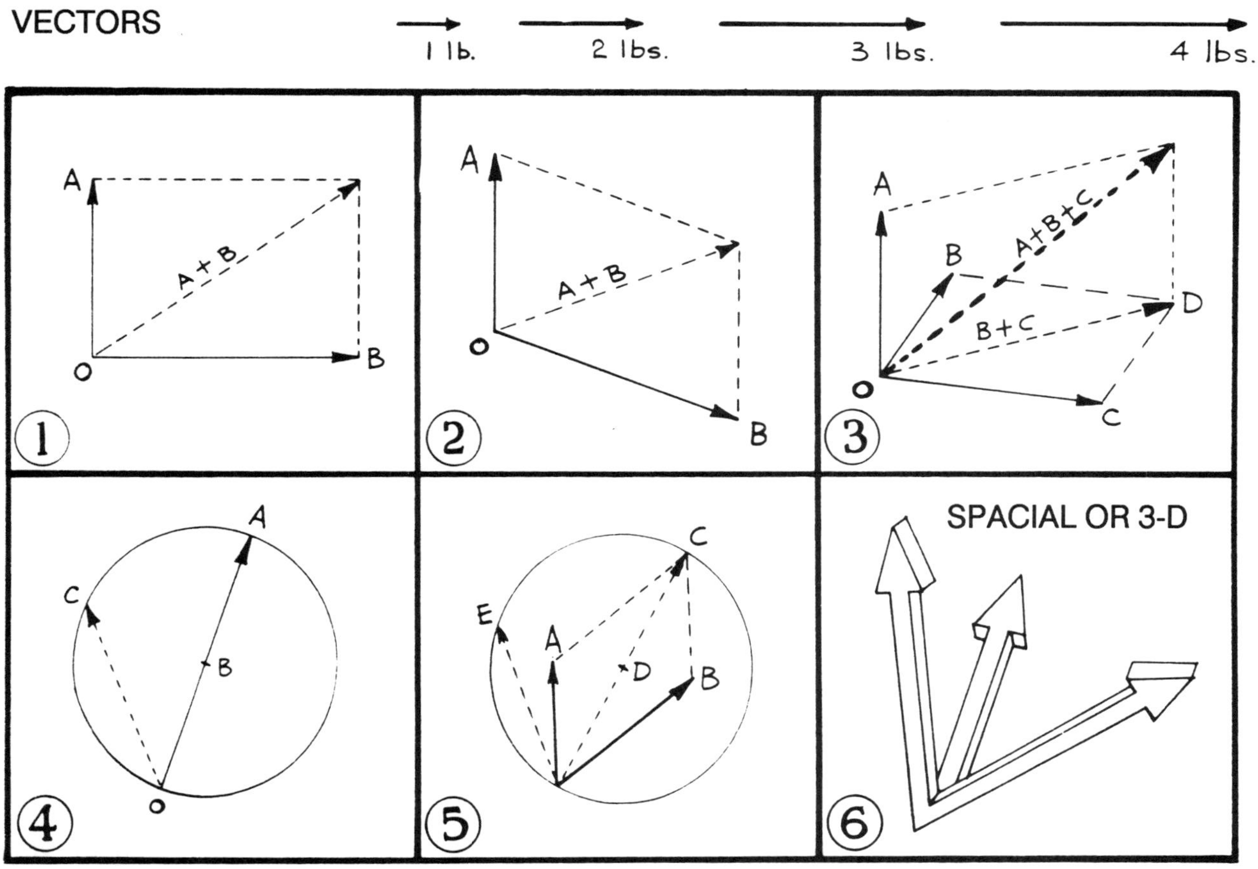

3-D VECTORS

In the real world, vectors function in space, not on a piece of drawing paper. Many problems involve forces in more than one plane, like the example in diagram 6, Figure J-2. To solve problems of this type, any pair of vectors can be chosen and added together. If you know the angle between two forces, you can plot them on drawing paper and make a parallelogram to add them togther. This process eliminates one of the forces. Continue until there are only two forces left. They can be drawn on paper and added to result in a final solution. Use a *protractor* to measure your angles carefully.

GEOMETRY

Geometry is a technique for using lines, edges, planes (flat surfaces), circles, and triangles, for measuring things and for describing things. When you measure something with a ruler you are using geometry. When you order the large pizza you are using a knowledge of the areas of circles. Ordinary geometry is actually called *plane geometry*, because a plane is a flat surface like your table top or drawing table.

An excellent habit to get into is to draw the things you are talking about, and write notes about things you want to remember. The first time you have a few dollars to spare, buy a drawing table, a T square, a set square, and several pencils. Then use the drawing table instead of a desk. Get in the habit of drawing what you are thinking about. It makes life a lot more interesting. Eventually, the drawing table will pay for itself over and over again.

TRIGONOMETRY

The word trigonometry means the study of triangles. So all you have to do is draw a triangle and look at it, and you are right into trigonometry. It is not complicated. Like other branches of study, it has a few simple rules. If you learn the rules, the rest is simple.

MEASURING ANGLES

When two straight lines run in different directions, the difference in direction is called an *angle.* An angle is measured in *degrees*. The largest angle you can imagine is one complete circle. A complete circle has 360 degrees. Degrees are indicated by placing a small O up in the right hand corner of the number of degrees, as shown. One degree is divided into 60 smaller pieces called *minutes*. Because minutes are hard to calculate, they are seldom used. Instead of minutes, regular decimal notation is used for degrees. In other words, instead of writing 6 degrees and 30 minutes, you can write 6.5 degrees. This method is simple and therefore sensible.

The circle of 360 degrees is usually cut into four pieces, called *quadrants*. Each quadrant has 90 degrees. An angle with 90 degrees is usually called a *right angle*. A triangle which has one angle of 90 degrees is called a *right triangle.* These are shown in the diagram, Figure J-4.

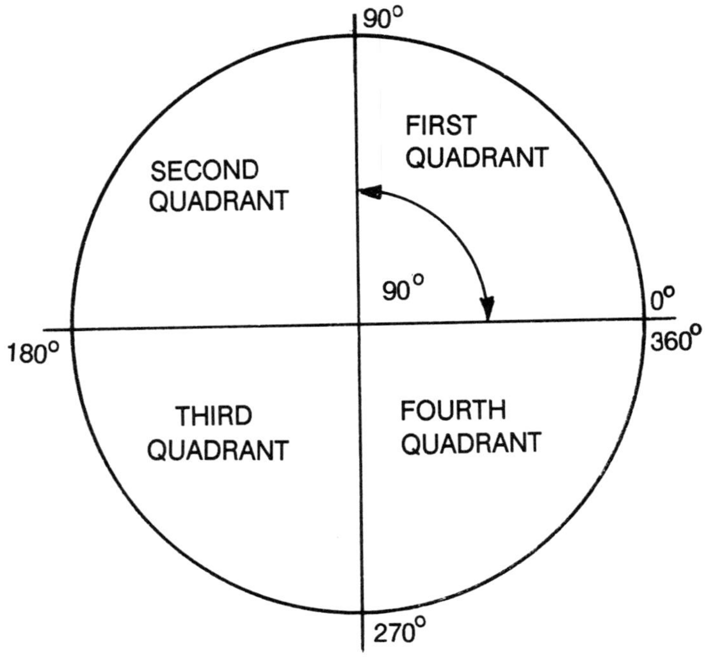

Figure J-3. A circle contains 360 degrees. It is usually cut into four pieces called quadrants. Each quadrant has 90 degrees. An angle of 90 degrees is called a right angle.

SINES AND COSINES

The SINE of an angle and the COSINE of an angle are names that have been given to relationships in triangles. When something has a name, it is easier to remember. Instead of SINE they could have called it JACK or JOHN.

The SINE of an angle is a fraction. If you carefully measure the length of one of the lines in a triangle, and then divide the number by the length of

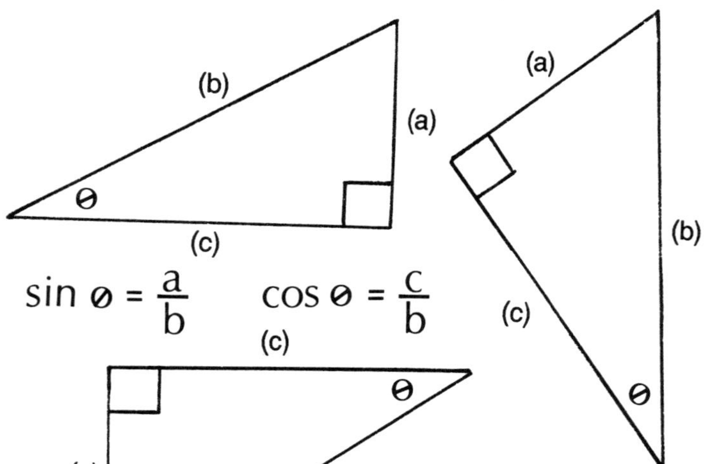

Figure J-4: Examples of right triangles. The 90 degree angle is marked with a small square.

another side of the triangle, you get a number that can be used in special ways.

In the three triangles shown in Figure J-4 each one has a right angle in one corner. This is indicated by a small square in that corner. The angle to which the SINE belongs is marked θ. Note that in each case the angle θ is opposite the right angle. The number that results when (a) side is divided by (b) side is the SINE of the angle.

The COSINE is also a fraction. When side (c) is divided by side (b), the resulting number is the COSINE of the angle θ. All you have to remember is that the side (a) is opposite the enclosed angle θ.

IMAGINARY RIGHT ANGLE

The right angle is necessary because it automatically defines the lengths of side (a) and side (b). However, a given angle is always going to have the same SINE. Therefore, we don't have to see the right angle. We don't even need a triangle. Since the SINE for a given angle will always be the same number, why not just write down the number?

When you want to know the SINE or the COSINE of an angle, all you have to do is look up the angle on a list of angles from zero to 90 degrees, and there it is. The lists of angles for SINES and COSINES are on page 168.

For convenience in using SINES and COSINES, they are shortened to three letters. The SINE is always shown as *sin*. The COSINE is always shown as *cos*. Now when you see these names, you will remember what they are for. They are used all the time by engineers, architects, surveyors and many other technicians.

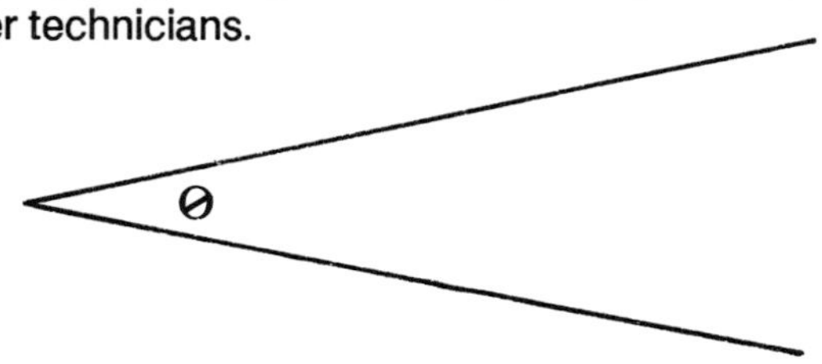

Figure J-5: In the diagram, it doesn't matter where you place the right angle. The sin will always be the same for that particular angle. This makes it possible to have a list of all possible angles with their sines included. To find the sin of an angle, it is only necessary to look up the angle on the list.

LAW OF SINES

Apparently, the right angle that generates the sin and cos relationships can be imaginary. This makes it possible to apply the *sin* relationship to any triangle without a right angle. The result is the law of sines, which is:

$$\frac{a}{\text{SIN A}} = \frac{b}{\text{SIN B}} = \frac{c}{\text{SIN C}}$$

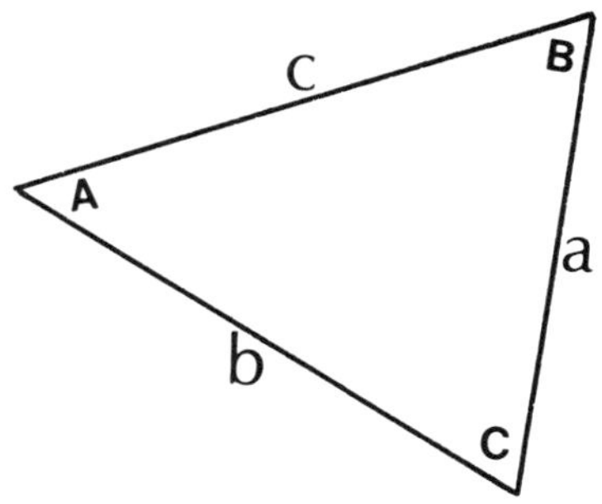

Figure J-6: In any triangle the sin of any angle is proportionate to the length of the side opposite.

USING EXPONENTIALS

When a number is multiplied by itself, the term that is used to describe that action is to square the number. Four time four (4 x 4) is sixteen (16). The number l6 is 4 squared. A quick way of writing four squared is to write down the number 4 and put a small 2 up in the right hand corner, like this, 4^2.

The number 100 is 10 squared, or 10^2 The number 1000 is 10^3 and 10,000 is 10^4 (called ten to the fourth). These little numbers are called *exponentials*. The use of exponentials provides a way to write really big numbers in a simple way. If we want to write 16 million, (16,000,000), it isn't necessary to write all those zeros. Instead you can just write down 16 x 10^6. Since 10^6 is a million, then 16 x 10^6 is 16,000,000.

It is also handy for writing very tiny numbers. For example, if you want to write a sixteen millionth of an inch, you can just write 16 x 10^{-6} inches. That is really a great time saver and it is easy to understand if you take a close look at the way it is done. The minus sign means that the exponential is inverted. Ten becomes a tenth. A hundred becomes a hundredth. That is easy, too. Try it.

EXAMPLES:

1.63×10^{12} is the same as 1,630,000,000,000

$10^2 = 100$ $\quad$ $3 \times 10^3 = 3{,}000$

$6 \times 10^6 = 6{,}000{,}000$

$10^{-2} = \frac{1}{100}$ $\quad$ $3 \times 10^{-3} = \frac{1}{3{,}000}$

$6 \times 10^{-6} = \frac{1}{6{,}000{,}000}$

SINES -- 0 TO 90 DEGREES BY HALF DEGREES			
0.0 0.00000	22.5 0.38268	45.0 0.70711	67.5 0.92388
0.5 0.00873	23.0 0.39073	45.5 0.71325	68.0 0.92718
1.0 0.01745	23.5 0.39875	46.0 0.71934	68.5 0.93042
1.5 0.02618	24.0 0.40674	46.5 0.72537	69.0 0.93358
2.0 0.03490	24.5 0.41469	47.0 0.73135	69.5 0.93667
2.5 0.04362	25.0 0.42262	47.5 0.73728	70.0 0.93969
3.0 0.05234	25.5 0.43051	48.0 0.74314	70.5 0.94264
3.5 0.06105	26.0 0.43837	48.5 0.74896	71.0 0.94552
4.0 0.06976	26.5 0.44620	49.0 0.75471	71.5 0.94832
4.5 0.07846	27.0 0.45399	49.5 0.76041	72.0 0.95106
5.0 0.08716	27.5 0.46175	50.0 0.76604	72.5 0.95372
5.5 0.09585	28.0 0.46947	50.5 0.77162	73.0 0.95630
6.0 0.10453	28.5 0.47716	51.0 0.77715	73.5 0.95882
6.5 0.11320	29.0 0.48481	51.5 0.78261	74.0 0.96126
7.0 0.12187	29.5 0.49242	52.0 0.78801	74.5 0.96363
7.5 0.13053	30.0 0.50000	52.5 0.79335	75.0 0.96593
8.0 0.13917	30.5 0.50754	53.0 0.79864	75.5 0.96815
8.5 0.14781	31.0 0.51504	53.5 0.80386	76.0 0.97030
9.0 0.15643	31.5 0.52250	54.0 0.80902	76.5 0.97237
9.5 0.16505	32.0 0.52992	54.5 0.81412	77.0 0.97437
10.0 0.17365	32.5 0.53730	55.0 0.81915	77.5 0.97630
10.5 0.18224	33.0 0.54464	55.5 0.82413	78.0 0.97815
11.0 0.19081	33.5 0.55194	56.0 0.82904	78.5 0.97992
11.5 0.19937	34.0 0.55919	56.5 0.83389	79.0 0.98163
12.0 0.20791	34.5 0.56641	57.0 0.83867	79.5 0.98325
12.5 0.21644	35.0 0.57358	57.5 0.84339	80.0 0.98481
13.0 0.22495	35.5 0.58070	58.0 0.84805	80.5 0.98629
13.5 0.23345	36.0 0.58778	58.5 0.85264	81.0 0.98769
14.0 0.24192	36.5 0.59482	59.0 0.85717	81.5 0.98902
14.5 0.25038	37.0 0.60181	59.5 0.86163	82.0 0.99027
15.0 0.25882	37.5 0.60876	60.0 0.86603	82.5 0.99144
15.5 0.26724	38.0 0.61566	60.5 0.87036	83.0 0.99255
16.0 0.27564	38.5 0.62251	61.0 0.87462	83.5 0.99357
16.5 0.28402	39.0 0.62932	61.5 0.87882	84.0 0.99452
17.0 0.29237	39.5 0.63608	62.0 0.88295	84.5 0.99540
17.5 0.30071	40.0 0.64279	62.5 0.88701	85.0 0.99619
18.0 0.30902	40.5 0.64945	63.0 0.89101	85.5 0.99692
18.5 0.31730	41.0 0.65606	63.5 0.89493	86.0 0.99756
19.0 0.32557	41.5 0.66262	64.0 0.89879	86.5 0.99813
19.5 0.33381	42.0 0.66913	64.5 0.90258	87.0 0.99863
20.0 0.34202	42.5 0.67559	65.0 0.90631	87.5 0.99905
20.5 0.35021	43.0 0.68200	65.5 0.90996	88.0 0.99939
21.0 0.35837	43.5 0.68835	66.0 0.91355	88.5 0.99966
21.5 0.36650	44.0 0.69466	66.5 0.91706	89.0 0.99985
22.0 0.37461	44.5 0.70091	67.0 0.92050	89.5 0.99996
22.5 0.38268	45.0 0.70711	67.5 0.92388	90.0 1.00000

COSINES--0 TO 90 DEGREES BY HALF DEGREES			
0.0 1.00000	22.5 0.92388	45.0 0.70711	67.5 0.38268
0.5 0.99996	23.0 0.92051	45.5 0.70091	68.0 0.37461
1.0 0.99985	23.5 0.91706	46.0 0.69466	68.5 0.36650
1.5 0.99966	24.0 0.91355	46.5 0.68836	69.0 0.35837
2.0 0.99939	24.5 0.90996	47.0 0.68200	69.5 0.35021
2.5 0.99905	25.0 0.90631	47.5 0.67559	70.0 0.34202
3.0 0.99863	25.5 0.90259	48.0 0.66913	70.5 0.33381
3.5 0.99813	26.0 0.89879	48.5 0.66262	71.0 0.32557
4.0 0.99756	26.5 0.89493	49.0 0.65606	71.5 0.31731
4.5 0.99692	27.0 0.89101	49.5 0.64945	72.0 0.30902
5.0 0.99619	27.5 0.88701	50.0 0.64279	72.5 0.30071
5.5 0.99540	28.0 0.88295	50.5 0.63608	73.0 0.29237
6.0 0.99452	28.5 0.87882	51.0 0.62932	73.5 0.28402
6.5 0.99357	29.0 0.87462	51.5 0.62252	74.0 0.27564
7.0 0.99255	29.5 0.87036	52.0 0.61566	74.5 0.26724
7.5 0.99144	30.0 0.86603	52.5 0.60876	75.0 0.25882
8.0 0.99027	30.5 0.86163	53.0 0.60182	75.5 0.25038
8.5 0.98902	31.0 0.85717	53.5 0.59482	76.0 0.24192
9.0 0.98769	31.5 0.85264	54.0 0.58779	76.5 0.23345
9.5 0.98629	32.0 0.84805	54.5 0.58070	77.0 0.22495
10.0 0.98481	32.5 0.84339	55.0 0.57358	77.5 0.21644
10.5 0.98325	33.0 0.83867	55.5 0.56641	78.0 0.20791
11.0 0.98163	33.5 0.83389	56.0 0.55919	78.5 0.19937
11.5 0.97992	34.0 0.82904	56.5 0.55194	79.0 0.19081
12.0 0.97815	34.5 0.82413	57.0 0.54464	79.5 0.18224
12.5 0.97630	35.0 0.81915	57.5 0.53730	80.0 0.17365
13.0 0.97437	35.5 0.81412	58.0 0.52992	80.5 0.16505
13.5 0.97237	36.0 0.80902	58.5 0.52250	81.0 0.15644
14.0 0.97030	36.5 0.80386	59.0 0.51504	81.5 0.14781
14.5 0.96815	37.0 0.79864	59.5 0.50754	82.0 0.13917
15.0 0.96593	37.5 0.79335	60.0 0.50000	82.5 0.13053
15.5 0.96363	38.0 0.78801	60.5 0.49242	83.0 0.12187
16.0 0.96126	38.5 0.78261	61.0 0.48481	83.5 0.11320
16.5 0.95882	39.0 0.77715	61.5 0.47716	84.0 0.10453
17.0 0.95630	39.5 0.77163	62.0 0.46947	84.5 0.09585
17.5 0.95372	40.0 0.76604	62.5 0.46175	85.0 0.08716
18.0 0.95106	40.5 0.76041	63.0 0.45399	85.5 0.07846
18.5 0.94832	41.0 0.75471	63.5 0.44620	86.0 0.06976
19.0 0.94552	41.5 0.74896	64.0 0.43837	86.5 0.06105
19.5 0.94264	42.0 0.74315	64.5 0.43051	87.0 0.05234
20.0 0.93969	42.5 0.73728	65.0 0.42262	87.5 0.04362
20.5 0.93667	43.0 0.73135	65.5 0.41469	88.0 0.03490
21.0 0.93358	43.5 0.72537	66.0 0.40674	88.5 0.02618
21.5 0.93042	44.0 0.71934	66.5 0.39875	89.0 0.01745
22.0 0.92718	44.5 0.71325	67.0 0.39073	89.5 0.00873
22.5 0.92388	45.0 0.70711	67.5 0.38268	90.0 0.00000

USING RECIPROCALS

Reciprocal is a long word with a simple intent. To make a reciprocal for any number, pick up your pocket calculator, press clear, then press the number one. Now press the division sign, and then enter your number, whatever it is. Finally, press the equals sign. The number that appears on the screen is a reciprocal.

Choose the number 649 (for example). Press 1. Press divide. Enter 649. Press equals. The answer you get is 0.0015408, and that is the reciprocal of 649.

Now find the reciprocal of 0.0015408. Press 1. Press divide. Press 0.0015408. Press equals. The answer is 649. In other words, the very same operation takes you back and forth over the same two numbers. That's called reciprocation. So the two numbers are called reciprocals.

$$\text{The reciprocal of 649} = \frac{1}{649} = 0.0015408$$

$$\text{The reciprocal of 0.0015408} = \frac{1}{0.0015408} = 649$$

THE GREEK ALPHABET

Α,	α	*Alpha*	Ν,	ν	*Nu*
Β,	β	*Beta*	Ξ,	ξ	*Xi*
Γ,	γ	*Gamma*	Ο,	ο	*Omicron*
Δ,	δ	*Delta*	Π,	π	*Pi*
Ε,	ϵ	*Epsilon*	Ρ,	ρ	*Rho*
Ζ,	ζ	*Zeta*	Σ,	σ	*Sigma*
Η,	η	*Eta*	Τ,	τ	*Tau*
Θ,	θ	*Theta*	Υ,	υ	*Upsilon*
Ι,	ι	*Iota*	Φ,	ϕ	*Phi*
Κ,	κ	*Kappa*	Χ,	χ	*Chi*
Λ,	λ	*Lambda*	Ψ,	ψ	*Psi*
Μ,	μ	*Mu*	Ω,	ω	*Omega*

SQUARE ROOTS FROM 0 TO 100 BY 0.5 INTERVALS

0.5	0.70711	17.5	4.18330	34.5	5.87367	51.5	7.17635	68.5	8.27647	85.5	9.24662
1.0	1.00000	18.0	4.24264	35.0	5.91608	52.0	7.21110	69.0	8.30662	86.0	9.27362
1.5	1.22474	18.5	4.30116	35.5	5.95819	52.5	7.24569	69.5	8.33666	86.5	9.30054
2.0	1.41421	19.0	4.35890	36.0	6.00000	53.0	7.28011	70.0	8.36660	87.0	9.32738
2.5	1.58114	19.5	4.41588	36.5	6.04152	53.5	7.31437	70.5	8.39643	87.5	9.35414
3.0	1.73205	20.0	4.47214	37.0	6.08276	54.0	7.34847	71.0	8.42615	88.0	9.38083
3.5	1.87083	20.5	4.52769	37.5	6.12373	54.5	7.38241	71.5	8.45577	88.5	9.40744
4.0	2.00000	21.0	4.58258	38.0	6.16441	55.0	7.41620	72.0	8.48528	89.0	9.43398
4.5	2.12132	21.5	4.63681	38.5	6.20484	55.5	7.44983	72.5	8.51469	89.5	9.46044
5.0	2.23607	22.0	4.69042	39.0	6.24500	56.0	7.48331	73.0	8.54400	90.0	9.48683
5.5	2.34521	22.5	4.74342	39.5	6.28490	56.5	7.51665	73.5	8.57321	90.5	9.51315
6.0	2.44949	23.0	4.79583	40.0	6.32456	57.0	7.54983	74.0	8.60232	91.0	9.53939
6.5	2.54951	23.5	4.84768	40.5	6.36396	57.5	7.58288	74.5	8.63134	91.5	9.56556
7.0	2.64575	24.0	4.89898	41.0	6.40312	58.0	7.61577	75.0	8.66025	92.0	9.59166
7.5	2.73861	24.5	4.94975	41.5	6.44205	58.5	7.64853	75.5	8.68907	92.5	9.61769
8.0	2.82843	25.0	5.00000	42.0	6.48074	59.0	7.68114	76.0	8.71780	93.0	9.64365
8.5	2.91548	25.5	5.04975	42.5	6.51920	59.5	7.71362	76.5	8.74643	93.5	9.66954
9.0	3.00000	26.0	5.09902	43.0	6.55744	60.0	7.74597	77.0	8.77496	94.0	9.69536
9.5	3.08221	26.5	5.14782	43.5	6.59545	60.5	7.77818	77.5	8.80341	94.5	9.72111
10.0	3.16228	27.0	5.19615	44.0	6.63325	61.0	7.81025	78.0	8.83176	95.0	9.74680
10.5	3.24037	27.5	5.24404	44.5	6.67083	61.5	7.84219	78.5	8.86002	95.5	9.77241
11.0	3.31663	28.0	5.29150	45.0	6.70820	62.0	7.87401	79.0	8.88819	96.0	9.79796
11.5	3.39117	28.5	5.33854	45.5	6.74537	62.5	7.90569	79.5	8.91628	96.5	9.82344
12.0	3.46410	29.0	5.38517	46.0	6.78233	63.0	7.93725	80.0	8.94427	97.0	9.84886
12.5	3.53553	29.5	5.43139	46.5	6.81909	63.5	7.96869	80.5	8.97218	97.5	9.87421
13.0	3.60555	30.0	5.47723	47.0	6.85565	64.0	8.00000	81.0	9.00000	98.0	9.89950
13.5	3.67423	30.5	5.52268	47.5	6.89202	64.5	8.03119	81.5	9.02773	98.5	9.92472
14.0	3.74166	31.0	5.56776	48.0	6.92820	65.0	8.06226	82.0	9.05538	99.0	9.94987
14.5	3.80789	31.5	5.61249	48.5	6.96419	65.5	8.09321	82.5	9.08295	99.5	9.97497
15.0	3.87298	32.0	5.65685	49.0	7.00000	66.0	8.12404	83.0	9.11043	100.0	10.00000
15.5	3.93700	32.5	5.70088	49.5	7.03562	66.5	8.15475	83.5	9.13783	121.0	11.00000
16.0	4.00000	33.0	5.74456	50.0	7.07107	67.0	8.18535	84.0	9.16515	144.0	12.00000
16.5	4.06202	33.5	5.78792	50.5	7.10634	67.5	8.21584	84.5	9.19239	169.0	13.00000
17.0	4.12311	34.0	5.83095	51.0	7.14143	68.0	8.24621	85.0	9.21954	196.0	14.00000

Tension (stretch)

Compression (squeeze)

Torsion (twist)

Deflection (bend)

Shear (sideways slip)

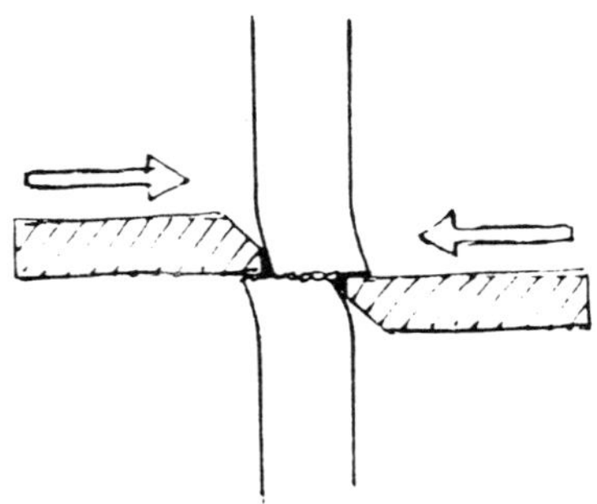

Shearing Action
(similar to pair of scissors)

ENGINEERING TERMS ILLUSTRATED

GLOSSARY OF WORDS FOR PATENT APPLICATIONS

TERMINOLOGY OF THE CLAIMS

The advice to the inventor from the experts is to have an expert prepare the patent application, and in particular, the claims. Often the inventor has been working on his invention for many months, or even years. Therefore, it is unlikely that anyone will perceive the working structure of the invention in the same way that the inventor sees it. For this reason, it is always a good idea for the inventor to make the first draft of the claims, because his intimate knowledge of the invention will be a factor. The inventor's written claims can then be edited and worked over by an expert in patent matters.

Wording used in the claims has the objective of clarity and avoidance of conflict in the actual meaning of words. It is a good idea to study the claims of as many inventions as possible and make notes of the words and combinations of words used to define structure and composition.

Some words commonly used in claims and in specifications are listed here alphabetically with brief descriptions of the meaning or intent:

Abandon: To give up or yield with the intent to never again claim a right or interest in.
Ablated: Removed due to erosion, evaporation or sublimation.
Abrasion: Grinding away by friction.
Absolute zero: 459.67 degrees F. below zero.
Abstract: To summarize.
Accelerate; To increase the speed of.
Accessible: Pertaining to availability....can be reached.
Accessory: A device that increases the effectiveness of the main object; in a secondary role.
Accordingly: Agrees and conforms.
Accurate: Freedom from error; exactly right.
Accrue, accumulate: To increase little by little.
Acetate: Cellulose acetate,
Acetone: $C3H6O$.
Acetylene: $HC\equiv CH$.
Achieve: To complete successfully.
Achromatic lens: A color-corrected lens.
Acoustic: Pertaining to sound and hearing.
Action: The way a mechanism operates.
Activate: To cause to change or move.
Actuate: To put into motion, to act on.
Actuator: Means for putting into motion.
Acquisition: Something got or gained.
Acrylic: Pertaining to acrylic acid (polymers).
Acrylic acid: $C3H4O$.
Act: To take action; to play a part.
Actual: Factual; real; not imaginary.
Acute: Sharp angle or pointed.
Adaptable: Can be made to function as described.
Adapted to: Made to accomplish the desired result.
Additive: A small amount of a substance added to change or improve the properties of the whole.
Adequate: Reasonably fulfilling the requirements.
Adjustable: Means of regulating or setting in position.
Adhere or cohere: To stick together firmly by adhesive means.
Adhesive: Substance used for fastening, like glue.
Adjacent: Near, beside or next to.
Adjustment: A small change in position.
Adsorption: Adhesion between layers of molecules at the interfaces of different substances, usually a layer of gas or liquid molecules adhering to the surface of a solid.
Adulterate: Mix with a foreign substance, usually an undesired result.
Advance: To move forward, to move by degrees or increments.
Aerate: To supply with air or oxygen.
Aesthetic: Pertaining to the artistic and beautiful; appreciation of beauty.
Affiant: One who takes an oath.
Affirm: To confirm the validity of.
Agent: Wielding the power of; instrumental representative.
Agitate: To impart a jerky motion to; to stir up.
Agreement: An understanding; a treaty or contract.
Aileron: The movable part of an aircraft wing.
Alacrity: Lively; prompt.

Alcohol: C_2H_6O.
Algebra: An extension from arithmetic with letters representing unknown numbers.
Al-Gol: A computer language in which algorithms are applied to provide solutions.
Alidade: A surveyor's transit or theodolite.
Alignment: The correct adjustment of a machine or vehicle.
Allegation: An unsupported assertion.
Almost: Nearly but not quite accurate.
Alnico: A powerful permanent magnet.
Alteration: Change; Modification.
Alternating current: Electric current that changes direction back and forth periodically.
Alternative: Pertaining to choice between possibilities.
Alternator: A device for generating alternating current.
Aluminum: Element 13, atomic weight 26.98. Silvery metal.
Amalgamate: To mix different substances together.
Ammeter: A meter for measuring electric current in amperes.
Ampere: The amount of electricity that flows during one second when the potential drop is one volt.
Ampersand: A symbol & as a short form of the word and.
Amplify: To enlarge on, to increase, to add power.
Amyl: Pentyl; C_5H_{11}.
Analogous: Similar; related.
Analogy: To show the resemblance or similarity.
Anemoscope: Wind direction indicator.
Aneroid: Containing no liquid.
Analyse: To resolve and reveal the factors affecting a situation.
Anneal: To heat and let cool slowly to relieve internal stresses.
Anchor: To fasten securely; a secure type of fastener.
Angular: With sharp angles; sharp corners.
Annular: Ring like, shaped like a ring.
Annunciator: Electrically controlled signal board or indicator board.
Anodize: To plate with a protective or beautifying layer of metal by electrolysis.
Antenna: Device for receiving radio waves.
Anti-: Meaning against; opposed.
Anticipate: To forsee the result by logic or intuition.
Anticatalyst: Retards a chemical reaction.
Antiparticle: A particle of matter which when combined with a known particle causes total annihilation of both particles.
Antiproton: The antiparticle of the proton.
Anvil: Any part designed to withstand repeated impact.
Aperture: An opening; a port or a hole.
Apogee: In an orbit, the point furthest from the center of the earth (or of the orbitted planet).
Apparatus: Parts and accessories usually made for a specific application.
Apparent: Visible to view; obvious.
Apply: To place one substance or object in contact with another object. To act to achieve a desired result.
Apprize: To give notice of.
Aptitude: Natural ability, tendency to learn quickly.
Aqueous: Watery, like water.
Arbiter: A person having the power to resolve a dispute.
Arch: A door or opening with a rounded top.
Arcade: A row of arches supported on pillars.
Area: An extent of surface, often with a boundary.
Argil: Potter's clay.
Argument: To display the facts in a reasonable way.
Armature: A soft iron core in an electromagnetic device.
Arsenic: Element 33; atomic weight 74.91. Poisonous substance.
Art: Craftsmanship; the methods of making anything manufactured or hand made.
Artifact: Something that has been made by hand or manufactured.
Artificial: Imitation; not natural.
Ascribe: To attribute to.
Assembly: A group of parts that go together.
Assign: To sign over rights to another.
Astatine: Element 85; atomic weight 210.
Astronomical unit: Distance from the center of the earth to the center of the sun; about 93 million miles.
Atom: Smallest elementary particle of matter.
Attachment: The means for joining one part to another.
Attain: Achieve or accomplish.
Attribute: The quality or character of anything.
Audio: Pertaining to sound.
Auto: A prefix meaning "self".
Automatic: Self-acting or self-regulating mechanism.
Awl: A sharp pointed tool for piercing holes.
Axis: The line about which it may turn or revolve.
Axle: A shaft or spindle on which a wheel turns.

Babbitt: A white, antifriction alloy of metals.
Back: Towards the rear; the rear part of.
Backlash: Loose action in a machine due to wear or poorly fitted gears or parts.
Baffle: A part designed to check or deflect the flow of a substance.
Bake: To heat in an oven, usually dry heat.
Bakelite: Hard, durable plastic compounded from synthetic resin.
Balance wheel: A wheel designed to regulate motion.
Ball-and-socket: A joint with universal motion.
Ball-bearing: A rotary bearing comprising a ring of steel balls enclosed within a containment race and the bearing surfaces. Converts sliding friction to rolling friction.
ball cock: A valve that opens and closes due to the floatation of a hollow ball.
Ballistics: The science of the trajectory of a projectile.

Band: A shackle, bracelet or bracket that supports, binds, encircles or restrains.
Bandwidth: The range of radio frequencies allocated to a modulated carrier wave.
Bar: Usually a structural member having a maximum length and a minimum thickness.
Bar code: A set of printed bars of varying thickness which represent coded prices on products.
Barb: A sharp point designed to snare or irritate, as on barbed-wire.
Base: The bottom of anything, especially a pedestal shaped part.
Basic: Fundamental; essential.
Basis: Fundamental principle.
Basin: An open vessel or natural depression with sloped sides.
Batch: The raw materials for a process, mixed ready for use or application.
Battery: Electric cell or cells.
Beacon: Signal light; a source of inspiration.
Beam: A principle horizontal member of a machine or structure.
Bearing: The supporting part within which a pin, spindle or axle rotates.
Beat: An audible frequency caused by two or more devices operating at different RPM or different frequencies.
Bedplate: The baseplate or foundation frame.
Bellcrank: A bell-shaped lever for changing the direction of motion.
Belljar: A jar in the shape of a bell, used upside down for vacuum experiments.
Bellows: Means for directing blasts of air into a forge.
Belt: Means for transmitting power from pulley to pulley.
Bend: To shape by applying pressure.
Benzene: C_6H_6.
Bessemer process: Forcing air through molten metal to oxidize and burn off impurities.
Betatron: A device for accelerating electrons.
Between: A position midway.
Bevel: The angled edge of a component to eliminate the sharp edge.
Bevel gear: Toothed gears bevelled to mesh non-parallel shafts.
Biconcave: Concave on both faces.
Biconvex: Convex on both faces.
Bifurcate: To divide into two branches, Y shaped.
Bilateral: Having two sides emphasized.
Billion: A thousand million.
Binary: Characterised by two parts.
Biodegradable: Can be broken down and assimilated by natural processes.
Bisect: Divide equally into two parts.
Blackout: Failure of an electrical system.
Block: A squarish, solid chunk of matter such as wood or stone.
Blow-off: A pressure relief valve.
Body: Main part of.
Boil: To increase heat of a liquid until cavitation due to vapor pressure is evident.
Bolt: A sliding cylinder to lock a door, a cylinder with a head on one end and threads or an eye on the other end. A fastener.
Bona fide: Genuine; not fraudulent or deceitful.
Border: The outer part, the edge or margin.
Bore: Diameter of a cylinder; to make a cylindrical opening with a rotating tool.
Boron: Element 5, atomic weight 10.82. Metaloid element.
Boss: A part that sticks out, enlarged part, a lug or stud that is protuberant.
Boundary: A border; the limit of.
Brace: To support with braces, to secure with brackets.
Bracket: An angle support.
Brake: Means for stopping motion.
Break: To fracture.
Bridge: Any bridgelike support or part.
Brittle: Easily broken, fragile.
Broad: Wide.
Broach: Ream.
Bromine: Element 35, atomic weight 79.9. Volatile liquid.
Bronze: An alloy of copper and tin.
Buffer: A cushion to lessen shock.
Bumper: A shock absorber.
Buoyancy: The tendency for light objects to rise in a fluid.
Burn: Consume by combustion.
Burst; Rupture suddenly.
Butterfly valve: A clack valve with a central axis.
Butylene: C_4H_8.
Bypass: An auxiliary duct or outlet.

Cable: Wire rope; several insulated conductors twisted spirally together around a central core.
Cadmium: Element 48; Atomic weight 112.41. Metalic, bluish-white.
Calcium: Element 20; atomic weight 40.08. Silvery-white, metalic.
Calculate: Careful reckoning.
Calibrate: To set an instrument to exact or accepted standards of operation.
Caliper: An adjustable feeler for measuring small diameters or thicknesses.
Calorie: Measurement of heat or energy; The amount of heat energy necessary to raise the temperature of a kilogram of water by one degree Celsius.
Cam: A rotating or sliding projection for imparting reciprocating motion to a roller or push-rod.
Camber: Slightly bent or curved.
Camphor: $C_{10}H_{16}O$
Camshaft: The shaft with one or more cams on it.
Cancel: To cross out, to erase, abolish.
Cant: An oblique or slanting surface.
Cantilever: A projecting beam or support attached at the opposite end.
Capacitance: The name given for the strength of the charge held by a capacitor.
Capacity: the power to contain or absorb a volume.
Capillary: a tube or duct as fine as a hair.
Carbide: A compound of carbon and calcium.

Carbon: An element with atomic weight 12.011.
Cardinal: Of essential importance.
Carriage: A moving frame or support in a machine.
Casette: A container for something easily damaged by handling; A container for magnetic recording tape..
Case: A box or covering; a set of arguments.
Cast: To form to shape, as in a mold.
Caster: A wheel mounted on a vertical off-center pivot so as to be self-steering.
Catalyst: a substance that increases the speed of a chemical reaction.
Category: Classification.
Catenary: The curved shape that a chain or heavy rope takes when suspended between two points of equal height; the curve of a suspension bridge.
Cathode: The electrode that emits electrons in a vacuum tube.
Caulk: To pack cracks and joints with a sealing cement.
Cation: A positive ion.
Cause: The action that produces a result.
Caustic: Corrosive chemical reaction.
Cavity: A hollow place; an opening closed at the opposite end.
Cease, Cessation: To stop the action of.
Celerity: Speed.
Cell: A container; receptacle.
Celsius: Name of an international thermometer in which the freezing point of water is zero degrees and the boiling point of water is 100 degrees, (centigrade degrees).
Cement: Any substance used for making objects adhere to each other, like a glue. The act of gluing, to cement.
Center: Mid-point, especially of rotary parts.
Center of gravity: The point at which an object is exactly balanced in any direction.
Centimeter: The hundredth part of a meter.
Centrifugal force: The force that tends to impel the parts outward from the center of a rotating body.
Centripetal force: The force that resists centrifugal force.
Ceramic: A baked material made from clay, earth or ground stone.
Certain: Thoroughly established.
Certificate: Document attesting to authorization.
Cesium: Element 55; Atomic weight 132.91. silvery-white substance.
Cetane: C16H34.
Chain pump: A chain arranged to run over two pulleys, one of which is below the surface of the liquid to be pumped.
Channel: A groove, rut, gutter or furrow. A conduit.
Charge: The amount of electricity stored by a capacitor, etc.
Chassis: The frame on which the parts of a machine are mounted.
Chase: To cut a groove in, to incise or engrave.
Chemistry: A science that reveals the composition of substances.
Chip, silicon chip: Microelectronics.
Chisel: A tool with a short cross blade used for working wood or stone.
Chlorine: Element 17; atomic weight 35.457. Greenish, yellow gas.
Chroma: Color and hue (intensity) defined.
Chronometry: The art of measuring time.
Chuck: A rotary vise for holding a drill-bit or tool.
Circa: About that time.
Circuit: A loop of wire for conducting an electric current.
Circulation: The movement of a fluid through a duct system.
Circum: A prefix meaning "around".
Circumference: The distance around a circle.
Cite: To quote, to mention, often as an authority.
Clack valve: A hinged plate or baffle to block an opening in one direction.
Clarify: Top make clear.
Clamp: To hold together by pressure.
Classify: To set into specie, rank or type.
Clearance: Space between objects or surfaces.
Cleat: A small projection in contact with the ground to prevent slippage.
Clevis: A "U" bracket with a removable cross-pin.
Clockwise: In the direction that clock hands turn.
Clutch: A devise that can be engaged temporarily to transmit rotary motion by friction.
Coaxial: On the same axis.
Cobalt: Element 27; Atomic weight 58.94. Metalic.
Co-bol: A business language for programming a computer.
Code: To put intelligence into the form of symbols.
Coexist: To exist together.
Coextensive: To have the same dimension, to the same extent.
Cogwheel: A wheel with cogs, bumps or teeth, like a gearwheel.
Cohere: Adhere.
Coil: Electrical, a wound spiral of wire.
Collar: A circular band.
Collateral: Side by side or parallel.
Collet chuck: A threaded ring or circular flange around three or more vise jaws, for holding small parts or bits.
Combination: Two or more things mixed together or made to function together.
Commutator: A rotating set of contacts used for distributing electricity in sequence to the coils of an electric motor.
Compact: Closely united or packed; a small version.
Compare: To reveal similarities or differences in two objects.
Complex: Not simple; not easily described.
Compound: Two or more elements mixed together in proportions.
Compress: To squeeze into a smaller space.
Computer: An electronic system for storing, processing and retrieving information.
Concave: A depression or hollow.
Concept: A thought; a mental im-

age.
Concise: Brief and to the point.
Conrete: Solid and rigid; unchanging.
Condense: Change a vapour into a liquid by pressure and cooling. To make more compact.
Condition: State of usefulness; a stipulation.
Conductive: Able to transmit heat or electricity.
Conduit: A duct, raceway or channel for conveying a substance; a duct for electric wires.
Cone: A shape that is circular at one end and reduces in size toward the other end.
Conformable: Can be brought into harmony with, made to fit.
Congeal: To change from a liquid to a semi-solid by freeezing or by chemical reaction.
Conic: Like a cone.
Connect: Join or link together.
Consequence: Chain of causes; as the result of.
Consistency: Degree of stiffness, as a liquid or paste.
Constant: Unchanging.
Contiguous: In actual contact with.
Continuous: Uninterupted.
Contour: The outline shape of an object or part.
Contra: A prefix meaning opposite to.
Control: Means to regulate or direct the action of.
Convection: The movement of a fluid due to variations in temperature and density.
Convex: Oblate, blister-shaped, like a magnifying lens.
Convoluted: Complicated; Intricate (twisting in various ways).
Cook: Immerse or treat in heat or fire.
Co-operate: To function together; operate jointly.
Co-ordinates: The numbers that specify locations of points or spacial elements.
Copper: Element 29; atomic weight 63.54. Redish-gold metal used for conducting electricity.
Core: The central part of anything; the center strand in a rope or cable; the inner part of a solenoid.
Cork: The tough, resilient bark of the cork oak.
Corroborate: To confirm as true.
Corrugated: With equally spaced parallel ridges.
Corrode: eat away; the action of a powerful acid.
Cotter pin: A pin with a split which permits the end to be spread out for a retaining method.
Coulomb: A measure of electric current in Amperes per second.
Counter: A prefix meaning to act against; to oppose.
Counterproductive: not productive.
Countersink: To shape the mouth of a screwhole so that the screwhead will neatly fit in.
Couple: linkage, join.
Crank: A spindle or handle offset from a shaft but parallel to it.
Create: To bring something new into existence.
Creosote: wood preservative.
Crimp: To fasten by pressure indentations with a crimping tool.
Critical: Indispensible component.
Crossfeed: A transverse adjustment of the cutting tool on a lathe.
Crucible: A refractory repository for melting steel; for high temperatures.
Cryogenic: Relating to very low temperatures.
Cryometer: Measuring low temperatures.
Crystal: Mineral in the solid state, usually transparent.
Crystalize: To become brittle from overworking a metal.
Cube: A solid figure, box-shape, with six identical square sides.
Cumbersome: Heavy and bulky; hard to handle.
Cupric: Pertaining to copper.
Current: A flow, as of water or electricity.
Cybernetics: The science of automatic control systems; robotics.
Cycle: A circle of events, returning to the starting point, in a given interval of time.
Cycloid: The path followed by a point near the edge of a rolling wheel.
Cylinder: The shape of a round tube. The piston chamber within an engine.
Cylindrical: Shaped like a cylinder.

Damper: For reducing vibration or shock; to hold down within proper bounds.
Darwinism: Relating to the theory of evolution.
Data: Plural of datum; information based on factual conditions.
Data processing: Converting information to computer language; storing, sorting, filing, the information in a computer.
Davit: A form of crane, often used in pairs to hoist life boats.
Debatable: Open to question and reason.
Decelerate: Retard the motion of.
Decibel: A unit of sound; the measurement of audio energy.
Declivity: Sloping down.
Decoration: Ornamentation.
Defective: Broken or not up to standard.
Definite: Distinct; not ambiguous.
Deflection: Not straight.
Degree: Measurement of temperature; measurement of angles. Relative intensity.
Degrees of freedom: The number of different ways a particle can move representing stored kinetic energy.
Delegate: To represent another; an agent for.
Delete: Erase, leave out.
Deliberation: Thoughtful consideration.
Delicate: Fine, dainty, exquisite, fragile.
Delineate: To draw an outline of.
Deltoid: triangular.
Demodulate: To extract the audio signal from a radio frequency carrier wave.
Demonstrate: To show by reasonable evidence.
Denominator: The lower number in a fraction.
Dense: compacted, heavy, opague.
Density: The ratio of mass to volume.
Depend: To hang from.
Depict: To illustrate with a picture.

Deplete: Used up the supply.
Depress: To press down on.
Depth: The concept of distance downward, or away from in perspective, degree of deepness.
Derivative: Procured from the source; to draw conclusions from a source (derive).
Design: Usually to plan out on drawing paper the way something is going to look, or how something is going to function.
Destabilize: To make something unstable, insecure.
Destructive distillation: A process in which wood-chips or other matter is enclosed in an evacuated chamber and then heated to such temperature that all volatile elements are vaporized, leaving only ash. The vapors are then condensed as in a still, and finally fractionated.
Detach: Remove from (opposite of attach); unjoin.
Determine: To know for certain, to find out.
Detail: To include all minor items.
Detonate: Loud explosion.
Deuterium: Heavy hydrogen.
Deuterium oxide: Heavy water; D2O.
Detector: An indicator.
Develop: To evolve and express all the possibilities of.
Device: Instrument or contrivance formed by design.
Dexterity: Mental skill; quickness of hand and mind.
Diagonal: Crosswise from corner to corner.
Diagram: A chart, graph or drawing made for explanatory purposes.
Diamond: Crystaline carbon, said to be the hardest of all materials.
Diameter: The distance across through the center.
Diaphram: A thin flexible partition.
Diazine: C4H4N2.
Dichroic mirror: Partially silvered mirror such that light rays are half reflected and half transmitted.
Die: A tool for cutting threads; the part in a punch press that provides the shape of the punched out product. To cut or shape with a die.
Dielectric: An insulating material; a non-conductor.
Diesel: An internal combustion engine in which the ignition is accomplished by the heat of compression.
Differential: Gearing; an arrangement of gears that permits the driven wheels of an automobile to rotate at different speeds when turning a corner.
Diffraction: The tendency for a light ray to change direction slightly as it bends around the edges of any objects in its path.
Diffusion: The mixing of substances by the motions of individual molecules or atoms.
Digital: Pertaining to the numbers from 1 to 9 and zero.
Dihedral: The angle of an aircraft wing from the fusilage to the tip of the wing, measured from the horizontal.
Dilute: To attenuate; to thin out with solvent.
Dint, Dent: An indentation.
Diode: A vacuum tube with two filaments; a heated cathode and a cold anode.
Diopside: Pyroxene.
Dioptase: Hydrous copper silicate.
Dioxane: C4H8O2.
Direct: By the quickest or shortest path.
Disintegrate: Decompose.
Disc: A flat circular plate.
Disc brake: A disc fixed to a wheel, which when squeezed by a caliper or vise, stops the wheel from turning.
Discontinuous: Interupted.
Discovery: To uncover for the first time.
Disintegrate: To crumble; to break down into the constituent elements.
Dissolve: To mix into or melt into a solution.
Disparity: Unequal, incongruous.
Distance: Measurement along a straight line.
Distillation: To vaporize volatile substances by heating, then condensing the vapors to liquid form.
Distinct: Definite; clear.
Distortion: Twisted; not like the original.
Divide: To separate into parts.
Document: Evidence or information put into written form.
Dolly: A small roller truck for moving heavy objects.
Dominant: Exerting control.
Doppler effect: The shift in frequency of sound or light by the motion of the emitter.
Dowel: A cyclindrical pin, often made of wood.
Draft: The act of drawing.
Drill: A bit for boring holes; also the act of boring holes.
Dry-ice: Solid carbon dioxide.
Ductile: Said of pliant metals, lead, gold, etc.
Dual: Pertaining to a pair of similar objects.
Duplicate: Two of; to make another the same.
Dynamic: Pertaining to physical force or energy.
Dynamotor: An electric motor that automatically becomes a generator when driven by the shaft.
Dyne: A unit of force.

Earnest: Serious state of mind.
Earth: Soil; name given to the whole world.
Eccentric: Off center; a device that imparts a back-and-forth motion by being off center.
Echo-location: Finding the direction and distance to an object by listening for its echo.
Economic: Pertaining to the management of finances and valuable commodities.
Eddy: Whirling fluid; a vortice.
Eddy currents: Vortices of electricity.
Effect: Result of a cause.
Elasticity: Quality of being springy and resilient.
Effort: Effective force.
Eject: Expel with force.
Elastic: Resilient; rubbery; stretchy.
Electrodynamics: Pertaining to the application of magnetic and electrostatic fields for energy conversions
Electromagnetic waves: Light; radio; infrared, ultraviolet, X-rays

and gamma rays.
Electrolyte: A liquid or paste capable of conducting electricity.
Electron: Elementary charge of electricity; small particle of matter with a negative charge.
Electroplate: To place a thin coat of metal on an object by means of electrolysis.
Element: One of 96 basic particles of matter that are immutable except for spontaneous decay of some heavy elements.
Elementary: Pertaining to the fundamentals of.
Elevator: A machine that lifts people or goods to different levels.
Eliminate: To remove; expel from a system.
Ellipse: Elongated circle; egg-shaped.
Elongate: To make longer.
Ellucidate: To enlighten by explanation.
Embodiment: The whole body of.
Emboss: To apply a raised device or design; opposite of engraving.
Emerge: To come out of.
Emit: To throw off; eject out of (usually pertains to fine particles, smoke for example).
Emphasize: To point out; to give special attention to a detail.
Empirical: From actual experience, proven by experiment.
Emulsion: Fine solids suspended in a liquid or gel.
Enamel: A impervious vitreous coating applied to metal; a paint that imitates enamel.
Endorsement: Written approval; sanction.
Energy: Capacity for doing work.
Engine: Any machine, but used mainly for the internal combustion engine.
Engage: Put in gear; interlock or intermesh.
Energy: The capacity to perform work.
Engrave: To incise; to cut a design or impression in any material.
Enhance: Improve on.
Ensue: To follow as a consequence.
Entropy: Availability of energy in a thermodynamic system.
Envision: To have a mental picture; in the mind's eye.
Epicycloid: The path followed by a point on the edge of a wheel as it rolls around the circumference of a larger wheel.
Equalibrium: A state of balance between opposing forces.
Equivalent: Having the same standards; interchangeable.
Eradicate: Exterminate; get rid of.
Erase: To rub out; to expunge.
Erect: Upright; also, to construct.
Erg: The work done by a force of one dyne over a distance of one centimeter.
Erronious: Containing errors.
Erupt: To burst forth.
Escapement: A device which permits a spring loaded gear or rack to move periodically, one tooth at a time.
Essential: The essence of; the intrinsic characteristic of; indispensable.
Ethane, Ethyl: $CH3CH3$.
Ether: $(C2H5)2O$.
Ethylene: $C2H4$.
Euclidian: Geometry based on local reality.
Evacuate: To create a vacuum within.
Evaporate: Change from liquid to vapor by application of heat energy.
Evidently: Clear exposition; obvious from the evidence.
Evolve: Process of change; develop naturally.
Exact: Agree with a standard of accuracy.
Examine: Scrutinize; investigate.
Example: A sample chosen for illustration or demonstration.
Exciter: A device or circuit that generates the initial phase of a process.
Exclude: Leave out; eliminate from participation.
Execute: Carry out; complete; apply a signature.
Existing: The thing we are discussing right now.
Expanded: Inflated; extended.
Expedient: Suitable to the end in view.
Experiment: To prove by actually doing it.
Expertise: All of the past experience relating to it.
Explicit: Distinct; clear; exact.
Explode: Detonate.
Extant: Presently in existence.
Extensive: Size; length; extent of dimension.
Exterior; External: On the outside of.
Extract: To derive from; to distill; that which is extracted.
Extract: To draw out of.
Extreme: At the utmost point; at the very edge of.
Extricate: To break free; to release from difficulty.
Extrude: Method of manufacturing elongated members by squeezing ductile material through shaped dies.
Eyelet: A small grommet.
Eyepiece: The part of an optical instrument placed against the eye of the observer.

Fabricate: To assemble; manufacture.
Facade: The front face of a structure.
Face: The front vertical or upright face of an object.
Face plate: A clamping disc or wheel turning with the spindle of a lathe.
Facilitate: To assist; to make easy or simplify.
Facing: The material used for the face of an object.
Factor: One of the main elements that contributes to a result.
Failsafe: An automatic feature that turns off a device if it malfunctions.
Fallacy: A false idea.
Fall-out: Undesirable things that might follow an accident or unplanned event.
Fan: Any device designed to produce an air current.
Farad: A unit of capacitance.
Faraday: A unit of electricity; 96,500 coulombs.
Fashion: Process of making; craftsmanship.
Fastener: Any means of fixing together, including nuts and bolts, rivets, eyehooks, etc.
Feasible: Capable of being done successfully.

Feature: Something that stands out: unusual or special attribute.
Feeler: Touch antenna, robotics.
Feedback: Recycle a fraction of the output and add it to the input; regenerative feedback.
Fender: Guard rail, splashboard, defender.
Ferro-: A prefix indicating iron as a constituent of a mineral.
Ferric-: Containing some iron.
Ferrule: A short tube or bushing, often forming a joint.
Fiberglass: Glass wool. etc.
Fiber-optics: Piping light through a fine transparent tube.
Field: A region; like a field; spacious; magnetic field for example.
Field of vision: The visual area encompassed by the eyesight.
Figure: A number; a character; drawing of a person or an object.
Filament: The fine wire used for electron tube electrodes and for incandescent lamps.
File: To arrange in a methodical order.
File: A small bar of hardened steel with cutting teeth or ridges on one or more of its surfaces.
File (rat tail): A tapering cylindrical file.
Film: Very thin membrane; or the appearance of.
Fillet: Triangular reinforcement at the inside corners of castings, supports, etc.
Filter: Any fine mesh device through which a fluid passes.
Filter: Radio; an inductance and a capacitance to attenuate unwanted frequencies.
Fin: Name given to any projecting metal rib for cooling an engine or any fin-shaped part. Triangular appendage.
Finance: The management of monetary affairs.
Fired: Baked in a kiln.
Finish: Degree of smoothness or of quality.
Firewall: Any fireproof partition designed to prevent the spread of fire.
Fireproof: Made non-combustible.
Fission: Natural splitting into two parts.
Flame: Source of heat and light from combustion.
Flange: A raised edge; turned edge.
Flat: Even surface; not rounded.
Flaw: A defect or fault.
Flecks: Visible flakes or particles.
Flexible: Mobile; pliant, resilient.
Flexure: A sharp bend or a fold.
Float: Lifted by the buoyancy of a liquid or gas.
Flock: Very short fibres, usually wool.
Floor: Bottom surface, usually strong enough to support a person.
Flow: To circulate as a running liquid.
Fluctuate: Waver; constantly changing.
Fluid: A liquid or a gas.
Fluorescence: Emits radiation when energized.
Fluorine: Element 9, a corrosive yellow-green gas.
Flush: To clean out with swishing solvent, usually water.
Flute: A long straight groove, usually in vertical groups of several side by side.
Flutter: A faster vibration than flicker.
Flux: The flow of energy through a boundary.
Flying: Airborne as a result of applied energy.
Flywheel: Any wheel designed to store energy due to its momentum.
Focus: In a lense system, the plane where a clear image occurs.
Foil: Metal worked extremely thin and applied as decoration. Thin aluminum.
Footage: Length of material expressed in feet.
Foot-pound: A unit of energy or work.
Force: The energy of pushing; that which causes acceleration.
Forceps: Small size pincers or tongs, used by doctors, jewellers, etc.
Force pump: To pump by pushing instead of suction; and therefore is not limited to atmospheric pressure.
Foreign matter: A substance contaminating a mixture.
Forge: A furnace where metal is heated and worked to shape.
Formaldehyde: CH2O.
Formula: Any general rule or principle expressed in symbols.
For-tran: A logic and algebraic computer language.
Forward: In the direction ahead.
Fossil: Any trace of living matter that has been petrified in the rocks from the distant geological past.
Foundry: The art of casting metals; the building and works where castings are made.
Foundation: A permanent solid base.
Fraction: A portion of the whole; a part of.
Fractionate: Separate a liquid into its components by distillation. The most volatile portion is the first fraction to condense out of the still. Repeating the process releases new fractions, each slightly less volatile.
Fracture: A break or crack in a hard object.
Fragile: Easily broken.
Fragmentize: Break into small pieces.
Frame: The main members of a structure or machine.
Franchise: To exercise a right.
Free: Not bound; easily moved.
Free-wheeling: A ratchet or similar means that permits the wheels to turn freely when the power is disengaged.
Freezing point: The temperature at which a liquid reverts
to a solid.
Freon: A fluorocarbon gas used as a refrigerant.
Frequency: The timing of a periodic event; usually beats per second.
Friction: The wearing of one object against another by rubbing or sliding.
Frigid: Intensely cold.
Frontal: On the front; the most forward part of anything.
Fruition: Brought to a state of realization.
Fuel: Any substance that is suitable for combustion.
Fulcrum: The point about which a

lever moves; the pin that a lever turns on.
Full circle: Completed cycle.
Fulminate: Explosive noise.
Fumes: Smoke; vapours; aromatic incense.
Functional: Able to act, operate, work, achieve what was intended.
Fundamental: A principle, law, or essential concept on which a system is based.
Funnel: A cone-shaped duct with a wide mouth and small output port.
Furnish: Equip or supply.
Fuse: To mix by melting together.
Fuselage; The main central body of an aeroplane.
Fusible: Can be fused; easily fused.
Fusion bomb: Hydrogen bomb.
Future: Time not yet expired.
Future shock: Rapid social and technilogical changes; distress suffered as a result of.

Gadget: A device; usually a hand-held contrivance for accomplishing a household task.
Gadolinium: Element 64, Atomic weight 157.3.
Galactic: Pertaining to the milky way.
Galaxy: The milky way. A cluster of stars.
Galena: Lead sulphide.
Gallium: Element 31; Atomic weight 69.72.
Galvanized iron: Steel coated with zinc.
Gamma ray: Hard radiant energy, harder than x-rays.
Gantry: A frame structure with tracks to support a travelling crane.
Gas: A vapor; also slang for gasoline.
Gasify: To convert to gas.
Gasohol: A mixture of gasoline and alcohol, mainly gasoline.
Gear: A toothed wheel in a machine; also equipment.
Gearing: A train of toothed wheels, intermeshed.
Gelation: Cooling and thickening to a jelly-like consistency.
Generator: A device for producing something; means for converting rotary energy to electricity.
Genuine: Authentic; original.
Geometric: Pertaining to the study of lines, angles, surfaces, planes, figures and their relationships.
Geoscience: Study of the earth.
Germanium: Element 32; atomic weight 72.60.
Getter: A particle of cesium imbedded within a vacuum tube to absorb traces of unwanted substances.
Gigawatt; A billion watts.
Gimbals: Compensating bearings which permit tipping in any direction. Universal joint supporting an instrument.
Gimlet: An awl with a threaded screw point for making holes.
Glare: Dazzling light.
Glass: Clear transparent vitreous material made by fusing silicates, borates and phosphates. Most glass contains sand (silicates), potash or soda (alkali) and lime or lead oxide (a base).
Globe: A ball or spherical in shape.
Gloss: Polished to a lustre.
Glowing: Heated to luminosity.
Gold: Element 72, atomic weight 197.2.
Governor: A device that controls the speed of an engine by metering the fuel.
Gradient: A slope; the rate of ascent or descent.
Gram: The unit of weight in the CGS system of measurement.
The CGS stands for Centimeter, Gram, Second.
Granite: Extremely hard igneous rock.
Granular: Consisting of small particles like grains of wheat.
Graphic: Pertaining to drawing, engraving, designing, etching, printing, etc.
Graphite: A form of carbon suitable for pencils, etc.
Grapho: A prefix meaning to write.
Grapple: Three or more metal hooks welded together to form a star, when attached to a rope, to be thrown so as to hook onto a projection.
Gravitate: To draw together slowly.
Gravitation: The force that holds objects to the earth; a universal force that tends to draw all matter together.
Grease: A lubricant in the form of a paste; to lubricate.
Grid: Equally spaced vertical and horizontal lines forming squares.
Grid: The filament between the emitter and the plate in a triode.
Grind: To sharpen; to shape or to smooth out by friction;
to reduce to a powder by crushing.
Grindstone: A bar or wheel of stone suitable for grinding; such as carborundum.
Grip: A shaped handle on a manual control.
Grit: Grindings or fine sand.
Grommet: A metal eyelet, especially used along the edges of a canvas sail, banner or tarpaulin.
Groove: A narrow elongated depression; a scratch or a furrow for example.
Gross: Bulky or massive.
Grounds: The basic principles on which belief is founded.
Grout: A finishing plaster or cement.
Guarantee: Assurance that a condition will be fulfilled.
Guard: A deflector or baffle to prevent injury by moving parts or by electric shock, chemical attack, etc.
Guide: A groove or device to direct motion.
Gum: An organic sticky substance used for adhesion, etc.
Gum elastic: Rubber.
Gussett: A fillet; a triangular piece welded or cast into a corner to strengthen it.
Gutter: An open drain, or a trough for drainage.
Gyrate: To whirl or spiral around; to rotate while moving laterally.
Gyro-: A prefix meaning spiral or circle, usually with motion.

Hack: Cut up quickly; chop & slash.
Hacksaw: A fine blade stretched

taut in a frame for cutting metal.
Hair; hairbreadth: The width or diameter of a hair.
Hairline: Cross hairs; the fine lines used in optical instruments.
Hairspring: A fine coil-spring for regulating a timepiece, etc.
Halogen: An element that forms a salt by direct union with a metal.
Hammer: A weighted tool for impacting on a driven fastener, or for crushing material, etc.
Handbook: A manual on a specific subject; a guidebook.
Hang: Suspend from; depend.
Hanger: A device made for supporting by hanging.
Hangar: Building for housing aircraft, etc.
Hard: Flinty, tough, unyielding.
Hardware: The necessary tools, parts, accessories for a project.
Hazard: A possible source of accident or danger.
Headstock: The part of a lathe that supports the spindle.
Hearing: The audio sense.
Heat: The energy of molecular motion; infrared rays.
Heavyduty: Able to withstand rigorous usage.
Height: Altitude, elevation.
Helical: In the form of a helix or spiral.
Helio-: Pertaining to the sun.
Heliocentric: With the sun as the center.
Hema-: Relating to blood.
Hematoid: Resembling blood.
Heptagon: A seven sided figure.
Hertz: Cycles per second.
Heterodyne: Generating a beat frequency by combining two frequencies (radio, superheterodyne).
Hexagon: Six-sided figure.
Hexane: C6H14.
Hinge: A door pin; a part with a pin or flexible joint that permits motion in a part or door.
Hexane: C6H14.
High-energy: The physics of particles travelling at relativistic velocities; particle acceleration.
High tension: High voltage.
Hinge joint: Turning point of mechanism with limited angle of turn.
Hitch: The connecting device between vehicles, farm implements, etc.
Holo-: A prefix meaning the whole of, the entire thing.
Hold-down; hold-fast: Fasteners of various types.
Hone: Sharpen to a fine cutting edge.
Honeycomb: A strong, lightweight cellular structure.
Horizontal: Level from side to side, like the horizon; straight and level.
Horsepower: A unit or work or energy, 550 footpounds per second.
Hose: A flexible duct or pipe for conveying liquids or gases.
Hue: Color density; chroma.
Hydrate: Combined with water.
Hydraulic: Any process where water is the prime mover or water is used to transmit power or pressure.
Hydric: Pertaining to or containing hydrogen.
Hydriodic acic: HI; An acid compounded of Iodine and Hydrogen.
Hydrocarbon: Any compound of Hydrogen and Carbon; acetylene or benzene for example.
Hydrodynamic: Water power.
Hydroelectric: Electricity generated by directing pressurized water through a water turbine to drive an electric generator.
Hydrogen: Element number 1; atomic weight 1.0078.
Hydrographics: The study and charting of water courses.
Hydroponics: Growing plants in chemical solutions.
Hyper-: A prefix meaning over, above, beyond.
Hyperon: A baryon greater in mass than a proton.
Hypothesis: A theory proposed to explain evidence.
Hypotenuse: The longest side of a right-triangle; the side opposite the right-angle in a right- triangle.

Ice: The crystals formed when water changes from a liquid to a solid.
Iconoscope: Electronic tube for converting the variations in a pictorial image to variations in a continuous electric signal suitable for transmission; a television camera.
Ideally: Conforming closely to a perfect state.
Identical: The same in every way; equivalent.
Idle wheel: A gear wheel that only transmits power between working gear wheels.
Ignition: Setting alight; to begin combustion; explode by heat or fire.
Illegal: Unlawful; Illicit.
Illuminate: To throw a light on; light up; to explain with lucidity.
Illustrate: To make clear with figures and examples; to draw a picture of.
Imagine: To form a mental image; to conceive or think about.
Imbue: Inherent attribute; to saturate.
Imitation: A copy of the real thing; to mimic; close resemblance.
Immiscible: Cannot be mixed together.
Immerse: To submerge into liquid.
Immobile: Cannot be moved.
Immune: Free from obligation; not subject to attack.
Impact: The meeting with force of two solids or of one solid and one liquid; to pack tightly by force.
Impair: To diminish in quality or strength.
Impart: To give a share; to disclose.
Impedance: A resistance to a flow of current in an alternating current circuit.
Impel: To push ahead; to give motion to.
Imperative: Obligatory; compulsory.
Impervious: Cannot be penetrated.
Impetus: Impulse; momentum.
Impinge: Strike against; encroach; infringe on.
Implode: To explode inward; to clap together.
Implied: Inferred.
Impotent: Powerless.
Impress: to press in; imprint.
Improvise: to successfully make

do with a substitute.
Impulse: A sudden force.
Impure: Adulterated.
Incandescent: Glowing hot.
Incendiary: For starting fires.
Incise: Cut into or engrave.
Incline: A slope.
Incognizant: Not aware of.
Incompatible: Cannot be made to work together.
Incongruent: Not conforming; inharmonious, not fitting.
Inconsistent: contradictory.
Increment: Minute increase or decrease.
Incursion: Unwanted entrance of; invasion.
Indefinite: Not limited; without clear attributes.
Indelible: Cannot be crossed out or erased.
Indentation: A recess or cusp; a notch.
Independent: Not hung up; not supported; free of guidance.
Index: To point or indicate; an organized directory.
India ink: Black ink of intense opacity.
Indicate: To show by pointing.
Indirectly: Action through a medium; round-about approach.
Indistinct: Undefined.
Induction: Specifically, the production of electromotive force in an electric circuit due to the change in the magnetic flux of a nearby circuit.
Innefficient: Cannot produce the effect or objective intended.
Industry: Systematic technology for the production and distribution of goods.
Inert: No power to move or react by itself: inactive.
Inertia: The ability of a body to resist acceleration or deceleration; the tendency of matter to remain at rest.
Inexplicable: Cannot be explained.
Infer: To draw conclusions about; to surmise.
Infiltrate: To filter into; enter into the pores of a permeable substance.
Infinite: Without end.
Infrared: Electromagnetic rays emitted by a hot body.
Inflatable: Capable of holding air by expanding; a balloon.
Influx: Inflow; pouring in.
Infringement: Trespass on a right or privilege.
Inherent: The essential character or attributes of anything.
Injector: A device for forcing liquid into a cavity under pressure.
Innovate: To add novelty to; to invent.
Innumerable: Without limit; cant be counted.
Inoperative: Producing no result; not working.
Insert: A part that is placed in or cast within a larger part.
Instantaneously: in an instant; in a flash of time; immediately following.
Instrumentation: The control devices in a system, such as indicators, meters, monitors, regulator dials, gauges, manometers, etc.
Insulator: That which prevents or controls the transmission of heat, sound or electricity.
Intact: Not cracked or broken; the whole thing.
Integer: Not a fraction; a whole number.
Integrity: Unimpaired; wholeness; unbroken.
Intentional: Done with intent; on purpose.
Intend: To have in mind as a decision to act.
Intensity: The degree of force, energy, pressure, etc.
Inter-: A prefix meaning among or between.
Interface: The boundary plane between two bodies or regions.
Intercept: To catch and hold before it reaches its destination.
Interfere: To claim the same priority in an identical invention.
Interior; internal: The inside of.
Intermittent: Periodically on and off.
Interpolate: To take the average between quantities; to compare by proportion.
Interpret: to explain the meaning of.
Interrupt: To stop the action.
Interstice: A small space between objects; the spaces between atoms in a crystal.
Interval: A short pause; a hiatus.
Intersect: Meet at a cross-point.
Intricate: Complicated; many small components operating together.
Intrinsic: Inherent.
Inverted: Inside out; reverse order.
Involved: Connected with by complication.
Ion: An electrically charged particle.
Iodine: Element 53, atomic weight 126.91.
Ion: An atom or molecule with an electric charge.
Ionize: To make conductive by the addition of ions.
Iron: Element 26, Atomic weight 55.85.
Irridescent: Shining with the colors of the rainbow.
Irradiate: To direct radiation upon.
Irrelevant: Not pertaining to the subject at hand.
Island universe: The galaxy.
Iso-: A prefix meaning alike.
Isometric drawing: A perspective drawing in which all lines Intersect at 60 degrees or 30 degrees.
Isolate: To separate from contact with other things.
Isothermal: To change the pressure without changing the temperature of a gas.

Jack: A device for lifting a heavy object by a small amount.
Jaws: The parts of a machine that are designed to grasp or crush.
Jet: A stream of fluid forced through a nozzle or small opening; the nozzle itself is also called a jet.
Jig: A device for guiding or supporting parts in metal working or woodworking.
Jimmy: A short pry bar.
Joint: The connection or device that connects different parts.
Joule: A unit of work or energy;
Journal: The part of a shaft, axel or spindle that turns within a bearing.

Joy stick: A manual control that can be moved in any direction from the vertical off position.
Juncture: Joint.
Juxtapose: To place side by side.

Kaleidoscope: An optical device for making rapidly changing patterns; also, patterns and designs of color and shape changing rapidly.
Kaolin: White clay used in ceramics.
Keen: Sharp.
Kelvin: A temperature scale with degrees the same size as Celsius degrees except that zero degrees K is at absolute zero.
Ketene: Poisonous gas C_2H_2O.
Key: The critical information needed; the basic principle that applies.
Key: Slang, a kilogram.
Kiln: Industrial oven for heat treating or drying materials.
Kilo: One thousand.
Kilolitre: 1000 litres; a cubic meter.
Kilowatt: 1000 watts; one watt is a flow of current of one ampere through a voltage drop of one volt.
Kinescope: Television camera.
Kinetics: Pertaining to motion caused by the application of a force.
Knuckle joint: A hinged joint like a knuckle.
Knurl: To machine a surface to give it a gripping texture for hand controls; to machine a surface with a design to make it attractive.
Krypton: Element 36, atomic weight 83.80. A colorless inert gas.

Label: Give a name to; designate.
Labyrinth: Many interconnecting small ducts or passageways.
Labyrinth gland: A method of rapidly reducing high pressure steam to zero pressure as it passes through a rotating journal.
Lacquer: A natural varnish usually made of selected tree saps dissolved into wood alcohol.
Lactose: $C_{12}H_{22}O_{11}$.
Lag: A short interval of time when the second event follows the first event.
Lambent: Softly radiant.
Lambert: Measure of illumination; one lumen per square centimeter of surface (emitted or reflected).
Laminate: Build up with thin layers.
Lampblack: Carbon black powder; pigment for black ink, shoe polish, etc.
Land: The flat space between grooves in a grooved device.
Lanthanum: Element 57; atomic weight 138.92.
Lap: Fitting sliding metal parts together by gentle abrasion of the touching surfaces with fine oiled emery cloth.
Lapidary: The art of polishing and shaping gemstones.
Laser: A device for generating coherent light or other electromagnetic radiation.
Latent: Possible talent; possible energy source.
Lateral: Pertaining to sidewards; crosswise; transverse.
Latex: The raw material from which rubber is made.
Lathe: A machine that rotates the workpiece while the cutting tool is operated by an automatic screw action.
Lattice: Strips of material usually crossed over in
a diamond pattern and fastened at the crossovers; the formation of crystals by regular patterns of atoms or molecules in straight rows.
Law: A statement of scientific fact; the practice of legal matters.
Lawyer: One who practices law; advises clients as to their legal rights and obligations.
Layout: Act of planning, usually drafting or drawing to scale.
Lead: Element 82; atomic weight 207.22.
Lead: To travel in front of in time or distance, leading.
Ledge: A projecting ridge.
Legible: Can be plainly seen and read.
Lengthwise: In the direction of greatest dimension.
Lens: A transparent disc with curved faces adapted to form a image when light rays are passed through it.
Lens-shaped: Shaped like a lentil; circular from the plan view and oval from any side view.
Lethal: Deadly.
Level: Flat and smooth; horizontal.
Lever: A bar with a spindle through it, on which it turns, with end pins used for transmitting a force, or for changing the direction of a force or for changing the strength of a force.
Levitate: To cause to float disregarding gravity.
Licensee: The person to whom a license is given.
Light: Luminous energy used by the eyes for sight.
Light-pen: A pen-shaped piece that interacts with the face of a computer monitor.
Likely: Probably true.
Lime: Quicklime; CaO.
Limitation: Restriction; retarded by a flaw or confined by a borderline.
Lineal: Pertaining to lines.
Linear measure: Measured by length alone.
Linkage: The means by which a motion is communicated through a machine.
Link: A connection, like the links in a chain.
Liquid: Freely flowing, like water.
Liquify: To convert a solid to liquid.
Liquid crystal: A liquid that exhibits some characteristics of a crystal.
Lip: The ridge or ringed edge of a cavity or port.
Lithium: Element 3, atomic weight 6.940.
Litre; Liter: 1000 cubic centimeters.
Lithoid: Like a stone.
Litigation: Dispute in a court of law.
Lobe: The protuberant portion of a cam; any rounded part that sticks out.
Localize: Confine to a definite space or area.
Locate: To specify the position of.

Locknut: Any of several types of nuts with means to prevent turning after it has been tightened.
Logarithmic Table: A list of the exponents of common numbers; (since exponents can be added or subtracted to effect multiplication or division, arithmetic is thereby simplified).
Logic: The science of what is the true cause and the true result; valid result of applied reason and intelligence.
Longitude: Pertaining to the length of; not transverse.
Loss: Power or energy not available for use.
Low pressure: The part of a cycle operating at a pressure less than normal.
Lubricant: Oil, grease, graphite; substance to reduce friction.
Lucid: Clearly written or spoken.
Lug: To haul a heavy load; a small projection on a wheel for gripping the ground to prevent slippage.
Lumen: A unit of light energy.
Luminosity: Degree of illumination.
Lunate: Crescent shaped.
Lux: A unit of light energy; one lumen per square meter.
Lymph: A clear body fluid containing white blood cells.

Macadam: Inventor of asphalt roadways; asphalt road.
Machine: A gadget or contrivance of a mechanical kind; a vehicle; a motorized shop tool. Turning, shaping, planing, milling metal.
Machine language: Binary code; computer language.
Mach number: The speed of sound being unity or one at a particular altitude and weather conditions. Mach 2 is twice the speed of sound at the point where the speed is determined.
Macro-: Large; large scale.
Magnesium: Element 12, atomic weight 24.32.
Magnet: Any object that has the power to attract iron or steel; electromagnet.
Magnetic field: The space near a magnetized object.
Magnify: To make objects appear larger than lifesize.
Magnetochemistry: The science that pertains to magnetic effects encountered in chemistry.
Magnitude: Relative greatness.
Maintain, maintenance: The act of keeping machinery, vehicles or property of any kind in good repair, including cleaning, lubrication, painting, etc.
Major: Dominant; opposite of minor.
Make public: Public disclosure of information.
Malleable: Ductile.
Mallet: Hammer with a cylindrical wooden head (usually).
Mandate: Authoritative command; imperative; social obligation. command, imperative.
Mandrel: A tapered shaft or spindle; the metal core in a an object cast of plastic.
Manganese steel: Steel with 12 to 14 percent manganese is extremely hard and durable.
Manganese: Element 25, atomic weight 54.94.
Manifest: Plain to see and understand; not obscure.
Manifold: The part of a duct or conduit where it branches off with several ports; in an automobile, the duct between the carburetor and the cylinders.
Manipulate: To control the action by hand.
Marginal: Near the edge; borderline situation.
Manufacture: To produce in quantity a finished product out of raw materials.
Marginal: Pertaining to an edge or border.
Marine: Pertaining to the navigation of the sea.
Market value: The amount of money charged for an item in the stores; the amount of money that someone is willing to pay for an item.
Marshal: To arrange in proper rank or order.
Mash: Crushed grain; a pulpy mass of anything.
Masonry: Concrete and brick structures in general.
Mass: The general term for matter in terms of its inertia.
Material: The real substance of an object or situation.
Maximize: To increase as much as possible; to take the greatest advantage of.
Mathematics: Science of numbers, arithmetic, algebra, trigonometry, calculus.
Matrix: Trellis; a rectangular array of numbers, etc.
Matter: Mass, the existing subject.
Mature: To complete the development of.
Maxim: Fundamental principle stated.
Maximum: The most possible, to the greatest degree.
Maxwell: The name given to magnetic flux; the density of magnetic flux per square centimeter.
Meaning: Intention, interpretation.
Means: What is available to accomplish a task or to effect a solution; what to use to make it work, suitable means.
Mean distance: The average of several measurements of distance over the same course.
Meaning: The interpretation intended.
Measure: To define distance in exact units.
Mechanical: Pertaining to machines; done by a machine.
Mechanism: An arrangement of parts of a machine adapted to produce a specific result.
Median: An average in the middle.
Medium: Average; middle state.
Medium: A substance or condition through which an effect is transmitted; intervening.
Meet: Touch together during assembly.
Mega: Large, million.
Melting point: Temperature at which a solid changes to a liquid.
Membrane: A very thin, pliable skin.
Mercury: Silver liquid, element 80, atomic weight 200.61.
Mercury arc: Mercury heated in a vacuum tube releases conductive vapor permitting the flow of a large current.

Merge: Combine; coalesce.
Mesh: To merge correctly and smoothly, as gear wheels.
Metacenter: A vertical line through the center of buoyancy.
Metallic: Resembling metal; made of metal.
Meter: The distance chosen to represent a yard in the metric system; 39.37 inches, one ten-millionth part of the distance from the equator to the pole.
Methane: CH_4.
Methanol: CH_3OH.
Methodical: Systematic.
Metric: Pertains to the science of measurement.
Micro-: One millionth part of; very small; infinitesimal.
Microelectronics: Silicon chip computer element.
Microfiche: Microfilm array.
Micrometer: A tool for measuring extremely small sizes.
Microscope: A device for examining very small items by magnification.
Microphone: Sensitive device for converting sound waves into electrical impulses.
Milky Way: The galaxy; the local island universe.
Million: One thousand thousand.
Mineral: Occurring naturally in the earth; produced by a mine.
Miniaturize: To represent on a small scale.
Minimum: The least amount; to the smallest degree.
Minute: Extremely small.
Mirror: Surface that forms images by reflection.
Mis-: Prefix meaning bad or wrong.
Misalign: Badly lined up.
Miscible: Easily mixed.
Missile: Designed to be thrown or projected with force; a projectile.
Mitigate: To lessen the severity of.
Mode: Method.
Modem: A device or desk-top unit for communicating over telephone lines from computer to computer.
Moderate: Not excessive.
Modern: Of recent origin; new.
Module: Standardized unit.
Modulus: A constant or coefficient.
Mold: The negative or reverse shaped cavity into which the desired object is cast.
Molecule: A particle comprising two or more elementary particles.
Molten: Liquified by the application of heat.
Molybdenum: Metalic element; number 42, atomic weight 96.
Moment: An instant of time.
Moment: The numerical value of a leverage.
Momentum: The inertia of a moving mass; the power to keep moving.
Mono-, mon-: Signifying one; or alone.
Monitor: A device that advises, warns or instructs; a computer screen.
Monochromatic: One color.
Monograph: A writing on a single subject.
Monoxide: An oxide with one atom of oxygen.
Morse Code: A signalling code based on dots and dashes.
Motion: The process of changing position by translation or rotation.
Mount: To install; to place in its proper position.
Multi-: Signifying many times; many things; repetition.
Mutable: Easily changed; variable.
Mutual: Operating together; cooperation.
Myopia: Lack of foresight; narrow view.
Myriad: Innumerable; a great many.

Nadir: The very lowest point; the point exactly opposite the Zenith.
Nail: To attach by driving slender metal fasteners in or through with a hammer.
Nano-: One billionth.
Napthalene: $C_{10}H_8$.
Necessitate: Unavoidable.
Needlelike: Any part that is slender and straight like a needle.
Nematic: Pertaining to liquid crystals.
Neon: Element 10, atomic weight 20.183.
Neptunium: Element 93, atomic weight 237.
Nerve center: Source of control and leadership.
Network: Many interconnected communications channels.
Neutral: Not engaged nor involved either way.
Neutron: An elementary particle about the same mass as a proton but with no electric charge.
Neutrino: A tiny particle with no mass and no charge.
Newton: Discovered the laws of motion.
Nexus: Link; connection.
Niccolite: Nickel arsenide.
Nickel: Element 28; atomic weight 58.69.
Niobium: Element 41, atomic weight 92.91.
Nippers: Small wire cutters.
Nitric Acid: HNO, oxidizing agent.
Nitrogen: Element 7; atomic weight 14.008.
Nodule: Small lump like a node; protuberance.
Non-: A prefix meaning not.
Nonviable: Not likely to succeed.
Normal: Not deviating from established standards.
Noticeable: Easily observed.
Notify: To advise; alert by memo.
Noxious: Harmful to humans.
Nucleate: To cluster together; to condense around a particle of dust or ice.
Nucleus: Central nut or seed; center of an atom.
Nullify: To make invalid; convert to no value.
Numerical: Expressed by numbers; in sequence by numbers.
Numerous: Plenty.
Nut: Any small hard object; A uniform block of metal with a threaded hole and four or six vertical sides, for threading onto a bolt as a fastening device.
Nutrition: Pertaining to nourishment; the science of food.

Object: Something that has dimension, and is tangible and visible.
Objective: Existing independently from the mind; the aim or goal of a

project.
Oblate: Slightly flattened spheroid.
Obligation: Duty; binding on the conscience; mandatory.
Oblique: Slanting or inclined.
Obliterate: Wipe out; cancel; eliminate.
Observer: One who participates only by seeing and recording what took place.
Obsolete: No longer used; outdated by technical advances.
Obverse: The opposite side of; reverse.
Occlude: To cover from sight.
Obvious: Easily understood.Occasional: Happens from time to time.
Octagonal: A figure with eight sides.
Octahedron: A solid figure with eight faces.
Octane: C8H18.
Ocular: Pertaining to the eye; pertaining to sight; an eye piece.
Odometer: An instrument with a dial showing the number of miles travelled by a vehicle, or the like.
Olfactory: Pertaining to the sense of smell.
Omit: To leave out; not included.
Opacity: State of being opague; non-transparent.
Opening: Aperture or port; orifice; hole.
Operate: To perform; to function as intended.
Opposite: Across from; on the other side; diametrically positioned.
Optical: Relating to lenses; pertaining to the eye.
Optimize: To make the most of; increase or improve to near perfection.
Optional: Having the choice of; discretionary.
Orbit: The path followed by a point, usually cyclical, often circular or elliptical.
Orbit: The path followed by a satellite of a planet.
Orbicular: Spherical.
Organize: To place in a systematic condition.
Orifice: A mouth-like opening; vent; aperture.
Origin: From where it started; the source or inception of.
Original: Not copied; having a known origin.
Originate: To innovate; to invent.
Ornament, decoration: Added for the sake of eye appeal, made to look at.
Oscilation: Movement back and forth; cyclical; vibrations.
Oscilator: Radio-frequency generator.
Osmium: Element 76, atomic weight 190.2
Osmosis: The passage of molecules through a semi-permeable membrane.
Ounce: 28.35 grams.
Outage: Disruption of electric service.
Outcome: The result.
Outer space: Outside the earth's atmosphere.
Oval: Elliptical.
Oven: Kiln; a chamber for heating, drying or baking.
Overlay: To superimpose.
Overrun: Excess production; more than ordered.
Oxidize: To combine with oxygen.
Oxygen: Element 8; atomic weight 16.
Ozonosphere: A layer of the atmosphere about 20 to 30 miles above the surface of the earth, with a high content of ozone.

Pacemaker: An implanted electrical device for controlling the rate of the heartbeat.
Paddle: A tool flat on both sides for stirring or mixing.
Palladium: Element 46; atomic weight 106.7.
Panacea: A universal remedy.
Panchromatic: Sensitive to all colors of visible light.
Pantograph: A device in the shape of a parallelogram for enlarging or reducing maps, etc.
Parabolic: The shape of a section through the center of a reflecting lens; a section of the reflecting lens in an automobile headlight; a conic section.
Paradox: Self-contradictory; opposed to common sense.
Paragraph: A subdivision in a written treatise or article, indicated by an indentation in the first line.
Parallel: Pertaining to lines that run in the same direction and remain the same distance apart.
Parallelogram: A figure with four sides, with opposite sides parallel, and therefore having the same length; diamond-shaped.
Parameter: One of two or more continuous factors; one of a set of coefficients, one of several criteria, one of a set of properties of the subject or object.
Paramount: Dominant.
Participate: To take part in, to share.
Particle: Smallest bit of matter or energy.
Particular: Separate or specific.
Partition: An interior dividing wall or panel.
Passive: Has no energy of its own.
Paste: Any composition of powder and liquid to form a squishy consistency; sludge; mud; clay and water mixture for making pottery, etc.
Pasteurization: To heat-up for the purpose of sterilizing or for killing germs.
Patent: An official document conferring a right or privilege, the exclusive right to an invention.
Patina: A coating or coloring in the surface of an object, usually accomplished by anodizing or by an acid bath.
Pattern: A design, an arrangement of shapes of different colors.
Pattern: A full size drafting of an object of manufacture; material cut to specific shape.
Pawl: A sliding pin or hinged lug that engages indentations to control the position of a machine part.
Peculiarity: Unusual characteristic.
Pedestal: Base; foundation; bed frame.
Pedal: Foot operated control.
Pellet: A Beebee.
Pending: Expected to happen.
Penetrate: To enter into.
Pentane: C5H12.
Pentagon: Five sided.
Percolate: Solvent passing

through solid particles.
Percussion: Sound waves created by sudden impact or explosion.
Periodic: Repeating in an interval of time; a repeating cycle.
Periscope: A device with mirrors for looking over the top of obstructions; An optical system in a submarine for looking all around.
Perimeter: The border around a plane figure.
Periphery: The outer border around a circular object.
Peristalic pump: A flexible duct mechanically squeezed at three places in sequence in order to press liquid through it.
Permeable: Can be penetrated by fluids or by magnetic fields.
Perpendicular: At right angles to.
Perpetual: Without end; everlasting.
Perspective view: A drawing that as near as possible mimics reality by showing how an object would look to the eye of an observer.
Pertain: Relates to.
Petroleum: Oil from the earth.
Pewter: An alloy of tin and lead.
Phase: A state or condition during a cyclic process.
Phenomenon: An event that can be observed and recorded; a scientific record of a unique event.
Phon-: Sound or voice. Audio.
Phonics: Acoustics.
Phosphorus: Element 15, atomic weight 30.975.
Photo-: Pertaining to radiant energy.
Photoemission: Release of electrons due to light rays falling on a photoelectric surface.
Photon: A unit of radiant energy.
Physical: Pertaining to real events.
Physics: The science that relates to matter and energy transformations.
Picture tube: Cathode ray tube; television monitor.
Pierce: To make a hole using a sharp point and pressure.
Piezo-electric: Crystals that produce an electric current when pressure is applied.
Pig: A metal ingot.
Pinch-off: Cut off a bit.
Pinion: The small gear in a gear train.
Pin: Any pin-shaped part.
Piston: The sliding part within the cylindrical chambers of an engine.
Pivot: A spindle or pin about which something rotates or swivels.
Pixel: A tiny picture element on a TV csreen.
Plan: To devise the steps of a process ahead of time.
Plane: A flat surface.
Planet gear: A gear wheel that turns around a central gear called the sun gear.
Plasma jet: A jet of extremely hot matter; matter so hot the electrons are freed from their atoms; plasma.
Plasticity: The quality of being changed in shape by heat and pressure; can be injection molded.
Plate: A flat slab of metal or glass, etc.
Plating: Surfacing a metal by an electrolytic process.
Platinum: Element 78, atomic weight 195.09.
Pliable: Supple; flexible.
Plutonium: Element 94, atomic weight 242.
Pneumatic: Pertaining to the use of air or a gas for inflation or for its operation.
Point: A position indicated by a dot fixed in space; a sharp object.
Point of view: An opinion; standpoint.
Polarity: The characteristics of a system that give it intrinsic direction.
Polarized light: Light ray in which the wave components have mainly only one direction of oscilation.
Pole: The end of a magnet from which the magnetic lines emerge.
Polonium: Element 84, atomic weight 210.
Poly: Pertaining to many.
Porcelain: A glassy, vitreous surface on pottery, insulators, etc.
Porous: Permeable.
Positive: Upbeat; opposite of negative.
Positron: The antiparticle of the electron.
Potential: Inherent power.
Power: Pertaining to control, authority, use of energy.
Practical: Can be put into practice.
Pre: Prefix meaning beforehand, in advance of.
Precise: Very accurate.
Prefabricate: To build the sections in advance for later assembly.
Preference: Having the opportunity to choose.
Primary: First in order of importance.
Print-out: A printed record produced by a computer.
Prior: Preceding.
Prism: A transparent bar of glass with a triangular cross-section; usually a right triangle.
Probability: The chance that something will occur; the odds.
Probe: A long thin knife blade or paddle for examining small articles within small spaces; exploratory electrode, for example.
Procedure: A step by step operation.
Process: The art of changing the attributes of raw materials to make a finished product, proceeding with a step by step operation.
Production: The output of an industry.
Profile: Contour.
Prognosis: To forecast what is likely to occur because of existing evidence.
Projectile: An object forcefully ejected.
Project: The job at hand.
Promote: To develop the marketability of.
Proof: Validity established.
Propane: $C3H8$.
Propellant: The fuel used in a reaction engine.
Proportion: Size relative to another size; a fraction or equation.
Proposal: Put forth for consideration.
Propulsion: Being propelled.
Proton: Nuclear particle with a positive charge equal to the

charge on the electron.
Prototype: A model; a first example; the original made to full size.
Provide: To make available.
Proximate: Close; near to.
Proximity: State of nearness.
Publish: To make known widely, usually by the printed word.
Pulsate: Beat.
Pulverize: To crush into very fine particles.
Pump: Mechanism for circulating or compressing a fluid.
Purify: Remove undesirable matter from.
Pylon: Any high pointed structure to identify an entrance or to mark a position; cone-shaped marker for race courses and airports.
Pyramid: A triangular structure that converges uniformly as it rises to a point.
Pyrex: Heat resistant glass.
Pyroelectricity: Crystals that produce an electric current when heated.
Pyrotechnic: Pertaining to fireworks.

Quadrant: An angle of ninety degrees at the center of a circle; a circle when cut horizontally and vertically through the center forms four quadrants.
Quad: Pertaining to four or a square.
Quadruplicate: To make four copies; four the same.
Qualified: Eligible; Up to standard.
Quality: An attribute that stands out.
Quantize: Subdivide into small elements.
Quartz: SiO2, silicon dioxide.
Quin: Pertaining to five.

Race: Part of the groove within which ball bearings circulate.
Rack: A straight bar notched with gear teeth along one side, made to engage a gear wheel (rack and pinion).
Radar: Determination of direction and distance by analysing reflections from a high frequency ray of electromagnetic waves.
Radial: From the center out; lines that co-incide with any radii of a circle.
Radiant energy: Usually electromagnetic radiation.
Radio beacon: Radio signals broadcast for the purpose of guiding aircraft to destinations.
Radio frequency: The frequencies between the highest audio frequency and the low end of infrared radiation, frequencies below 30,000 megahertz.
Radium: Element 88, atomic weight 226.05.
Radius: In a circle, the distance from the center to the edge.
Raise: To elevate; to give more height to.
Ramjet: A jet engine that uses its forward velocity for compressing the air it encounters in flight.
Random: Events that occur without plan; hit-and-miss.
Range: Distance; scope.
Rapid: Speedy; quick, fast.
Rapport: Good relationship.
Rasp: A course file.
Ratchet: A pawl designed to engage one tooth at a time in a rack or toothed wheel.
Ray: The iine of sight to a radiating body.
Reaction engine: A jet engine.
Ream: To smooth or enlarge a hole in metal, wood, etc.
To remove material by reaming.
Receptor: A sensing device.
Recess: Indentation or hollow, alcove.
Reciprocation: Regular movement back and forth.
Recoil: Jump backwards; Sudden reaction.
Recording: A means of saving the information produced during an event.
Rectangle: Squarish figure, longer than wide. Like the top of a shoebox.
Rectifier: Changes alternating current into direct current.
Reduction gear: A gear box with a low RPM, high torque output.
Reel: Bobbin or spool.
Refraction: The change in direction of a light ray when it enters a transparent surface.
Refrigerate: Preserve by keeping it cool.
Regard: Pay attention to.
Register: To make a record of.
Regular: Steady; normal, typical.
Regulator: A governor (machine).
Reinforce: Strengthen with additional factors or structure.
Reiterate: Repeat more than once.
Related: Connected by a common origin, type, method, or by fitting together in an operational mode.
Release: Trip lever on a trap door or the like.
Relevant: Has a significance to the project on hand.
Renovate: To restore; renew.
Replica: Duplicate reproduction.
Requisite: A necessary item.
Research: To investigate thoroughly.
Resistance: Opposition to.
Resonance: A vibrating element is stimulated at its natural frequency.
Resourceful: Able to improvise and make do.
Responsive: Sensitive to directives, reacts quickly.
Result: The effect of an action.
Reticle: The cross-hairs in an optical instrument.
Retro-: Pertaining to moving backwards or reversing.
Reverberant: Resound with an echo.
Rheostat: A variable resistor.
Rhodium: Element 45, atomic weight 102.91.
Rhombus: A parallelogram with four equal sides and no right angles.
Ridge: A raised elongated part.
Rigidity: Stiff; state of being hard and unyielding.
Robotics: Science of automatic control, especially adapted to imitating human beings.
Rocker: A lever or part that has a rocking action.
Rocket: A device operated by a reaction engine.
Roentgen ray: X ray.
Roller bearing: Hard steel cylindrical rollers that circulate within a

container; converts sliding friction to rolling friction.
Rotation: Moving around a central axis; turning on a point.
Rotary engine: An engine with no reciprocating parts.
Router: A machine for milling out a shaped groove.
Run: Pertaining to continuous operation.

Saber-saw: A portable saw in which the reciprocating saw blade protrudes like a knife blade or saber.
Sabotage: Deliberate malpractice or secret destruction of tools and equipment.
Safety valve: Automatic relief valve.
Salient: Conspicuous.
Sanitize: Cleaning antiseptically.
Saturated: The most concentrated liquid solution possible.
Saw: A cutting blade with shaped teeth.
Scale: The proportions of a drawing as it relates to full size, ie., quarter size, etc.
Scan: To make a record point by point in orderly lines.
Scanning microscope: Electron beam scanning of an object results in ray scattering. Focussing the scattered ray produces an image on a fluorescent screen.
Schematic: A drawing or diagram in which everything is left out except the essential feature or features.
Science: The knowledge of the universe collected by mankind.
Scientific method: (1) Collecting evidence, (2) forming a theory and (3) testing the theory.
Scope: Range; width or depth of interest and intent.
Screen mesh: Wire formed into a square pattern with the spaces small enough to prevent insects from passing through.
Larger mesh screens are used for sorting gravel, etc.
Screw: A metal fastener with a spiral thread for use in wood or metal, especially sheetmetal.
Screw, machine: A metal fastener with a spiral thread onto which a nut with a female thread is matched and installed.
Scrutinize: Inspect carefully.
Sculpture: To create a three-dimensional image in a likeness of something or somebody.
Sealed beam: A light with a parabolic reflector sealed inside the evacuated bulb of the lamp.
Secrecy: Keep the information to one's self; preserve privacy; keep under cover.
Section: How it appears when cut through cleanly.
Sector: One of two or more sections or sectors.
Secure: Safely fastened.
Segment: A portion, usually one of several the same.
Seismic: Pertaining to examining the earth at depth by analysing vibrations that have been reflected from deep strata.
Seismology: The science that relates to vibrations in the earth such as earthquakes.
Selenium: Element 34, atomic weight 78.96.
Self-starter: Automatic starter.
Semi-: Prefix meaning half.
Semitransparent: Partly transparent; diffused or cloudy.
Sensor: A device that perceives motion, sound or light variations, or electrical or magnetic changes.
Sensitive: Responsive to minimal stimulus.
Separate: Set apart, slice in two.
Sequence: The order of occurrence.
Sequential: Arranged in order.
Serpentine: Curved back and forth like a snake.
Serrated: Toothed edge.
Service module: Unit with backup materials and gear.
Servo-mechanism: Automatic control with low voltage of large high voltage equipment.
Servomotor: An electric motor used in a servo-mechanism; a motor that automatically duplicates the motions of a control motor.
Set-screw: A screw that can be tightened to make related parts rigid with respect to each other, such as to tighten a wheel on a shaft.
Shaft: A spear-shaped weapon; a spindle or axle for mounting wheels and gears.
Shank: The straight piece of a machine part.
Sharpen: To grind or hone to a cutting edge.
Shear: An automatic machine for cutting large flat stock such as paper or metal.
Sheath: A covering or case, usually close fitting.
Sheer: Semi-transparent.
Shield: Guard or fender.
Short circuit: Causing electric current to flow accidentally; not intended. not desired and not planned.
Shunt winding: A winding temporarily energized to magnetize field coils in an alternator; permanent field coils wired in parallel in some types of electric generators.
Siemens: German electrical engineer and inventor.
Signal: (Communications) Information transmitted electronically or by radio and television.
Silencer: Muffler.
Significant: Of some importance.
Silicon: Element 14, atomic weight 28.09.
Silver: White metal element 47, atomic weight 107.88.
Silver iodide: Compound AgI turns black on exposure to radiant energy.
Similarity: Resemblance, alike, comparable.
Simplify: To remove complexities.
Simultaneous: Occurring at the same instant of time.
Single phase: Standard household current.
Siphon: A duct connecting the liquids in two vessels, one of which is higher than the other. The liquid will run into the lower vessel even when part of the path it takes is uphill.
Sixth sense: Sharp intuition.
Skeleton: The basic framework.
Skim: To remove floating material or scum from the surface of a liquid.
Skin effect: Pertaining to the fact that high frequency current tends

to reside near the surfaces of the conductors.
Socket: A shaped depression or metal pocket into which a part fits exactly.
Sodium: Element 11, atomic weight 22.991.
Solar: Pertaining to the sun.
Solder: An alloy of lead and tin melted for fastening light metalwork and tin-smithing.
Solid: Being of hard matter throughout.
Solute: Dissolved matter.
Solution: Answer to a problem.
Solvent: A liquid capable of taking solids into liquid solution.
Sonar: Determining the distance and direction to a submerged object by sound reflection.
Sonic: Pertaining to audio.
Span: The distance from pier to pier in a bridge.
Specific: Exact details provided.
Specification: The written description of an invention.
Specific gravity: The weight of a given mass of a substance as compared to the same volume of distilled water.
Spectrum: The colors achieved when white light is passed through a prism.
Speculate: Theorize.
Sphere: Shaped like a globe or ball.
Spindle: The rod-shaped part of any machine; axle.
Spiral: Helix shaped.
Spline: The grooves in a sliding joint in a driven shaft.
Spontaneous: Action on a natural impulse.
Spontaneous ignition: Fire caused by waste material becoming hot from confinement and decomposition or chemical action.
Springy: Resilient; elastic.
Sprocket: A gear wheel designed to engage a chain link drive.
Standard: Conforming to what is acceptable by law.
State of the art: The most recent developments in any field of technical innovation or manufacture.
Static: Without motion; at rest.
Statute: Regulation.
Steel: Commercial iron with a carbon content to make it tough and hard.
Stereoscopic: Pertaining to a device for displaying photographs in three dimensions.
Stirling engine: An external combustion engine invented in 1896.
Strain: To place under a strong force or pressure.
Streamline flow: The movement of a fluid around an object following paths of least resistance.
Stress: The counter forces that resist an applied force or pressure. Resisting strain.
Striated: Marked with many parallel grooves.
Stretch: Deform by pulling.
Stroboscope: A light that flashes on an off with a set frequency. The effect is to freeze the motion of a rotating or vibrating body with the same frequency. Also a light that is flashed on and off at high speed by spinning a slotted disc in front of it.
Strontium: Element 38, atomic weight 87.63.
Structural: Relating to the main members used in construction.
Stud: A bolt with no head, and with threads on both ends.
Also a small metal pin or cleat.
Sub-: Prefix meaning under or below.
Sublimate: To evaporate a solid directly to a vapor or gas without passing through the liquid phase.
Submerged; submersed: Placed under the surface of a liquid.
Substantiate: To establish the proof of; to show the reality of.
Substitute: To use one thing in place of another.
Subterranean: Under the surface of the earth.
Suction: A force resulting from creating a partial vacuum.
Suitable: Fulfills the requirements; satisfies the criteria.
Sulfur: Element 16, atomic weight 32.066.
Summary: Concise orderly review.
Sump: Sink. A depression for the collection of liquids, often including a pump to dispose of the collected liquids.
Super-: A prefix meaning over or above.
Supercharger: A fan or blower for compressing the air entering an internal combustion engine.
Superconductor: A conductor that permits the flow of electricity with practically no resistance.
Supersonic: Faster than the speed of sound.
Surge: Sudden increase in electric current (sometimes due to lightning).
Suspension: Pertaining to hanging from supports.
Sustain: To continue to support.
Sweeping: Anything with a sweeping motion, windshield wipers for example.
Swivel: An object that can turn freely on a pin or pivot.
Symmetrical: Having the same shape each side of an imaginary center line.
Symptom: Evidence that something is wrong or malfunctioning.
Synoptic: Having an overall viewpoint.
Synthetic: Pertaining to something created by man rather than by nature.
Systematic: Organized; well planned.

Tabbed: Designated.
Tachometer: Gauge measures RPM.
Tactile: Pertaining to the sense of touch.
Tangent: A line on the circumference of a circle at right angles to the radius.
Tantalum: Element 73, atomic weight 180.95.
Tap: To cut threads into a metal nut or bolt, or stud hole.
Taper: Slender cone-shaped cylinder.
Taper off: To become less and less.
Taut: Tight; under tension.
Teardrop: Streamlined shaped that is spherical at the front and tapers to a point at the back.
Technical: Pertaining to applied scientific knowledge; engineering expertise.

Technician: One who is proficient in a particular technical occupation.
Tele: Pertaining to distance communications.
Telescope: Device designed to see at a distance.
Tellurium: Element 52, atomic weight 127.61.
Temper: The degree of hardness, in steel for example.
Tenacity: Ability to stick it out through duress.
Tensile strength: Ability of a material to withstand being pulled apart.
Tension: Pertaining to pulling force or stretching force.
Tentative: Not completed entirely.
Tenuous: Of little force; temporary condition.
Terminal: End point; battery post.
Terrestrial: Pertaining to earth.
Tetra-: Pertaining to four.
Tetrode: Vacuum tube with four electrodes.
Texture: The surface feel of an object; The tactile attribute of a surface.
Thalium: Element 81, atomic weight 204.39.
Thaw: Melt.
Theodolite: Surveyors instrument like a telescope mounted on a tripod.
Theoretical: A supposition not supported by evidence; speculative.
Thermal: Pertaining to heat; temperature.
Thermion: A charged particle emitted by an incandescent filament.
Thermodynamics: The science of thermal energy sources and their application.
Thermostat: Automatic temperature control device.
Thio: Pertaining to sulfur content.
Thorium: Element 90, atomic weight 232.05.
Three-dimensional: Illusion of depth in a flat picture.
Throttle: Controlling valve; butterfly valve.
Thyristor: Semi-conductor used for rectifier or switch.
Tie-rod: Usually a steel rod with a connector on each end.
Time machine: An invention that permits travel to and from the past or future (imaginary).
Tin: Element 50, atomic weight 118.70.
Titanium: Element 22, atomic weight 47.90.
Tolerance: Allowable deviation from standard.
Tone: The musical effect of sound.
Toolmaker: Machine-shop specialist.
Topography: The science of surfaces, particularly the earth.
Toroidal: Doughnut shaped.
Torque: Twisting force; the twist applied by a screwdriver.
Torsion: The twisting force applied to a screw by a screwdriver (example).
Touch: The tactile sense.
Toxic: Poisonous.
Trace: To copy onto semi-transparent paper overlaid on the original.
Tractor: Prime-mover designed for pulling.
Traction: The quality of the grip between a drive wheel and the road.
Trailing edge: The rear edge of aircraft wing.
Trans: Pertaining to across, or from side to side.
Transcript: A printed or taped copy of a written record.
Transduce: To convey energy or information from one system to another system.
Transistor: A solid-state device for controlling information signals.
Transit: Theodolite.
Transition: Gradual change from one phase to another; energy conversion.
Transitory: Temporary.
Translation: Steady motion of an object in a straight path.
Translucent: Permitting light to pass through in a diffused way.
Transmission: The process of transporting energy from one point to another.
Transmute: To change to a different kind; transform.
Transparent: Permitting light to pass through including images.
Transverse: Crosswise.
Travel: The distance moved by a reciprocating part.
Trellis: Latticework usually made of crossed slats to be used as a screen or for climbing plants.
Trial and error: To try different ways of accomplishing something until the correct way is found.
Triangulation: Using triangles to accurately locate a point on an uneven or curved surface.
Tri-: Pertaining to three.
Trigonometry: The study of the properties of triangles.
Trisect: To cut into three pieces.
Tritium: Heavy hydrogen with three nucleons instead of one.
Triton: The nucleus of tritium.
Truss: A horizontal roof support made up from criss-cross members and an upper and lower longitudinal member.
Tube: A cylindrical duct or pipe.
Turbine: An energy converter which uses rotating blades to absorb the energy from steam or water.
Turbocharge: Supercharger with a turbine-type compressor.
Turboelectric: A turbine that converts hydraulic pressure or steam to electricity.
Turbulent flow: Departure from smooth flow.
Turret lathe: A tool holder that turns to several positions with a different tool in each position.
Twist: To apply torque.
Typical: Conforming to a type.
Tyro: Novice.

Ultimate: Absolute maximum.
Ultra-: The ultimate; superior without equal; beyond the limits of the ordinary.
Ultrasonic: Pertaining to sound waves above the range of human hearing.
Unambiguous: Clearly shown.
Uncharged: Free of electrical bias.
Undercarriage: The framework around the base of a moving machine.
Understand: Comprehend.
Undulation: Wavy; wave motion.

Uni-: Pertaining to one.
Universal: Known everywhere; common.
Universal motor: An electric motor that runs on D-C or A-C at any speed.
Uranium: Element 92. atomic weight 238.07.
Utensil: Implement.
Utilize: Apply to a useful purpose.

Vacuum: Absence of air or other matter.
Valence electrons: The electrons in an atom that are the basis of chemical reactions.
Validate: Confirm as true or legal.
Valve: The movable part of a faucet ; anything with an on-and-off, or open-and-closed, action.
Vane: A fin-shaped paddle intended to be affected by flowing air; turns so as to present the least resistance to air flow.
Variable: Pertaining to changing over a range of conditions.
Vector: A diagram in the likeness of an arrow in which the length of the arrow represents the force of the vector and the arrow head points in the direction of the vector.
Vehicle: Usually a wheeled conveyance, but can be intended to mean a fluid within which a medication or a pigment is suspended.
Velocity: Usually speed or rate of change of position.
Venturi: A constriction in a flow-tube used for measuring flow-rate; also used for accelerating a fluid; a nozzle.
Verify: To validate.
Vernier engines: Small rocket engines used for making fine adjustments to the position of a spacecraft in space.
Vertex: The highest point.
Vertical: Straight up and down.
Vestige: A small piece of evidence showing that something once existed.
Viable: Having a good chance to succeed.
Vibration: Oscilation; reciprocation.
Vicinity: Proximity; near.
Video: Pertaining to television images.
Viscid: Adhesive; sticky.
Viscosity: Degree of resistance to the flow of a fluid over a surface, lubricating quality.
Vitreous: Like glass.
Volatile: Easily vaporized.
Voltage: Electrical potential; electrical pressure.
Voltmeter: A device for measuring electrical pressure.
Volume: Cubic contents.
Vortex: Eddy or whirlpool.

Wafer: A thin slice of a suitable semi-conductor such as silicon used for micro-circuits; also silicon chip.
Waive: To withdraw from enforcing a lawful claim or right.
Wankel: Inventor of rotary engine.
Warp: Twist that is usually not wanted.
Warranty: A guarantee.
Watt: A rate of work or energy expenditure equal to one ampere per second over a potential difference of one volt.
Wave length: The distance from the peak of one wave to the peak of the next.
Wax: A hydrocarbon by-product that is solid at room temperature.
Ways: Grooves or guides within which a reciprocating part slides.
Wear: The change in the dimensions of a surface caused by friction and use.
Weld: To fuse together by melting.
White heat: Temperature at which some metals become incandescent (1600 degrees C).
Whorl: Shape of a spiral or vortex.
Wind farm: Groups of wind turbines generating electricity.
Winding: Coil of wire in an electrical device.
Wireway: A conduit for electrical conductors.
Wound coil: Ready-made electrical winding.

X-axis: One of the three axes used to designate three dimensional space, the others being the Y-axis and the Z-axis.
Xenon: Element 54, atomic weight 131.30.
Xerography: A process for making rapid copies of graphic matter.
Xray: Radiant energy of very short wave length, less than 100 angstroms.
X-section: Abbreviation for cross-section.
Xylene: C8H10.
Y-intercept: Point where a curve intersects the Y-axis; also X-intercept and Z-intercept.

Zenith: High point; the highest point in the heavens.
Zeolyte: A hydrous silicate used as an ion exchanger in water softeners.
Zinc: Element 30, atomic weight 65.38.
Zone: A designated space; a specified region.
Z-particle: A neutral basic particle that is 90 times heavier than a neutron.

INDEX